Lecture Notes in Computer Science 1203

Edited by G. Goos, J. Hartmanis and J. van Leeuwen

Springer
Berlin
Heidelberg
New York
Barcelona
Budapest
Hong Kong
London
Milan
Paris
Santa Clara
Singapore
Tokyo

Giancarlo Bongiovanni Daniel Pierre Bovet
Giuseppe Di Battista (Eds.)

Algorithms and Complexity

Third Italian Conference, CIAC '97
Rome, Italy, March 12-14, 1997
Proceedings

Springer

Volume Editors

Giancarlo Bongiovanni
Daniel Pierre Bovet
Giuseppe Di Battista
University of Rome "La Sapienza"
Via Salaria 113, I-00198 Rome, Italy
E-mail: ciac97@dsi.uniroma1.it

Cataloging-in-Publication data applied for

Die Deutsche Bibliothek - CIP-Einheitsaufnahme

Algorithms and complexity : third Italian conference ; proceedings / CIAC, '97, Rome, Italy, March 12 - 14, 1997. Giancarlo Bongiovanni ... (ed.). - Berlin ; Heidelberg ; New York ; Barcelona ; Budapest ; Hong Kong ; London ; Milan ; Paris ; Santa Clara ; Singapore ; Tokyo : Springer, 1997
(Lecture notes in computer science ; 1203)
ISBN 3-540-62592-5

NE: CIAC <3, 1997, Roma>; Bongiovanni, Giancarlo [Hrsg.]; GT

CR Subject Classification (1991): F.1-2, E.1, I.3.5, G.2

ISSN 0302-9743
ISBN 3-540-62592-5 Springer-Verlag Berlin Heidelberg New York

Typesetting: Camera-ready by author
SPIN 10549543 06/3142 – 5 4 3 2 1 0 Printed on acid-free paper

Preface

The papers in this volume were presented at the Third Italian Conference on Algorithms and Complexity (CIAC '97). The Conference took place on March 12-14, 1997 in Rome (Italy).

International CIAC Conferences present research contributions in theory and applications of sequential, parallel, and distributed algorithms, data structures, and computational complexity, and are held at the Conference Center of the University of Rome "La Sapienza"

In response to the Call for Papers, 74 were submitted. From these submissions, the Program Committee selected 25 for presentation at the Conference. In addition to the selected papers, Gianfranco Bilardi, Shimon Even, and Christos Papadimitriou were invited to give plenary lectures.

We wish to thank all the members of the Program Committee and of the Organizing Committee, the plenary lecturers who accepted our invitation to speak, all those who submitted papers for consideration, and all the referees.

Rome, January 1997

Giancarlo Bongiovanni
Daniel Pierre Bovet
Giuseppe Di Battista

Organizing Committee:

D.P. Bovet (Rome, chair)
G. Bongiovanni (Rome)
A. Clementi (Rome)
S. De Agostino (Rome)
R. Silvestri (Rome)

Program Committee:

G. Di Battista (Rome, chair)
H. Bodlaender (Utrecht)
D. Breslauer (Aarhus)
P. Crescenzi (Rome)
T. Hagerup (Saarbrücken)
G. Italiano (Venice)
L. Pagli (Pisa)
G. Persiano (Salerno)
R. Petreschi (Rome)
J. Rolim (Geneva)
S. Whitesides (Montreal)

Additional Referees:

A. Andreev
V. Auletta
P. Bertolazzi
T. Biedl
M.A. Bonuccelli
A. Brogi
T. Calamoneri
B. Carpentieri
M. Cesati
J. Cheriyan
B. Chopard
P. Cignoni
A. Clementi
B. Codenotti
A. Cortesi
A. Cresti
S. De Agostino
M. de Berg
M. Di Ianni
D. Dubhashi
P. Favati
P. Ferragina
A. Ferreira
R. Fleischer
P. Flocchini
A. Galluccio
N. Garg
L. Gargano
L. Gasieniec
R. Galvadà
J. Gergov
D. Giammarresi
R. Grossi
J.L. Träff
H.P. Lenhof
G. Lenzini
G. Liotta
E. Lodi
G. Lotti
F. Luccio
L. Margara
A. Massini
K. Mehlhorn
R. Milanese
A. Monti
N. Nishimura
J. O'Rourke
S. Orlando
M. Papatriantafilou
M. Parente
A. Pedrotti
M. Pelillo
C. Petrioli
A. Piperno
V. Priebe
T. Przytycka
A. Rescigno
F. Romani
C. Rüb
A. Salibra
V. Scarano
S. Schirra
U. Schöning
R. Seidel
J. Sibeyn
R. Silvestri
M. Smid
D. Sosnowska
G. Tel
D. Thérien
L. Toniolo
G. Toussaint
L. Trevisan
P. Tsigas
E. Urland
U. Vaccaro
M. van Kreveld
M. Veldhorst
R. Veltkamp
T. Warnow
S. Wismath
M. Yung
A. Zemach

Sponsored by:

Dipartimento di Scienze dell'Informazione , Università di Roma "La Sapienza"
Università degli Studi di Roma "La Sapienza"

Table of Contents

Invited Presentations

Regular Presentations

Algorithms and Data Structures for Control Dependence and Related Compiler Problems

Gianfranco Bilardi

DEI, Università di Padova, 35131 Padova, Italy, and
EECS, University of Illinois, Chicago, IL 60607

Abstract

The *control dependence* relation plays a fundamental role in program restructuring and optimization. It is used in many phases of modern compilers, such as dataflow analysis, loop transformations, and code scheduling. This talk will survey recent results concerning the efficient computation of control dependence information for a given program. The concrete application of these results to the problem of obtaining the static single assignment form for a program will be discussed. Finally, a framework will be outlined that generalizes the classical notion of control dependence to include other interesting variants. The talk is largely based on joint work with Keshav Pingali [1, 2, 3, 4].

Technically, control dependence is a relation between nodes and edges of the *control flow graph* (CFG) of a program, a directed graph in which nodes represent statements, and an edge $u \rightarrow v$ represents possible flow of control from u to v. A node $w \in V$ is said to be *control dependent* on edge $(u \rightarrow v) \in E$ if w postdominates v, and if $w \neq u$, then w does not postdominate u.

Several applications of control dependence require the computation of sets of the following type: $\texttt{cd}(e)$, which is the set of statements control dependent on control-flow edge e; $\texttt{conds}(w)$, which is the set of edges on which statement w is dependent, and $\texttt{cdequiv}(w)$, which is the set of statements having the same control dependences as w.

A novel data structure is introduced, the *augmented postdominator tree* ($\mathcal{APT}$), which can be constructed in space and time proportional to the size of the program, and which supports enumeration of the above mentioned control-dependence sets in time proportional to their size. Hence, $\mathcal{APT}$ provides an *optimal representation of control dependence.*

By exploiting the well known relation between control dependence and dominance frontiers, the $\mathcal{APT}$ is used to develop an algorithm for the *single static assignment* form of a program, which takes linear time per variable.

Finally, a notion of *generalized dominance* is introduced, based on a parametric set of paths in the control flow graph. This new definition leads to a generalized notion of control dependence, which includes *standard control dependence* and *weak control dependence* as special cases. To illustrate the utility of the framework, it is shown how the $\mathcal{APT}$ can be used for the optimal computation of weak control dependence sets.

References

[1] K. Pingali and G. Bilardi. APT: A data structure for optimal control dependence computation. In *Proc. ACM SIGPLAN '95 Conference on Programming Language Design and Implementation (PLDI)*, pages 32–46, La Jolla, California, June 1995.

[2] G. Bilardi and K. Pingali, A generalized framework for control dependence, In *Proc. ACM SIGPLAN '96 Conference on Programming Language Design and Implementation (PLDI)*, pages 291–300, Philadelphia, Pennsylvania, May 1996.

[3] K. Pingali and G. Bilardi. Optimal control dependence computation and the Roman chariots problem. *ACM Transactions on Programming Languages and Systems.* To appear.

[4] G. Bilardi and K. Pingali. Computing the single static assignment form of programs. Manuscript. October 1996.

Embedding Interconnection Networks in Grids via the Layered Cross Product

Guy Even[1] and Shimon Even[2]

[1] Universitaet des Saarlandes, FB 14 Informatik, Lehrstuhl Prof. Paul, Postfach 15 11 50, 66041 Saarbrücken, Germany. e-mail: guy@cs.uni-sb.de
URL: http://www-wjp.cs.uni-sb.de/~guy/

[2] Bell-Labs, Lucent Technologies, 700 Mountain Ave., Murray Hill, NJ 07974. On leave from the Computer Sci. Dept., Technion - Israel Inst. of Tech., Haifa, Israel 32000. Research was partly supported by the United States -Israel Binational Science Foundation, grant No. 94-00266/1. e-mail: even@research.bell-labs.com

Abstract. A technique for automatically producing a rectilinear planar drawings of interconnection networks is described. It is based on the Layered Cross Product suggested by Even and Litman. The technique is demonstrated on the Butterfly network, the binary tree and the Mesh-of-Trees of Leighton.

1 The Layered Cross Product

Many layered graphs are known to be useful as interconnection networks. Examples are the Butterfly network, including all its different looking representations, meshes of trees, Fat-trees, the Beneš network, multibutterflies.

In [4], Even and Litman introduced the concept of *Layered Cross Product*, LCP, of layered graphs and showed that several of the important and well known networks are LCP-s of simple layered graphs, such as trees. Following is the definition of LCP from [4].

A *layered graph*, of $\ell+1$ layers, $G=(V_0, V_1, \ldots, V_\ell, E)$, consists of:

- $\ell+1$ *layers* of vertices; V_i is the (nonempty) set of vertices in layer i.
- E is a set of directed edges. Every edge $\langle u,v\rangle$ connects two vertices of two adjacent layers; i.e. if $u\in V_i$ then $v\in V_{i+1}$.

Let G^1, G^2 be layered graphs of $\ell+1$ layers each; i.e., for $j\in\{1,2\}$, $G^j = (V_0^j, V_1^j, \ldots, V_\ell^j, E^j)$. Their *Layered Cross Product, LCP*, $G^1\times G^2$ is a layered graph, $G=(V_0, V_1, \ldots, V_\ell, E)$, where:

- For every $0 \leq i \leq \ell$, $V_i = V_i^1 \times V_i^2$.
- There is an edge $\langle u, v \rangle$ in G, connecting vertices $u = (u^1, u^2)$ and $v = (v^1, v^2)$, if and only if $\langle u^1, v^1 \rangle$ and $\langle u^2, v^2 \rangle$ are edges in G^1 and G^2, respectively.

In [4], no regard is paid to the labels or names assigned to vertices, or to the order in which vertices on the same layer are drawn; i.e. two layered graphs are considered *equal* if they are isomorphic via a level preserving mapping. Under this assumption, the LCP operation is commutative and associative. Thus, we may consider the LCP of more than two layered graphs, all with the same number of layers, without regard to the order in which they are written, or the order in which the binary operations are applied.

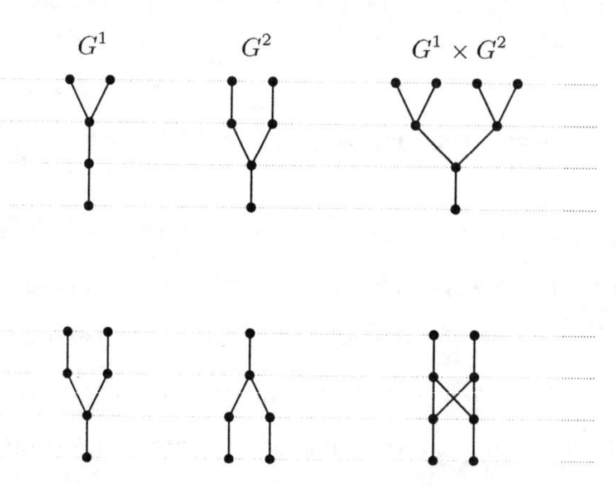

Fig. 1. Two examples of Layered Cross Products

Two examples are depicted in Fig. 1. Some of the results shown in [4] are:

- The Butterfly network is the LCP of two binary trees. The mesh of trees is the LCP of two binary trees with paths attached to their leaves. A particular Fat-tree is the LCP of a binary tree and a quad-tree.
- Several of the important properties of these networks were shown to be trivial consequences, once a network is presented as an LCP of simpler graphs.

2 Creating Layouts via Layered Cross Products

2.1 Objectives

An abstraction of a VLSI layout problem is to model the circuit as a graph, and look for an embedding of the graph in a rectilinear grid, using the following rules.

- Vertices of the graph are mapped to grid-points, at most one vertex per grid-point.
- Edges of the graph are to be routed along grid paths by an edge disjoint mapping. Namely, an edge of the grid may not belong to more than one routing path. Two paths may share an intermediate grid-point, but they must cross at this point.
- If a vertex is mapped to a grid-point, then all paths representing edges incident on this vertex must end at that grid-point, and no other path is allowed to pass through that point.

A layout which obeys these rules is called *rectilinear*. The *area* of a layout is defined to be the smallest area of a rectangle, sides parallel to the grid lines, which contains the entire layout. (The height of a rectangle is measured by the number of horizontal grid-lines in it; similarly for the width. The area is the product of the two.)

Another desirable property, when a grid layout represents a circuit layout, is to have the input/output (I/O) terminals on the boundary of the region of the grid that contains the layout [2, 5]. The terminals in the layouts proposed in this paper are not placed along the boundaries. Note however, that this may not be an important constraint if the terminals denote the processors to be connected, and these are realized on the same chip or printed board.

2.2 The Projection Methodology

In this section we describe a *projection methodology* (PM) for layout construction, which grows from the LCP. Consider the situation illustrated in Fig. 2. The butterfly network is the LCP of two binary trees, one drawn upwards and one drawn downwards. (It is helpful to dedicate a column to each vertex; this will prevent vertices of the layout from colliding.) Draw one tree on the x-y plane (the north wall) and the other on the y-z plane (the west wall). Construct their LCP in three dimensions inside the cube, in such a way that the two trees are the projections of the resulting butterfly on the x-y and y-z planes. The projection of this three-dimensional figure on the floor, depicted as the shadow cast by the butterfly, is a planar layout of the butterfly in area $4n^2$. This layout of the butterfly is symmetric under rotation; however, it is not rectilinear since it uses diagonal lines. Note that the I/O-terminals of the butterfly are on the central north-south and east-west symmetry axes.

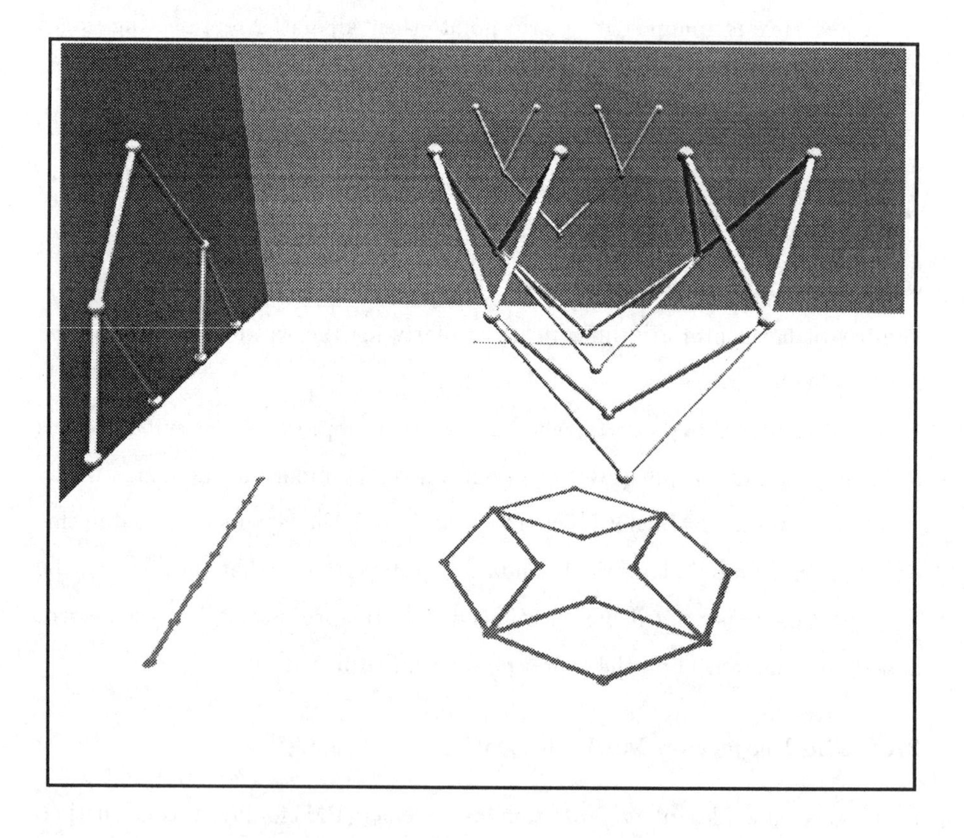

Fig. 2. Butterfly layout with diagonal lines

As the reader can observe in Fig. 3, we use standard techniques of projective geometry to represent the *vertical* projection on the x-y plane, the *horizontal* projection on the y-z plane, and the floor projection. Vertices belonging to layer V_0 of the multiplicand graphs and the product graphs are drawn as circles, whereas vertices belonging to layer V_1 are drawn as diamonds. All edges are drawn as straight line segments.

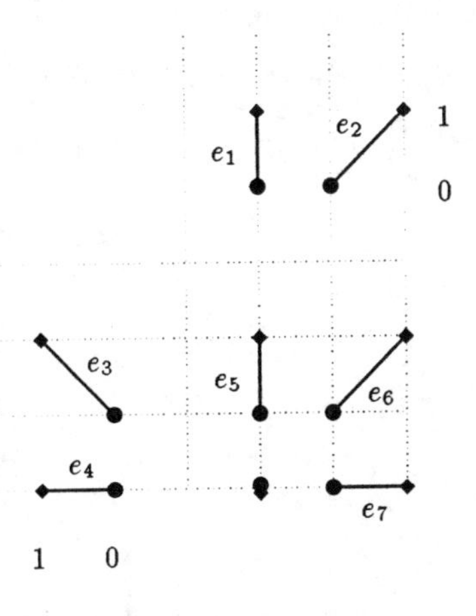

Fig. 3. An example of PM

Figure 3 depicts three types of edges in the product graph: (a) The product of the edges e_3 and e_2 (with their endpoints) yields the edge e_6. Note, that when both multiplied edges are diagonal, then their product edge is also diagonal. (b) The product of the edges e_3 and e_1 yields the edge e_5. Similarly, the product of the edges e_4 and e_2 yields the edge e_7. Note that if exactly one of the edges we multiply is diagonal, then the product edge is rectilinear. (c) The product of the edges e_4 and e_1 yields a product edge. However, the endpoints of this product edge are mapped to the same point in the floor projection, thus violating the layout condition that vertices must be mapped to different grid-vertices.

Figure 3 leads to the following idea for avoiding the diagonal edges in the layout of the butterfly as suggested in Fig. 2. Let us double the number of edge-levels and stretch each edge of the two multiplicands to become a path of two edges, where the primary and secondary vertices alternate. Each such path consists of one diagonal edge and one rectified edge (which is vertical in the vertical projection and horizontal in the horizontal projection). Observe that in the floor projection, the product of two such paths creates a rectilinear path; i.e. only right angles are obtained. Figure 4 depicts a rectilinear embedding of a single product path, and Fig. 5 depicts a rectilinear layout of the butterfly in area $4n^2$. (Smaller rectilinear layouts of the butterfly are known. Dinitz [3] showed that area $\frac{11}{6}n^2$ suffices. Avior, et.al. [1] showed that $n^2 + o(n^2)$ area is necessary and sufficient. However, these layouts are not nearly as esthetic.)

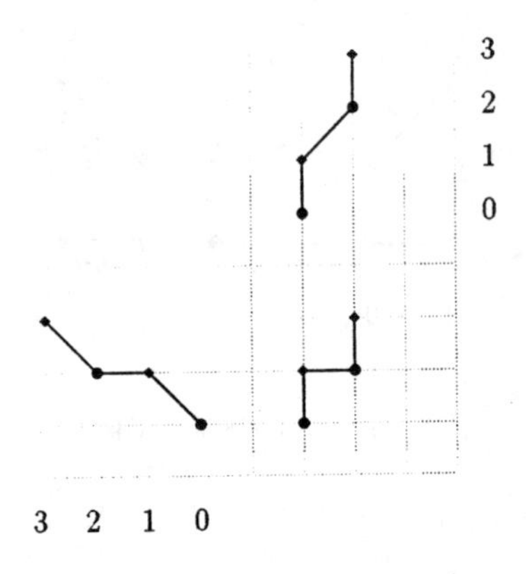

Fig. 4. A rectilinear path obtained by alternating diagonal levels

Let ℓ be an even integer. The binary tree with $2^\ell = n$ leaves is the LCP of two stretched binary trees, each having $\sqrt{n}$ leaves. We can get a rectilinear drawing of the binary tree by a slightly modified PM "stretching" technique. In the vertical projection put a stretched tree, with a vertical extra edge attached to its root, and trim the leaves to be of one edge, instead of the (stretched) two. In the horizontal projection, we use a stretched tree, as described in the previous paragraph. Observe that here too, A diagonal edge is only multiplied by a rectified edge to obtain a rectilinear layout, as depicted in Fig. 6. In fact,

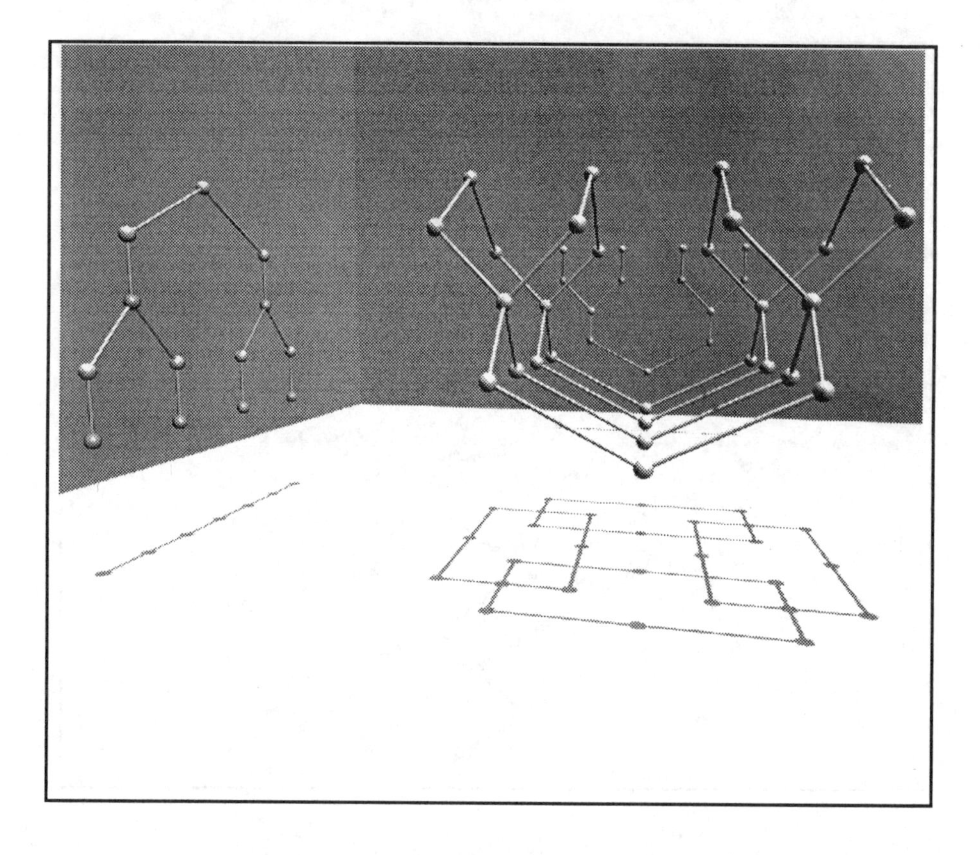

Fig. 5. A rectilinear layout of the Butterfly

the result is the famous H-tree, of area $4n - 4\sqrt{n} + 1$, invented by Shiloach [7].

Finally, let us demonstrate how PM can be used to generate a rectilinear layout of the Mesh-of-Trees, invented by Leighton [6], in area $(n(1+\log n)-1)^2$. This network has n input and n output terminals, and it is nonblocking in the strict sense. i.e. If a pair of input and output terminals are free, then there is a unique path of length $2\log n$ for connecting them, and this path is vertex disjoint from any other path which may be in use. (In fact we realize a slightly modified version of this network, in which n^2 vertices and edges are saved, and the terminal connecting paths are of length $2\log n - 1$.)

Even and Litman [4] showed that the (modified) Mesh-of-Trees is the LCP of two binary, with paths attached to their leaves, one with the root up, and one

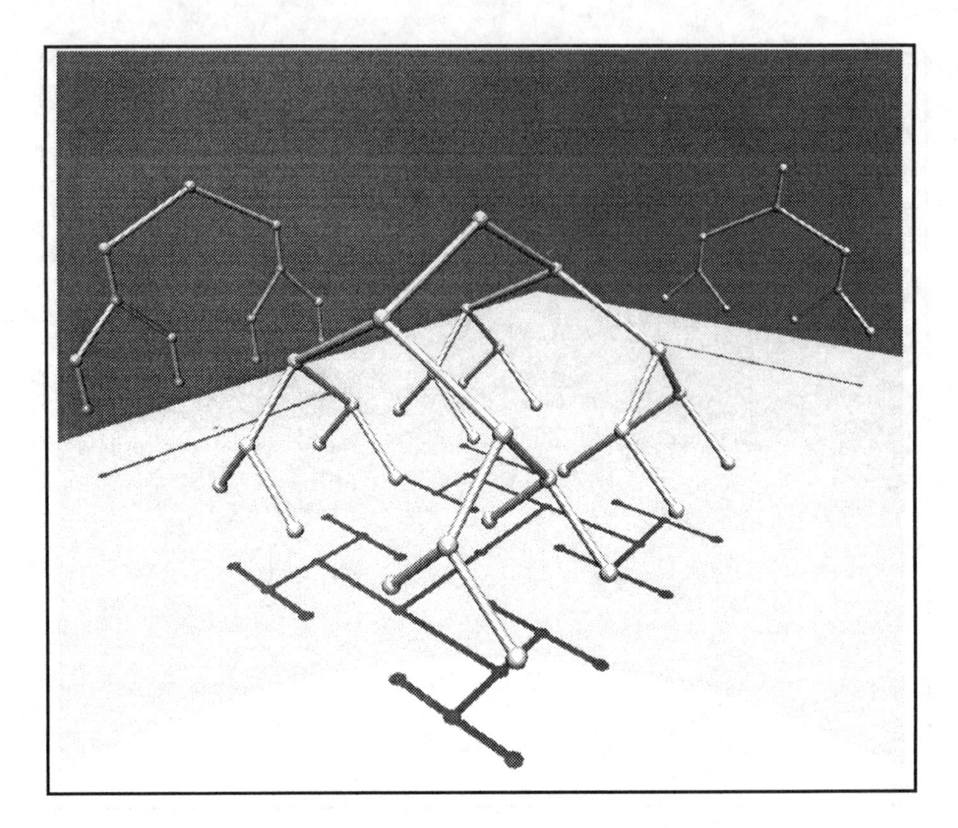

Fig. 6. H-tree : a rectilinear layout of a binary tree

with the root down. For $2^\ell = n$, each tree has n leaves on the ℓ-th level, and to each leaf, a path of length ℓ it attached. Thus, the total number of layers (of vertices) in each tree is $2\ell + 1$. The trees and their LCP, for the case $l = 2$ are shown in Fig. 7.

To make sure that vertices of the floor projection will not coincide, we use a dedicated column (row) for each of the vertices of the vertical (horizontal) projection. Thus, the number of columns (rows) used is $(\ell+1)n-1$, and therefore, the area of the floor projection is $((\log n+1)n-1)^2$. The next step, which makes it rectilinear, does not affect the area.

Now, stretch every edge of the vertical and horizontal projections into two, and alternate the levels in which the edges are diagonal and rectified. The level

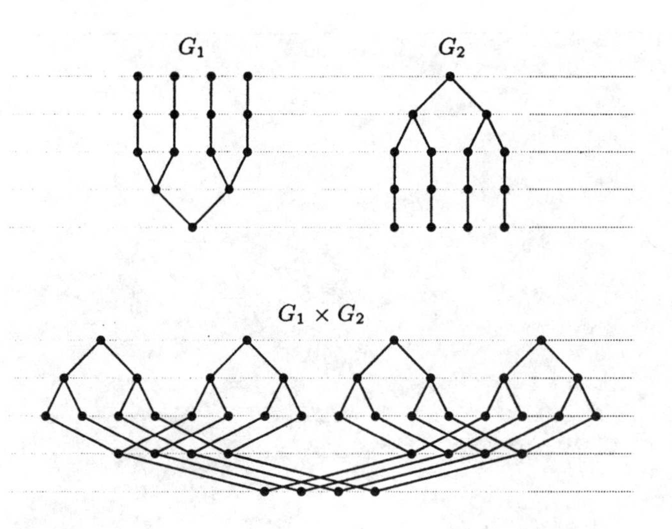

Fig. 7. A Mesh-of-Tress is the LCP of two binary trees with attached paths

of the root and its two neighbors is a diagonal level, the next is rectified, etc. The result for $\ell = 2$ is shown in Fig. 8.

Acknowledgments

We thank Volker Hofmeyer for his invaluable help with producing Figures 2, 5, and 6. We thank Arno Formella for helping produce Fig. 8. Arno's fast ray-tracing program, "Rayo", saved us many hours.

References

1. A. Avior, T. Calamoneri, S. Even, A. Litman and A.L. Rosenberg (1996): "A Tight Layout of the Butterfly Network", *8th Annual ACM Symposium on Parallel Algorithms and Architectures*, 170-175.
2. R.P. Brent and H.T. Kung (1980): "On the Area of Binary Tree Layouts." *Inf. Proc. Let. 11*, 44-46.
3. Ye. Dinitz: "A Compact layout of Butterfly on the Square Grid", Technical Report No. 873, Dept. of Comp. Sci., Technion, 1995, 7p.
4. S. Even and A. Litman (1992): "Layered Cross Product - A Technique to Construct Interconnection Networks", *4th ACM Symp. on Parallel Algorithms and Architectures*, 60-69. To appear in Networks.
5. M.J. Fischer and M.S. Paterson (1980): "Optimal Tree Layout." *12th ACM Symp. on Theory of Computing*, 177-189.

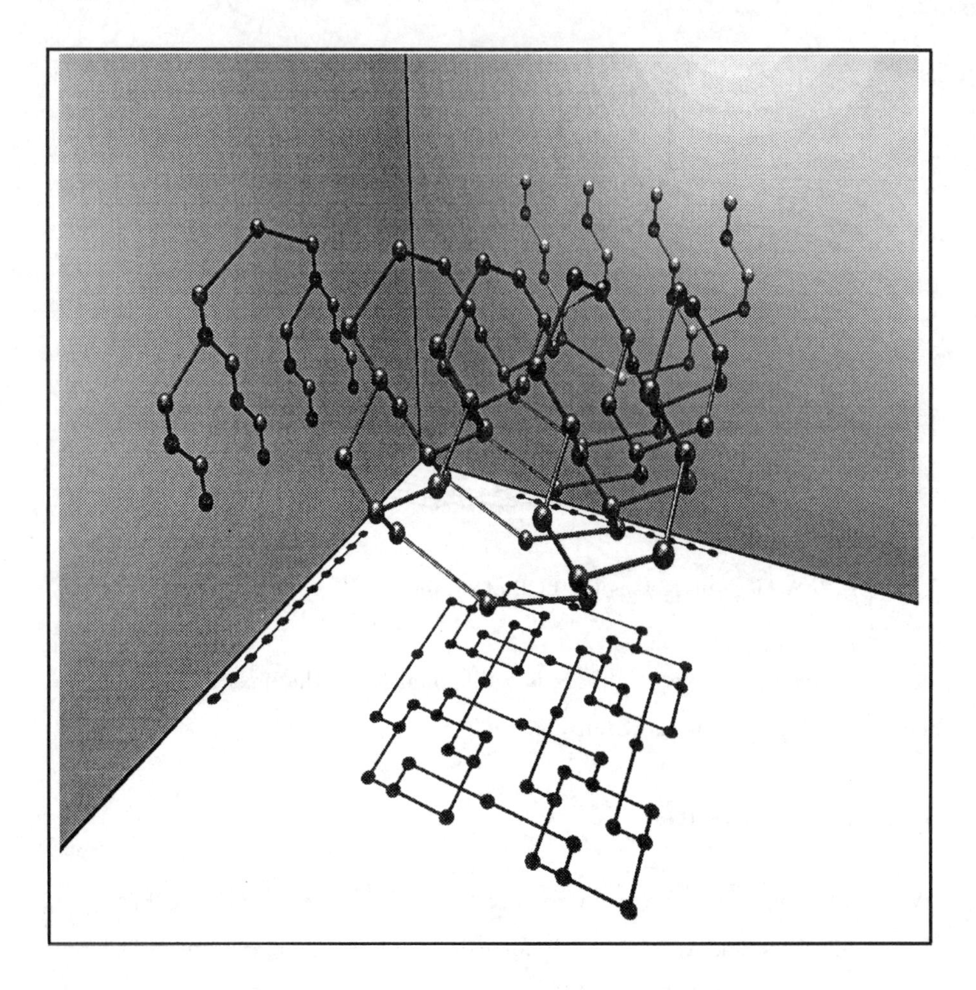

Fig. 8. A rectified layout of the Mesh-of-Trees

6. T. Leighton (1984): "Parallel Computation Using Meshes of Trees", *Proc. WG'83, International Workshop on Graphtheoretic Concepts in Computer Science*, (M. Nagl and J. Perl eds.), Trauner Verlag, 200-218.
7. Y. Shiloach (1976): "Linear and Planar Arrangements of Graphs", Doctoral thesis, Dept. of Appl. Math., The Weizmann Inst. of Sci., Rehovot, Israel.

Finding Optimum k-vertex Connected Spanning Subgraphs: Improved Approximation Algorithms for $k = 3, 4, 5$

Yefim Dinitz[1] * and Zeev Nutov[2]

[1] Dept. of Computer Science, Technion, Haifa 32000, Israel
e-mail: dinitz@cs.technion.ac.il
[2] Dept. of Mathematics, Technion, Haifa 32000, Israel
e-mail: nutov@tx.technion.ac.il

Abstract. The problem of finding a minimum weight k-vertex connected spanning subgraph is considered. For $k > 1$, this problem is known to be NP-hard. Combining properties of inclusion-minimal k-vertex connected graphs and of k-out-connected graphs (i.e., graphs which contain a vertex from which there are k internally vertex-disjoint paths to every other vertex), we derive polynomial time approximation algorithms for several values of k.

(i) For $k = 3$, we give an algorithm with approximation factor 2. This improves the best previously known factor 3.

(ii) For $k = 4$ and $k = 5$, we give an algorithm with approximation factor 3. This improves the best previously known factors $4\frac{1}{6}$ and $4\frac{17}{30}$, respectively.

1 Introduction

Connectivity is a fundamental property of graphs, which has important applications in network reliability analysis and network design problems. Recently, much effort has been devoted to problems of finding minimum cost subgraphs of a given weighted graph (network) that satisfy given connectivity requirements (see [3] for a survey). A particular important class of problems are the uniform connectivity problems, where the aim is to find a cheapest spanning subgraph which remains connected in the presence of up to $k-1$ arbitrary edge or vertex failures (i.e., a minimum cost k-edge- or k-vertex-connected spanning subgraph, respectively). For the practical importance of the problem see, for example, Grötschel, Monma and Stoer [9]. In this paper we consider the vertex version[3] (henceforth we omit the prefix "vertex"), that is, the following problem:

Minimum weight k-connected subgraph problem: given an integer k and a k-connected graph with a nonnegative weight function on its edges, find its minimum weight k-connected spanning subgraph.

---

* Up to 1990, E. A. Dinic, Moscow.

[3] For a survey on results concerning edge-connectivity see, for example, [11].

An equivalent problem is the *minimum weight k-augmentation problem* which is to increase the connectivity of a given graph to k by adding a minimum weight augmenting edge set. The reduction of each of these problems to the other is straightforward (see [3]).

Throughout this paper we assume that $k \geq 2$, since the case $k = 1$ is reduced to the problem of finding a minimum weight spanning tree. For $k \geq 2$, the minimum weight k-connected subgraph problem is known to be NP-hard even when the weights are all the same [8].

A few approximation algorithms are known for solving minimum weight k-connected subgraph problems (see [11] for a survey). An approximation algorithm is called *α-approximation*, or is said to achieve *factor α*, if it is a polynomial time algorithm that produces a solution of weight no more than α times the value of an optimal solution. For an arbitrary k, the best known approximation algorithm is due to Ravi and Williamson [18] which achieves the factor $2H(k)$, where $H(k) = 1+\frac{1}{2}+\ldots+\frac{1}{k}$ is the kth Harmonic number. Note that, for the cases $k = 2, 3, 4, 5$, this algorithm achieves factors $3, 3\frac{2}{3}, 4\frac{1}{6}, 4\frac{17}{30}$, respectively. Khuller and Thurimella [13] simplified Frederickson and Já Já [5] 2-approximation algorithm to increase the connectivity from 1 to 2. Khuller and Raghavachari [12] developed a $(2 + \frac{1}{n})$-approximation algorithm for $k = 2$ (improved to factor 2 in [17]), and, for the case when edge weights satisfy the triangle inequality, a $(2 + \frac{2(k-1)}{n})$-approximation algorithm for an arbitrary k. Penn and Shasha-Krupnik [17] were the first to introduce a 3-approximation algorithm for the case $k = 3$. A simpler and faster 3-approximation algorithm for $k = 3$ was developed by Nutov and Penn [15], where also the factor 2 was achieved for the problem of increasing the connectivity from 2 to 3.

The main results of this paper are a 2-approximation algorithm for the minimum weight 3-connected subgraph problem, and a 3-approximation algorithm for the minimum weight 4- and 5-connected subgraph problems. This improves the best previously known performance guarantees 3 [17, 15], and $4\frac{1}{6}$ and $4\frac{17}{30}$ [18], respectively. This is done by combining certain properties of minimally k-connected graphs, certain techniques from recent approximation algorithms [12, 17, 15], and some new ideas and techniques.

After this paper was written, it became known to us that Parente and Auletta [1] obtained independently results for $k = 3, 4$ similar to those of our work. They suggest the same main approach of considering minimally k-connected graphs. However, we use additional ideas, which lead to simpler analysis and algorithm for $k = 4$, enable us to derive a 3-approximation algorithm for $k = 5$, and provide simplification and reduction of the time complexity of algorithms.

This paper is organized as follows. In Sect. 2 we give some notations and describe known results used in the paper. Sect. 3 studies k-out-connected graphs (i.e., graphs that have a vertex from which there are k internally disjoint paths to any other vertex). In Sect. 4 we use properties of minimally k-connected graphs to derive a 2-approximation algorithm for the minimum weight 3-connected subgraph problem. Sect. 5 presents a deeper analysis of k-out-connected graphs and introduces our 3-approximation algorithm for the minimum weight 4- and 5-connected subgraph problems.

2 Preliminaries and Notations

Let $G = (V, E)$ be an undirected simple graph (i.e., without loops and multiple edges) with vertex set V and edge set E, where $|V| = n$ and $|E| = m$. For any set of edges and vertices $U = E' \cup V'$, we denote by $G \setminus U$ (resp., $G \cup U$) the graph obtained from G by deleting U (resp., adding U), where deletion of a vertex implies also deletion of all the edges incident to it. For $S, T \subset V$ we denote by $E(S, T)$ the set of edges with one end in S and the other end in T. We abuse the notations for singleton sets, e.g., use v for $\{v\}$. For a vertex v of a graph (or digraph) G we denote by $N_G(v)$ the set of neighbors of v in G, and by $\delta_G(v)$ the set of edges incident to v in G; $d_G(v) = |\delta_G(v)|$ denotes the *degree* of v in G. In the case G is understood, we omit the subscript "G" in these notations.

A graph G with a nonnegative weight (cost) function w on its edges is referred to as a *weighted graph* and is denoted by (G, w), or simply by G if w is understood. For a weight function w and $E' \subseteq E$, we use the notation $w(E') = \sum\{w(e) : e \in E'\}$. For a subgraph $G' = (V', E')$ of a weighted graph (G, w), $w(G')$ is defined to be $w(E')$. A subgraph $G' = (V', E')$ is called *spanning* if $V' = V$; in this paper, we use only spanning subgraphs and, thus, sometimes omit the word "spanning". Similar notations are used for digraphs.

A graph G is *connected* if for any two vertices of G there is a path connecting them. A subset $C \subseteq V$ is a *(vertex) cut* of G if $G \setminus C$ is disconnected; if $|C| = k$ then C is called a *k-cut*. A *side* of a cut C is the vertex set of a connected component of $G \setminus C$. Given a cut C and two nonempty disjoint subsets $S, T \subset V \setminus C$, we say that C *separates S from T* if $G \setminus C$ contains no path with one end in S and the other end in T. Clearly, if C is a cut and $\{C, S, T\}$ is a partition of V, then C separates S from T if and only if $E(S, T) = \emptyset$. A cut that separates a vertex s from a vertex t is called an *$\{s, t\}$-cut*. An $\{s, t\}$-cut C is called *$\{s, t\}$-minimum cut* if C has the minimum cardinality among all $\{s, t\}$-cuts. A graph G is *k-connected* if it is a complete graph on $k + 1$ vertices or if it has at least $k + 2$ vertices and contains no l-cut with $l < k$. The *connectivity* of G, denoted by $\kappa(G)$, is defined to be the maximum k for which G is k-connected. In what follows we assume that $n \geq k + 2$; thus $\kappa(G)$ is the cardinality of a minimum cut of G. It is well known that $\kappa(G) \leq \min\{d(v) : v \in V\}$.

A set of paths is said to be *internally disjoint* if no two of them have an internal vertex in common. Following [3], a graph (resp., digraph) which has a vertex r such that there are k internally disjoint paths from r to any other vertex is called *k-out-connected (from r)*. Observe that, by the famous Menger Theorem, a graph is k-out-connected from r if and only if it has no l-cut with $l < k$ separating r from any other vertex. Thus, in a graph which is k-out-connected from r, any l-cut with $l < k$, if exists, must contain r. As another consequence, a graph G is k-connected if and only if it is k-out-connected from every vertex of G. This implies that, for any vertex r of a k-connected weighted graph, the minimum weight of a k-out-connected spanning subgraph from r is less or equal to the minimum weight of a k-connected spanning subgraph.

A graph G is called *minimally k-connected* if $\kappa(G) = k$, but for any $e \in E$, $\kappa(G \setminus e) < k$. Observe that every k-connected graph contains a minimally k-connected spanning subgraph. Thus, among the subgraphs which are optimal solutions for the minimum weight k-connected subgraph problem, there always exists a minimally k-connected one (henceforth, called *minimally k-connected optimal solution*).

The *underlying graph* of a digraph D is the simple graph $U(D)$ obtained from D by replacing, for every $u, v \in V$, the set of arcs with endnodes u, v, if nonempty, by an edge (u, v). The *directed version* of a weighted graph (G, w) is the weighted digraph $D(G) = (D, w_D)$ obtained from (G, w) by replacing every undirected edge of G by the two antiparallel directed edges with the same ends and of the same weight. (For simplicity of notations, we usually use w instead of w_D.)

Frank and Tardos [4] showed that for a directed graph, the problem of finding a minimum weight k-out-connected subdigraph from a given vertex r is solvable in polynomial time; a faster algorithm is due to Gabow [6]. This polynomial solvability was used as a basis for deriving approximation algorithms for several augmentation problems (see, for example, [12, 17, 15]). The main idea behind most of these algorithms is as follows. First, to add a new "external" vertex r and connect it by edges to certain k vertices of the original graph. Then, to find a minimum weight k-out-connected subdigraph from r in the directed version. It is shown in [12] that the underlying graph of thus obtained k-out-connected subdigraph, after deleting r, is $\lceil \frac{k}{2} \rceil$-connected and its weight is at most twice the minimum weight of a k-connected subgraph.[4] For $k = 2$, a slight modification of this technique gives a 2-connected subgraph [12, 17], while for $k = 3$, an additional set of edges is added to make thus obtained subgraph 3-connected [17, 15].

In our algorithms, we show a method to choose such r as a vertex of the original graph. This guarantees that the resulting subgraph is $(\lceil \frac{k}{2} \rceil + 1)$-connected. For small values of k considered in this paper this is a crucial improvement on the $\lceil \frac{k}{2} \rceil$-connectivity guaranteed in [12].

Roughly, our algorithms work as follows. Among all k-out-connected spanning subgraphs from a vertex of degree k (the degree w.r.t. this subgraph), the algorithm finds one of weight at most twice the value of an optimal solution to our problem. For $k = 3$, such a subgraph is 3-connected, and it is the output of the algorithm. In the case $k \in \{4, 5\}$, such a subgraph is only $(k-1)$-connected, and the algorithm finds an additional set of edges to be added to make the resulting graph k-connected.

---

[4] In the case of edge connectivity, the underlying graph of *any* k-edge-out-connected subgraph is k-edge-connected. This observation was used in [13] to derive a fast and simple 2-approximation algorithm for the minimum weight k-*edge*-connected subgraph problem, k arbitrary.

3 Properties of k-out-connected graphs

In this section we study k-out-connected graphs. In particular, we show that if a graph is k-out-connected from a vertex of degree k, then it is $(\lceil \frac{k}{2} \rceil + 1)$-connected.

Our motivation to study k-out-connected graphs is that, in this paper, we choose to approximate a minimum weight k-connected spanning subgraph by a certain k-out-connected spanning subgraph. Observe, however, that an arbitrary k-out-connected graph is not necessarily even 2-connected. Indeed, let us take two complete graphs on at least k vertices each and connect an additional vertex r to some $t \geq k$ vertices in each of these two graphs. The resulting graph is k-out-connected from r, but not 2-connected (since r is a 1-cut). Observe that the degree of r in this example is at least $2k$. One may ask whether lower degree of r guarantees higher connectivity. The following Lemma establishes a lower bound on the connectivity of a k-out-connected graph from r relatively to the degree of r.

Lemma 1. *Let G be a k-out-connected graph from a vertex r, and let C be an l-cut of G with $l < k$. Then $r \in C$, and for every side S of C holds:*

$$l \geq k - |S \cap N(r)| + 1.$$

Thus $\kappa(G) \geq k - \lfloor \frac{d(r)}{2} \rfloor + 1$.

Proof. The fact that r is in C was already established in Sect. 2,

Let now S be a side of C. If $k \leq |S \cap N(r)|$, then the statement is trivial, so assume $k > |S \cap N(r)|$. Let us choose a vertex $v \in S$ and consider a set of k internally disjoint paths between r and v. Since those paths begin with distinct edges, at most $|S \cap N(r)|$ of them may not contain a vertex from $C \setminus r$. This implies that every one of the other at least $k - |S \cap N(r)|$ paths must contain each at least one vertex from $C \setminus r$. Thus $l - 1 \geq k - |S \cap N(r)|$, as required. To see that $\kappa(G) \geq k - \lfloor \frac{d(r)}{2} \rfloor + 1$, observe that every cut of G has a side S for which $|S \cap N(r)| \leq \lfloor \frac{d(r)}{2} \rfloor$. □

The highest connectivity that can be guaranteed by Lemma 1 for a k-out-connected graph from r corresponds to the lowest possible degree of r, which is k.[5] Hence, in this paper, we are interested in k-out-connected subgraphs from vertices of degree k. For such graphs, Lemma 1 implies the following statement.

Corollary 2. *Let G be a k-out-connected graph from a vertex r of degree k. Then G is $(\lceil \frac{k}{2} \rceil + 1)$-connected, and every side of an l-cut of G with $l < k$ contains at least $(k - l) + 1 \geq 2$ neighbors of r.*

[5] In fact, the bounds in Lemma 1 are tight in the following sense. For every $k \geq 2$ and $k \leq d \leq 2k$, there exists a graph which is k-out-connected from its vertex r of degree d and has connectivity exactly $k - \lfloor \frac{d}{2} \rfloor + 1$. Such a graph can be obtained by a generalization of the construction given in the beginning of this section, as follows: we identify $k - \lfloor \frac{d}{2} \rfloor$ vertices of the two complete graphs and connect the additional vertex r to one common vertex (if d is odd) and to at least $\lfloor \frac{d}{2} \rfloor$ noncommon vertices of each complete graph.

Proof. By Lemma 1, $\kappa(G) \geq k - \lfloor\frac{k}{2}\rfloor + 1 = \lceil\frac{k}{2}\rceil + 1$. Let S be a side of an l-cut of G, $l < k$. Then, by Lemma 1, $|S \cap N_G(r)| \geq k - l + 1 \geq 2$. □

For the cases $k = 3, 4, 5$, considered in this paper, we deduce:

Corollary 3. *Let G be a k-out connected graph from a vertex r of degree k.*

(i) *If $k = 3$ then G is 3-connected.*
(ii) *If $k \in \{4, 5\}$ then G is $(k-1)$-connected. Moreover, for every $(k-1)$-cut C of G holds:*
  - *C has exactly two sides;*
  - *If $k = 4$ then $|C \cap N(r)| = 0$ and each side of C contains exactly 2 vertices from $N(r)$;*
  - *If $k = 5$ then $|C \cap N(r)| \leq 1$ and each side of C contains 2 or 3 vertices from $N(r)$; thus at least one side contains exactly two vertices from $N(r)$.*

4 Minimally k-connected graphs and the minimum weight 3-subgraph problem

In this section we show a 2-approximation algorithm for finding a k-out-connected subgraph from a vertex of degree k (the degree is w.r.t. this subgraph). Combining this with Corollary 3(i), we derive a 2-approximation algorithm for the minimum weight 3-connected subgraph problem.

Our first aim is to establish that among optimal solutions to the minimum weight k-connected subgraph problem there always exists one which has a vertex of degree k (recall that its k-connectivity implies that it is k-out-connected from that vertex). This is straightforward by combining existence of a minimally k-connected optimal solution and the following theorem of Halin [10] (see also [2]).

Theorem 4 [10]. *Any minimally k-connected graph has a vertex of degree k.*

Remark. Let us define a *minimally k-out-connected graph* as a k-out-connected graph G such that, for every its edge e, $G \setminus e$ is not k-out-connected. Then, combining Theorem 4 with Corollary 3, we obtain an interesting characterization of minimally 3-connected graphs: *A graph is minimally 3-connected if and only if it is minimally 3-out-connected from a vertex of degree 3.*

Let us denote by w^* the weight of an optimal k-connected subgraph and by $\tilde{w}^*$ that of an optimal k-out-connected subgraph from a vertex of degree k. By Theorem 4, $\tilde{w}^* \leq w^*$. We now present an algorithm which finds a k-out-connected subgraph from a vertex of degree k with weight at most $2\tilde{w}^*$.

Out-Connected Subgraph Algorithm (OCSA)

Input: A weighted graph (G, w), $G = (V, E)$, and an integer k.
Output: A k-connected spanning subgraph $\tilde{G}$ of G and a vertex $\tilde{r}$.

Construct $D = D(G)$;
Set $\tilde{G}, \tilde{r}$ undefined, $\tilde{w} = \infty$, $M = 2w(G) + 1$;
For every vertex $r \in V$ *do*:
(1) Set $w_r(e) = w(e) + M$ if $e \in \delta(r)$ and $w_r(e) = w(e)$ otherwise;
(2) Find a minimum weight k-out-connected from r subdigraph $\tilde{D}_r$ of (D, w_r), if such exists, by the algorithm [6];
(3) If the degree of r in $U(\tilde{D}_r)$ is k and if $w(U(\tilde{D}_r)) < \tilde{w}$, then set $\tilde{G} = U(\tilde{D}_r)$, $\tilde{r} = r$, and $\tilde{w} = w(U(\tilde{D}_r))$;
end for
If $\tilde{w} < \infty$ then output $\tilde{G}$, $\tilde{r}$;
If $\tilde{w} = \infty$ then declare "G has no k-out-connected spanning subgraph from a vertex of degree k";

Lemma 5. *The algorithm OCSA outputs a k-out-connected spanning subgraph from a vertex of degree k, if there exists any such subgraph, and its weight is at most $2\tilde{w}^*$. The complexity of the algorithm is $O(k^2n^3m)$.*

Proof. In this proof we use the following simple observation:

Fact 6. *A graph G is k-out-connected from a vertex r if and only if $D(G)$ is k-out-connected from r.*

First, observe that, by Fact 6, any subgraph $\tilde{G}$ chosen at step (3) is a k-out-connected spanning subgraph from a vertex $\tilde{r}$ of degree k. Therefore, (i) the output is always of this type and (ii) if G has no such subgraphs, the algorithm always declares on this.

Second, let G have subgraphs of this type, and let G' be anyone of them of the minimum weight $\tilde{w}^*$, say, from a vertex r'. Let D' denote the subdigraph $D(G')$ with all edges entering r' removed. By Fact 6, D' is k-out-connected from r' (the removal of entering edges cannot affect this property); its total degree at r' is k. Therefore, the subgraph $\tilde{D}_{r'}$ is well defined.

We claim that the total degree of r' in $\tilde{D}_{r'}$ and, thus, its degree in $U(\tilde{D}_{r'})$ are exactly k. If this is not so, then this degree is at least $k+1$ (since $\tilde{D}_{r'}$ is k-out-connected from r'). However, this implies that $w_{r'}(\tilde{D}_{r'}) \geq (k+1)M + w(\tilde{D}_{r'}) > kM + w(D') = w_{r'}(D')$, a contradiction to optimality of $\tilde{D}_{r'}$. Thus, our claim is valid, which implies that the output of OCSA is well defined.

To see the approximation ratio, observe that $w_{r'}(\tilde{D}_{r'}) \leq w_{r'}(D')$ implies $w(\tilde{D}_{r'}) \leq w(D') \leq 2\tilde{w}^*$. Since, clearly, $w(\tilde{G}) \leq w(U(\tilde{D}_{r'})) \leq w(\tilde{D}_{r'})$, the output $\tilde{G}$ is a 2-approximation required.

Let us analyze the time complexity. The dominating time is spent for finding n subdigraphs $\tilde{D}_r$. The time complexity of the algorithm [6] is $O(k^2n^2m)$, and the bound follows. □

Combining Theorem 4, Corollary 3(i), and Lemma 5, we deduce:

Theorem 7. *For $k = 3$, OCSA is a 2-approximation algorithm for the minimum weight 3-connected subgraph problem. The time complexity of the algorithm is $O(n^3m)$.*

Now we suggest a randomized version of OCSA (of Monte-Carlo type) with the complexity $O(k^2n^2m\log_2 n)$. Let us consider a modification of OCSA (henceforth called ROCSA) in which the input graph is k-connected and the sequence of examined vertices r is chosen at random. Let G^* be a minimally k-connected optimal solution and r^* be any its vertex of degree k. It can be shown similarly to the proof of Lemma 5 that the degree of r^* in $U(\tilde{D}_{r^*})$ is exactly k and $w(U(\tilde{D}_{r^*}))$ at most $2w^*$. Mader [14] shows that, for $k \geq 2$, any minimally k-connected graph has at least $\frac{(k-1)n+2}{2k-1} > \frac{1}{3}n$ vertices of degree k. This implies that with probability $1 - \frac{1}{n}$ at least one of the first $\lceil \log_{\frac{3}{2}} n \rceil$ vertices examined in ROCSA has degree k in G^*. Hence, with probability $1 - \frac{1}{n}$, ROCSA finds a k-out-connected subgraph from a vertex of degree k and of the weight at most $2w^*$ after examining $\lceil \log_{\frac{3}{2}} n \rceil$ vertices of G. Thus, the complexity of ROCSA is as required.

5 Minimum Weight 4- and 5-Connected Subgraph Problems

In this section we present a 3-approximation algorithm for the minimum weight 4- and 5-connected subgraph problems. The idea is to execute Out-Connected Subgraph Algorithm as a first phase, which produces, for $k \in \{4, 5\}$, a $(k-1)$-connected graph, and then to augment this graph by an appropriate set of edges. Such a set of edges is produced as follows. We show that, for $k \in \{4, 5\}$, the subgraph produced by the OCSA contains two vertices, such that every $(k-1)$-cut separates one from the other. In the second phase of our algorithm, we destroy all those cuts by adding the edge set of a minimum weight k internally disjoint paths between these two vertices; such a set of paths is found by using a minimum cost k-flow algorithm. (The idea of such a destroying all small cuts separating two fixed vertices was used before by Nutov and Penn [15] in their 3-augmentation algorithm.)

Existence of two vertices as above is showed based on the following two Lemmas.

Lemma 8. *Let C, C' be two $\{s,t\}$-minimum cuts of a graph G such that each of C and C' has exactly two sides, say S, T and S', T', respectively, where $s \in S, S'$ and $t \in T, T'$. Then $R_s = (C \cup C') \setminus (T \cup T')$ separates $S \cap S'$ from $T \cup T'$, $R_t = (C \cup C') \setminus (S \cup S')$ separates $T \cap T'$ from $S \cup S'$, and both R_s, R_t are $\{s,t\}$-minimum cuts as well.*

Proof. Observe that each of $\{R_s, S \cap S', T \cup T'\}$ and $\{R_t, S \cup S', T \cap T'\}$ is a partition of V. Also, $E(S \cap S', T \cup T') = E(S \cup S', T \cap T') = \emptyset$, since $E(S,T) = E(S',T') = \emptyset$. This implies that R_s separates $S \cap S'$ from $T \cup T'$, and R_t separates $T \cap T'$ from $S \cup S'$ (for illustration see Fig. 1a). In particular, we obtain that R_s and R_t are both $\{s,t\}$-cuts. To see that R_s and R_t are both $\{s,t\}$-minimum cuts, observe that $|R_s| + |R_t| = |C| + |C'|$ and that $|R_s|, |R_t| \geq |C| = |C'|$. □

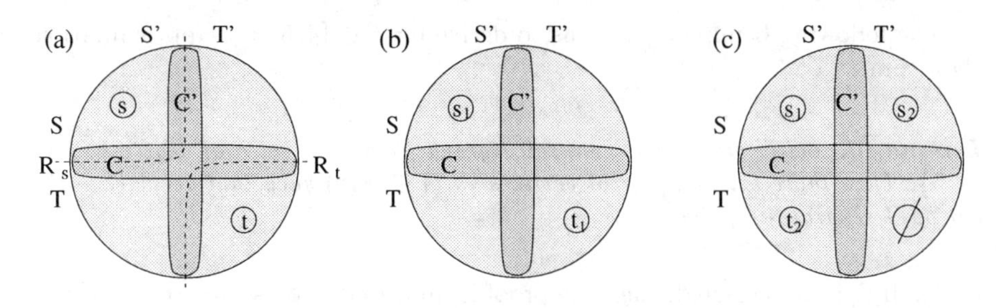

Fig. 1. Illustrations for the proofs of Lemmas 5.1 and 5.2.

Lemma 9. *Let G be a k-out-connected graph from a vertex r of degree k, $k \in \{4,5\}$. Let C be a $(k-1)$-cut of G and S a side of C that contains exactly two vertices from $N(r)$, say s_1 and s_2. Then, for any $(k-1)$-cut of G, both s_1 and s_2 are contained in the same its side.*

Proof. Recall that, by Corollary 3(ii), G is $(k-1)$-connected, and every $(k-1)$-cut of G contains at most one vertex from $N(r)$ and has exactly two sides, containing 2 or 3 vertices from $N(r)$ each. Let T be the side of C that is distinct from S. Assume in negation that there is a $(k-1)$-cut C' with sides S' and T' such that s_1 and s_2 are not contained in the same side of C'. Since C' contains at most one of s_1, s_2, we assume w.l.o.g. that $s_1 \in S'$, $s_2 \notin S'$. Let us consider two cases:

$T \cap T' \neq \emptyset$: Let $t_1 \in T \cap T'$ (see Fig. 1b). Then, clearly, C and C' are both $\{s_1, t_1\}$-minimum cuts. Thus, by Lemma 8, the cut $R_{s_1} = (C \cup C') \setminus (T \cup T')$ is a $(k-1)$-cut with sides $S \cap S', T \cup T'$. But $(S \cap S') \cap N_r = \{s_1\}$, a contradiction to Corollary 3(ii).

$T \cap T' = \emptyset$: In this case $s_2 \in S \cap T'$, since $|T' \cap N(r)| \geq 2$, $|C \cap N(r)| \leq 1$, and $(S \setminus S') \cap N(r) = s_2$. Moreover, $T \cap S' \neq \emptyset$, since $|T \cap N(r)| \geq 2$ and $|C' \cap N(r)| \leq 1$. Let $t_2 \in T \cap S'$ (see Fig. 1c). Then the contradiction is obtained by using similar arguments for s_2, t_2 instead of s_1, t_1.

□

Remark. Observe that Lemma 9, together with Corollary 3(ii), imply that a k-out-connected graph from a vertex r of degree k, $k \in \{4,5\}$, has a vertex s neighboring to r which does not belong to any cut smaller than k. Such a vertex s of $\tilde{G}$ can be easily recognized as a neighbor of $\tilde{r}$ in $\tilde{G}$ for which $\kappa(\tilde{G} \setminus s) \geq k-1$. This observation, together with Lemma 1, already imply a 4-approximation algorithm for the minimum 4- and 5-connected subgraph problems, as follows. Let $\tilde{G}, \tilde{r}$ be the output of the OCSA for the input (G, w) and k, and let, for s as above, $\tilde{G}_s$ be the underlying graph of the minimum weight k-out-connected from s subdigraph of $D(G)$. Then, for $k \in \{4,5\}$, the graph $\tilde{G} \cup \tilde{G}_s$ is k-connected. Similarly to the proof of Lemma 5, $w(\tilde{G}_s) \leq 2w^*$. Thus,

$$w(\tilde{G} \cup \tilde{G}_s) \leq w(\tilde{G}) + w(\tilde{G}_s) \leq 2w^* + 2w^* = 4w^*.$$

The following Lemma enables us to derive for $k \in \{4,5\}$ a 3-approximation algorithm.

Lemma 10. *Let G be a k-out-connected graph from a vertex r of degree k, $k \in \{4,5\}$. Then there exists a pair of vertices $\{s,t\} \subset N(r)$ such that every $(k-1)$-cut of G separates s from t.*

Proof. If G is k-connected, then the proof is immediate, so assume that $\kappa(G) = k-1$. Let us assume first that there is a $(k-1)$-cut C which has exactly two vertices from $N(r)$ in each its side, say s_1, s_2 and t_1, t_2, respectively. Then, by Lemma 9, any $(k-1)$-cut separates s from t for any $s \in \{s_1, s_2\}$ and $t \in \{t_1, t_2\}$. Otherwise, $k = 5$ and every $(k-1)$-cut has exactly two vertices from $N(r)$ in its one side and three vertices from $N(r)$ in its other side. Let C be a $(k-1)$-cut, and let $N(r)$ be partitioned into $\{s_1, s_2\}$ and $\{t_1, t_2, t_3\}$ by the two sides of C. Suppose now that there is a $(k-1)$-cut C' which does not separate $\{s_1, s_2\}$ from $\{t_1, t_2, t_3\}$. Then Lemma 9 implies that C' has s_1, s_2 and one of t_1, t_2, t_3, say, t_3, in its one side, and t_1, t_2 in its other side. But then, by Lemma 9, every $(k-1)$-cut must have s_1, s_2 in its one side and t_1, t_2 in its other side. In each of the cases, any pair $\{s_i, t_j\}$, $i, j \in \{1,2\}$, is appropriate. □

Let a subgraph G' contain a pair of vertices $\{s,t\}$ such that every its l-cut with $l < k$ has exactly two sides and separates s from t. Let F be the edge set of k internally disjoint paths between s and t. Then, clearly, $G' \cup F$ is k-connected. Following [3], the problem of finding a minimum weight set of k internally vertex-disjoint paths between two vertices is easily reduced to the problem of finding a minimum cost flow of value k between them (the vertex version).

We now present a 3-approximation algorithm for the minimum weight 4- and 5-connected subgraph problems. In this algorithm, let $OCSA(G,w,k)$ denote the output graph $\tilde{G}$ of OCSA for the input (G,w) and k, and let $\tilde{r}$ be the corresponding vertex of degree k from which $\tilde{G}$ is k-out-connected.

4 and 5-Connected Subgraph Algorithm

Input: A k-connected weighted graph (G,w) and an integer k, $k \in \{4,5\}$.
Output: A k-connected spanning subgraph of G.

1: Calculate $\tilde{G} = OCSA(G,w,k)$.
2: *For every* pair of vertices $\{s,t\} \subset N_{\tilde{G}}(\tilde{r})$ do:
  1) Using the minimum cost k-flow algorithm [16], find in G a minimum weight edge set $F(s,t)$ of k internally disjoint paths between s and t;
  2) Check whether $\tilde{G} \cup F(s,t)$ is k-connected by algorithm [7];
  If it is, then terminate with $\tilde{G} \cup F(s,t)$.
end for

Remarks. 1) In fact, at phase 2 of the algorithm it is sufficient to examine any two vertices $s_1, s_2 \in N_{\tilde{G}}(\tilde{r})$ for s, and any two other vertices $t_1, t_2 \in_{\tilde{G}} (\tilde{r})$ for t (see the proof of Lemma 10).

2) Note that one can probably get better solutions if after phase 1 the weight function is modified to be zero for all edges in $\tilde{G}$.

Theorem 11. *The 4- and 5-Connected Subgraph Algorithm is a 3-approximation algorithm for the minimum weight k-connected subgraph problems, $k \in \{4,5\}$. The time complexity of the algorithm is $O(n^3 m)$.*

Proof. The correctness of the algorithm follows from Corollary 3(ii), Lemma 10, and the discussion after the latter.

We turn to show the approximation ratio. Let $k \in \{4,5\}$ and let G^* be a minimum weight k-connected subgraph of an input graph G. At phase 2 the algorithm chooses a minimum weight set $F(s,t)$ of k internally disjoint paths between certain two vertices s and t. By Menger Theorem, G^* contains a set $F^*(s,t)$ of k internally disjoint paths between s and t. Clearly, $w(F(s,t)) \leq w(F^*(s,t)) \leq w(G^*) = w^*$. By Lemma 5, $w(\tilde{G}) \leq 2w^*$. Thus we have:

$$w(\tilde{G} \cup F(s,t)) \leq w(\tilde{G}) + w(F(s,t)) \leq 2w^* + w^* = 3w^*.$$

Let us discuss the time complexity. At phase 2 the algorithm finds a minimum cost k-flow and checks k-connectivity $O(k^2) = O(1)$ times. Orlin's [16] minimum cost k-flow algorithm takes $O(m \log n(m + n \log n))$ time, and k-connectivity is checked in $O(k^2 n^2 + k^3 n^{1.5})$ time by [7]. Thus, the dominating time is spent for executing phase 1, and the complexity of the algorithm is the same as the complexity of the OCSA given in Lemma 5. □

Remark. Observe that in this case, as in the case $k = 3$, we can use at phase 1 ROCSA for finding $\tilde{G}$, reducing the time complexity to $O(n^2 m \log_2 n)$.

Acknowledgment. The authors are very grateful to J. Cheriyan for drawing their attention to a method that enabled them to improve the time complexity of the algorithms.

References

1. V. Auletta, D. Parente: Better Algorithms for Minimum Weight Vertex-Connectivity Problems. Tech. Rep. Universita' di Salerno 6/96 (July 1996) (accepted to STACS'97)
2. B. Bollobás: Extremal graph theory, Chapter I. Academic Press, London (1978)
3. A. Frank: Connectivity augmentation problems in network design. *Mathematical Programming*, State of the Art, Ed. J. R. Birge and K. G. Murty (1994) 34–63
4. A. Frank and E. Tardos: An application of submodular flows. *Linear Algebra and its Application*, **114/115** (1989) 329–348
5. G. N. Frederickson and J. Já Já: Approximation algorithms for several graph augmentation problems. *SIAM J. Comput.*, **10(2)** (1981) 270–283

6. H. N. Gabow: A representation for crossing set families with application to submodular flow problems. *Proc. 4th Annual ACM-SIAM Symp. on Discrete Algorithms* (1993) 202–211
7. Z. Galil: Finding the vertex connectivity of graphs. *SIAM J. of computing* **9** (1980) 197–199
8. M. R. Garey and D. S. Johnson: Computers and intractability: A guide to the theory of NP-completeness. Freeman, San Francisco (1979)
9. M. Grötschel, C. Monma and M. Stoer: Design of survivable networks. Handbook in Operation Research and Management Science Vol. 7, Network Models, ed. by M. O. Ball, T. L. Magnanti, C. L. Monma and G. L. Nemhauser (1993)
10. R. Halin: A Theorem on n-connected graphs. *J. Combinatorial Theory* **7** (1969) 150–154
11. S. Khuller: Approximation algorithms for finding highly connected subgraphs. CS-TR-3398, University of Maryland (Feb. 1995)
12. S. Khuller and B. Raghavachari: Improved approximation algorithms for uniform connectivity problems. *Proc. 27th Annual ACM Symposium on the Theory of Computing* (1995) 1-10
13. S. Khuller and R. Thurimella: Approximation algorithms for graph augmentation. *Journal of Algorithms* **14** (1993) 214–225
14. W. Mader: Ecken vom Grad n in minimalen n-fach zusammenhängenden Graphen. *Archiv. Math. (Basel)* **23** (1972) 219–224
15. Z. Nutov and M. Penn: Improved approximation algorithms for weighted triconnectivity augmentation problems. TR-96-IEM/OR-1, Technion, Haifa, Israel (Feb. 1996)
16. J. B. Orlin: A faster strongly polynomial minimum cost flow algorithm. *Operations Research* **41** (1993) 338–350
17. M. Penn and H. Shasha-Krupnik: Improved approximation Algorithms for Weighted 2 & 3 Vertex Connectivity Augmentation Problems. to appear in *Journal of Algorithms*
18. R. Ravi and D. P. Williamson: An Approximation Algorithm for Minimum-Cost Vertex-Connectivity Problems. *Proc. 6th. Annual ACM-SIAM Symposium on Discrete Algorithms*, San Francisco, CA (1995) 332–341

The Optimum Cost Chromatic Partition Problem

Klaus Jansen[1]

Fachbereich IV - Mathematik, Universität Trier, 54 286 Trier, Germany, email: jansen@dm3.uni-trier.de

Abstract. In this paper, we study the optimum cost chromatic partition (OCCP) problem for several graph classes. The OCCP problem is the problem of coloring the vertices of a graph such that adjacent vertices get different colors and that the total coloring costs are minimum. First, we prove that the OCCP problem graphs with constant treewidth k can be solved in $O(|V| \cdot (\log |V|)^{k+1})$ time, respectively. Next, we study an ILP formulation of the OCCP problem given by Sen et al. [9]. We show that the corresponding polyhedron contains only integral 0/1 extrema if and only if the graph G is a diamond - free chordal graph. Furthermore, we prove that the OCCP problem is NP-complete for bipartite graphs. Finally, we show that the precoloring extension and the OCCP problem are NP-complete for permutation graphs.

1 Introduction

In this paper, we study the optimum cost chromatic partition (OCCP) problem for several graph classes. The OCCP problem can be described as follows: Given a graph $G = (V, E)$ with n vertices and a sequence of coloring costs $(k_1, \ldots, k_n)$, find a feasible coloring $f(v)$ for each vertex $v \in V$ such that the total coloring costs $\sum_{v \in V} k_{f(v)}$ are minimum. A coloring $f : V \rightarrow \{1, \ldots, n\}$ is feasible if adjacent vertices have different colors. Alternatively, the OCCP problem can be formulated as follows: Given a graph $G = (V, E)$ with n vertices and a sequence of coloring costs $(k_1, \ldots, k_n)$, find a partition into independent sets $U_1, \ldots, U_s$ such that $\sum_{c=1}^{s} k_c \cdot |U_c|$ is minimum. We may assume that $k_c \leq k_d$ whenever $c < d$.

The OCCP problem restricted to circle and permutation graphs (introduced by Supowit [10]) corresponds to a VLSI layout problem (see also [9]). Another application is given by Kroon et al. [7]. The OCCP problem for interval graphs is equivalent to the Fixed Interval Scheduling Problem (FISP) with machine dependent processing costs. In this scheduling problem each job $j \in J$ must be executed during a given time interval (s_j, f_j). We assume that a sufficient number of machines is available and that each job must be executed by one of the machines. If job j is executed by machine c, then the associated processing costs are k_c. The objective is to find a feasible schedule for all jobs with minimum total processing costs.

A generalization of the OCCP problem contains a graph $G = (V, E)$ and a $(n \times m)$ - cost-matrix $(k_{i,c})$ with unrelated costs $k_{i,c}$ to execute job i on machine c. The general optimum cost chromatic partition problem (GOCCP) can be formulated as follows: Find a partition of the graph G into independent sets $U_1, \ldots, U_s$ such that $\sum_{c=1}^{s} \sum_{i \in U_c} k_{i,c}$ is minimum.

It is not difficult to see that the OCCP problem is NP-complete for arbitrary graphs. Sen et al. [9] proved that the OCCP problem for circle graphs is NP-complete. Moreover, they considered an integer linear program formulation of the OCCP problem. The polytope corresponding to the constraints of this ILP contains only integral (0/1) vertices, if the cartesian product $G \times K_{|V|}$ is a perfect graph. Sen et al. [9] proved that $G \times K_{|V|}$ is perfect if G is a tree and that there exists a perfect graph G such that $G \times K_{|V|}$ is not perfect.

Kroon et al. [7] studied the OCCP problem for interval graphs and trees. They showed that the problem restricted to trees can be solved in linear time and that the problem restricted to interval graphs is NP-complete even if there are only four different values for the coloring costs. For interval graphs G, they proved that the zero-one matrix corresponding to the constraints of the ILP is perfect, if and only if $G \times K_{|V|}$ does not contain an odd cycle of size 7 or more as induced subgraph.

In this paper, we study the complexity of the OCCP problem for several graph classes. In Section 2 we propose an algorithm for graphs with constant treewidth k that runs in $O(|V|(\log |V|)^{k+1})$ time. Furthermore, we prove that an optimum solution of the OCCP problem restricted to graphs with constant treewidth needs at most $O(\log |V|)$ different colors. Moreover, we give an example that there exists a tree and a sequence of coloring costs such that each optimum solution needs $\Omega(\log |V|)$ different colors. The GOCCP problem can be solved in $O(m^{k+1}|V|)$ time for graphs with constant treewidth k.

In Section 3, we study the integer program formulation given by Sen et al. [9]. We show that the cartesian product $G \times K_m$ for $m \geq 3$ is perfect, if and only if G is a diamond - free chordal graph (a generalization of forests). Next, we show that these graphs have a special tree decomposition. Furthermore, we prove that the polyhedron corresponding to the constraints of the ILP contains only integral extrema, if and only if G is a diamond - free chordal graph. For diamond - free chordal graphs, the GOCCP problem can be solved in polynomial time using minimum weighted matchings in bipartite graphs.

In Section 4, we prove that the OCCP problem for bipartite graphs is NP-complete. For the reduction we use the 1 - precoloring extension problem (for short: 1-PrExt) that is NP-complete for bipartite graphs [3]. Finally, in Section 5, we show that the OCCP and the 1 - precoloring extension problem are NP-complete for permutation graphs. The first result solves an open question in [9] and the second an open question in [5].

In the full paper, we show that the OCCP problem for cographs can be solved in linear time and that the GOCCP problem for cographs is NP-complete. We note that the OCCP problem for complements of bipartite graphs or triangle free graphs can be solved in polynomial time. Furthermore, the GOCCP problem is NP-complete even for an union of two complete graphs.

2 Graphs with bounded treewidth

The notion of the treewidth of a graph was introduced by Robertson and Seymour [8], and is equivalent to several other interesting graph theoretic notions, for instance the notion of partial k-trees (see e.g., [1]).

Definition 1. A tree-decomposition of a graph $G = (V, E)$ is a pair $(\{X_i \mid i \in I\}, T = (I, F))$, where $\{X_i \mid i \in I\}$ is a collection of subsets of V, and $T = (I, F)$ is a tree, such that the following conditions hold:

1. $\bigcup_{i \in I} X_i = V$.
2. For all edges $(v, w) \in E$, there exists a node $i \in I$, with $v, w \in X_i$.
3. For every vertex $v \in V$, the subgraph of T, induced by the nodes $\{i \in I \mid v \in X_i\}$ is connected.

The treewidth of a tree-decomposition $(\{X_i \mid i \in I\}, T = (I, F))$ is $\max_{i \in I} |X_i| - 1$. The treewidth of a graph is the minimum treewidth over all possible tree-decompositions of the graph.

Note that a tree-decomposition of G with treewidth $\leq k$ can be found, if it exists, in $O(n)$ time [2]. Since each chordal graph has at most $|V|$ maximal cliques, we may assume that the decomposition tree has at most $|V|$ nodes.

First, we consider an optimum solution $(U_1, \ldots, U_m)$ of the OCCP problem with cost vector $(k_1, \ldots, k_n)$. Since $k_1 \leq \ldots \leq k_n$, we may assume that U_i is a maximal independent set in $G[V \setminus (U_1 \cup \ldots \cup U_{i-1})]$. Let $d(v)$ be the degree of vertex $v \in V$. Using a similar argument, we can generalize a result of Kroon et al. [7] for trees.

Lemma 2. *For a graph $G = (V, E)$ there exists a coloring f of minimum costs such that $f(v) \leq d(v) + 1$ for all vertices $v \in V$.*

Next, we prove that the length of the cost vector can be bounded by $O(\log |V|)$ for graphs with constant treewidth.

Lemma 3. *For a graph $G = (V, E)$ with constant treewidth, there exists a coloring f with minimum costs such that the color $f(v) \leq O(\log |V|)$ for each vertex $v \in V$.*

Proof. For a graph with constant treewidth k, there exists a tree decomposition T with at most $|V|$ nodes. The number of edges in G is bounded by $k|V|$, since every graph with treewidth $\leq k$ is a subgraph of a k-tree and a k-tree has maximal $k|V|$ edges [6]. If we choose one maximal independent set U, then the degree $d(v)$ of each remaining vertex v is decreased by at least one (otherwise U is not maximal).

Next, we prove the following assertion: If we choose $2k\ell$ maximal independent sets $U_1, \ldots, U_{2k\ell}$, then the number of edges in the remaining graph $G[V \setminus (U_1 \cup \ldots \cup U_{2k\ell})]$ is at most $\frac{k|V|}{2^\ell}$. This can be proved by induction on ℓ

and by considering two cases $\bar{n} \leq \frac{|V|}{2^{\ell+1}}$ and $\bar{n} > \frac{|V|}{2^{\ell+1}}$ (where $\bar{n}$ is the number of remaining vertices).

The assertion above implies that the number of edges is at most $2k-1$ after $2k\lfloor \log |V| \rfloor$ steps. After at most $2k-1$ further steps, the number of remaining vertices is zero. Since k is constant, an optimum solution $(U_1, \ldots, U_m)$ consists of at most $O(\log |V|)$ maximal independent sets. □

For every node $i \in I$, let Y_i be the set of all vertices in a set X_j with $j = i$ or j is a descendant of i in the rooted tree T. Our algorithm is based upon computing for every node $i \in I$ a table $minc_i$. Let m be the number of allowed colors. For each coloring $f : X_i \to \{1, \ldots, m\}$, there is an entry in the table $minc_i$, fulfilling

$$minc_i(f) = \min_{\bar{f}:Y_i \to \{1,\ldots,m\}, \bar{f}(x)=f(x) \forall x \in X_i} \sum_{j=1}^{m} k_j |\{y | y \in Y_i, \bar{f}(y) = j\}|.$$

In other words, for each coloring f of X_i, $minc_i(f)$ denotes the minimum costs over all colorings $\bar{f}$ of Y_i where $\bar{f}$ and f have the same colors for each vertex $x \in X_i$. The tables are computed in a bottom-up manner: start with computing the tables for the leaves, then always compute the table for an internal node later than the tables of its child or children are computed. In the full paper, we give the formulas to compute the table. For each node $i \in I$, at most $O(m^{k+1}k^2)$ operations are executed to compute the table $minc_i$. In total, the algorithm runs in $O(m^{k+1}k^2|V|)$ time if m different colors are allowed. We note that it is possible to modify the algorithm, such that it also yields a coloring with the minimum costs.

Theorem 4. *The problem OCCP for graphs $G = (V, E)$ with constant treewidth k can be solved in time $O((\log |V|)^{k+1}|V|)$. If the maximum degree $\Delta(G)$ is bounded by a constant, then the problem OCCP can be solved in linear time $O(|V|)$.*

Proof. Using Lemma 3 we set $m = \bar{c}\lfloor \log |V| \rfloor$ with constant $\bar{c}$. Next, we use the algorithm above that runs in $O(m^{k+1}|V|) = O((\log |V|)^{k+1}|V|)$ time.

If $\Delta(G)$ is bounded by a constant, then using Lemma 2 there exists a coloring f of minimum costs such that $f(v) \leq d(v) + 1 \leq \Delta(G) + 1$ for all $v \in V$. In this case, we define m as the constant $\Delta + 1$ and obtain an algorithm that runs in $O(|V|)$ time. □

We note that the definition of the table $minc_i$ and the algorithm can be modified such that we obtain the following result:

Theorem 5. *The GOCCP problem for graphs $G = (V, E)$ of constant treewidth k and $(n \times m)$ cost matrix can be solved in $O(m^{k+1}|V|)$ time.*

Next, we show that an optimum solution for a tree and also for a graph with constant treewidth can have a logarithmic number of colors.

Theorem 6. *There exists a tree $G = (V, E)$ and a cost vector $(k_1, \ldots, k_{n+1})$ such that the optimum solution for the OCCP problem is a partition of G into $n + 1 = \Omega(\log |V|)$ independent sets.*

Proof. Starting with a single vertex graph $G^0 = (\{v\}, \emptyset)$, we construct recursively a tree $G^{n+1} = (V^{n+1}, E^{n+1})$ as follows:

$$V^{n+1} = V^n \cup \{(v,1),(v,2) | v \in V^n\}$$
$$E^{n+1} = E^n \cup \{\{v,(v,1)\},\{v,(v,2)\} | v \in V^n\}$$

Since each vertex $v \in V^n$ gets two new neighbours, the number of vertices in G^n is equal to 3^n. Furthermore, for $n \geq 1$ the cardinality $\alpha(G^n)$ of a maximum independent set is $2 \cdot 3^{n-1}$. A maximum independent set of this size is the set $\{(v,1),(v,2) | v \in V^{n-1}\}$. If we choose iteratively maximum independent sets in G^n, then we obtain a partition into independent sets $U_1, \ldots, U_{n+1}$ such that $U_i = \{(v,1),(v,2) | v \in V^{n-i}\}$ for $1 \leq i \leq n$ and $U_{n+1} = \{v\}$. This implies that $|U_i| = 2 \cdot 3^{n-i}$ for $1 \leq i \leq n$ and $|U_{n+1}| = 1$.

Next, we define a cost vector and can prove that this partition is the optimum solution of the OCCP problem for G^n. This shows us that the optimum solution consists of $n+1 = \log_3 |V^n| + 1 = \Omega(\log |V^n|)$ independent sets. As cost vector we use $(k_1, \ldots, k_{n+1})$ with $k_1 = 1$ and

$$k_{i+1} = \sum_{j=0}^{i} \left(\frac{1}{3^n}\right)^j$$

for $1 \leq i \leq n$. We can prove that $(U_1, \ldots, U_{n+1})$ is the uniquely determined optimum solution of the OCCP problem (see also our full paper). □

3 Perfect matrices

In this section, we study the integer linear program formulation of the OCCP problem proposed in [7, 9]. Let I be an instance of the OCCP problem containing a graph $G = (V, E)$ with n vertices and a sequence of coloring costs $(k_1, \ldots, k_m)$. The objective function and the constraints of the problem can be described as follows:

$$\begin{array}{lll} min \sum_{i=1}^{n} \sum_{c=1}^{m} k_c \, x_{i,c} & & (0) \\ \sum_{c=1}^{m} x_{i,c} = 1 & \text{for } 1 \leq i \leq n & (1) \\ \sum_{i \in C} x_{i,c} \leq 1 & \text{for each clique C in G}, 1 \leq c \leq m & (2) \\ x_{i,c} \in \{0,1\} & \text{for } 1 \leq i \leq n, 1 \leq c \leq m & (3) \end{array}$$

The total coloring costs are minimized by the objective function (0). The constraints (1) specify that each vertex is colored exactly once and the constraints (2) guarantee that vertices that are connected by an edge are colored differently. We note that the GOCCP problem with cost matrix $(k_{i,c})$ can be solved with the same approach using the objective function

$$min \sum_{i=1}^{n} \sum_{c=1}^{m} k_{i,c} \, x_{i,c}.$$

The coefficient matrix corresponding to the restrictions (1) – (2) is called M. This matrix M is a zero - one matrix. A zero - one matrix M is called *perfect* if the polyhedron $P(M) = \{x | Mx \leq 1, x \geq 0\}$ has only integral extremal points. It follows that the OCCP problem can be solved by applying a linear programming algorithm, if the matrix M is perfect. We define $P_I(M)$ as the convex hull of the set $\{x | Mx \leq 1, x \in \{0,1\}^{mn}\}$. The submatrix of M that contains the first n rows of M is denoted by M_1, the submatrix that contains the other rows is denoted by M_2. Furthermore, we define $P^*(M) = \{x | M_1 x = 1, M_2 x \leq 1, x \geq 0\}$ and $P_I^*(M)$ as the convex hull of $\{x | M_1 x = 1, M_2 x \leq 1, x \in \{0,1\}^{mn}\}$.

The first goal of this section is to find necessary and sufficient conditions for the matrix M to be perfect in terms of the associated intersection graph $G(M)$. The second and main goal of this section is a classification of the graphs such that the polyhedron $P^*(M)$ contains only integral extrema. The intersection graph $G(M)$ of a zero - one matrix M is a graph with one vertex for each column of M. Two vertices v and v' are adjacent in $G(M)$ if and only if $M_{rv} = M_{rv'} = 1$ for at least one row r of M. The vertex set of $G(M)$ for our matrix M is equal to $\{(j,c) | 1 \leq j \leq n, 1 \leq c \leq m\}$. Two different vertices (j,c) and (j',c') are adjacent if $j = j'$ and $c \neq c'$ or if $c = c'$ and $\{j,j'\}$ is an edge in the original graph G of the OCCP problem. Notice that the subgraph $G_c(M)$ induced by the vertex set $\{(j,c) | 1 \leq j \leq n\}$ is equivalent to the original graph G, for each $1 \leq c \leq m$.

The *cartesian product* $G_1 \times G_2 = (V_1 \times V_2, E)$ of two graphs $G_1 = (V_1, E_1)$ and $G_2 = (V_2, E_2)$ is defined by the edgeset

$$E = \{\{(u_1,u_2),(v_1,v_2)\} | [u_1 = v_1 \wedge \{u_2,v_2\} \in E_2] \vee [u_2 = v_2 \wedge \{u_1,v_1\} \in E_1]\}.$$

It follows that the associated intersection graph $G(M)$ is the cartesian product $G \times K_m$ of the original graph G and a complete graph K_m with m vertices. Since the zero - one matrix M is a clique matrix of its associated intersection graph $G(M)$, the matrix M is perfect if and only if the graph $G(M)$ is perfect (Chvatal [4]).

First, we note that the cycle C_4 and the diamond (see Figure 1) generates a cycle C_7 in $G \times K_3$. Furthermore, each cycle C_k $(k \geq 4)$ generates a cycle of length $k+3$ in $G \times K_3$. This implies that each cycle with even length creates an odd cycle in $G \times K_3$. We may assume that G is a diamond - free chordal graph; otherwise $G \times K_m$ cannot be perfect. In the following we characterize the graphs G such that $G \times K_m$ remains perfect for each m.

Lemma 7. *If G is a diamond - free chordal graph, then for each maximal clique C in G it holds:*

(1) *each vertex $x \in V \setminus C$ is adjacent to at most one vertex $x_c \in C$,*
(2) *for vertices $x, y \in V \setminus C$ with $\{x, x_c\}, \{y, y_c\} \in E$ and $x_c, y_c \in C$, $x_c \neq y_c$ there exists no path in $G[V \setminus C]$ that connects x with y,*
(3) *for each vertex $c \in C$ the set of neighbours $\Gamma(c) \cap (V \setminus C)$ is a disjoint set of cliques $C_1, \ldots, C_k$ such that*

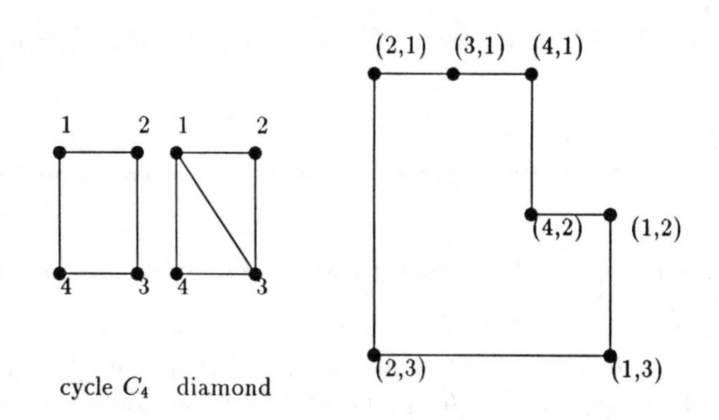

Fig. 1. A cycle C_4 and a diamond generate a C_7 in $G \times K_3$.

(3.1) *$C_1, \ldots, C_k$ are separated by C [there are no paths in $G[V \setminus C]$ that connect vertices in C_i with vertices in C_j for $i \neq j$],*
(3.2) *$C_i \cup \{c\}$ is a maximal clique in G.*

Proof. The proof of this Lemma is given in the full paper. □

Next, we define the graph class *tree - chordal* and show a classification of the perfect graphs $G \times K_m$ for $m \geq 3$.

Definition 8. A graph $G = (V, E)$ is tree-chordal if and only if there exists a directed tree $T = (U, W)$, an edge labelling $L : W \to V$ and sets $X_u \subset V$ for each $u \in U$ such that

(1) each vertex $v \in V$ is in exactly one set X_u,
(2) $L((u, v)) \in X_u$ for each edge $(u, v) \in W$,
(3) the edgeset E is given by the union of
  (3.1) the clique X_r where r is the root of T and
  (3.2) the cliques $X_v \cup \{L((u, v))\}$ for each edge $(u, v) \in W$.

A graph $G = (V, E)$ is forest-chordal if G is the disjoint union of tree - chordal graphs.

We notice that each forest-chordal graph is chordal. In the following we prove our main result.

Theorem 9. *The following statements are equivalent:*

(1) *G is a diamond - free chordal graph.*
(2) *G is a forest chordal graph.*
(3) *$G \times K_m$ (for $m \geq 3$) is a perfect graph.*

We have proved already the implication (3) ⇒ (1). If we apply Lemma 7 to each connected component of $G[V \setminus C]$, we get the implication (1) ⇒ (2).

Lemma 10. *If G is a diamond free chordal graph, then G is a forest - chordal graph.*

Proof. see our full paper. □

The last implication (2) ⇒ (3) for our main theorem is proved by the following Lemma.

Lemma 11. *If $G = (V, E)$ is a forest-chordal graph, then $G \times K_m$ is a perfect graph for each m.*

Proof. We prove this result only for $m \geq |V|$. For $m \leq |V|$, $G \times K_m$ is an induced subgraph of $G \times K_{|V|}$ and, therefore, also perfect. Clearly in $G \times K_m$ with $m \geq |V|$, $\omega(G \times K_m) = \chi(G \times K_m) = m$. Suppose that $H = (V_H, E_H)$ is an induced subgraph of $G \times K_m$ with $\omega(H) = k$. We may assume that G is connected; otherwise we compute a k-coloring parallel for each connected component of G. Let $T = (U, W)$ be the corresponding directed tree with edge labelling $L : W \to V$ and subsets X_u for $u \in U$. Recursively, we determine a k-coloring of H as follows.

First, we consider the set $X_r \subset V$ for the root r of T. Let $H' = (V_{H'}, E_{H'})$ be the subgraph of H induced by $\{(x,i)|x \in X_r, 1 \leq i \leq m\} \cap V_H$. We can describe H' by a bipartite graph $B = (V_B, E_B)$ with

$$V_B = X_r \cup \{1, \ldots, m\},$$
$$E_B = \{\{x,i\}|x \in X_r, 1 \leq i \leq m, (x,i) \in V_{H'}\}.$$

A vertex coloring of H' in $G \times K_m$ corresponds to an edge coloring of the bipartite graph B. The minimum number of colors for an edge coloring of a bipartite graph B can be computed in polynomial time and is equal to the maximum degree of a vertex v in V_B. The edges incident to v correspond to a clique in H' and, therefore, the chromatic number $\chi(H') \leq k$.

Next, we consider a set X_v for a node $v \in U$ with parent node $p(v) = u$. Suppose that the set $\{(x,i)|x \in X_u, 1 \leq i \leq m\} \cap V_H$ has been colored in a previous step with at most k colors. We define H' as the subgraph of H induced by $\{(x,i)|x \in X_v \cup \{L((u,v))\}, 1 \leq i \leq m\} \cap V_H$ and can show that H' can be colored also with at most k colors.

By induction on the height of T, it follows that $\omega(H) = \chi(H)$ and, therefore, that $G \times K_m$ is perfect. □

Now we are ready for our main classification.

Theorem 12. *Let I be an instance of the OCCP problem containing a graph G and a sequence of coloring costs $(k_1, \ldots, k_{|V|})$. Then, we have the following equivalence: G is a diamond - free chordal graph if and only if the polyhedron $P^*(M)$ contains only integral extrema.*

Proof. Let n be the number of vertices. First, suppose that G is diamond - free chordal. Then, the cartesian product $G \times K_n$ is perfect. Using the theorem of Chvatal [4], the matrix M corresponding to the restrictions $(1)-(2)$ is a perfect zero-one matrix. This implies that the polyhedron $P(M) = \{x|Mx \leq 1, x \geq 0\}$ has only integral extrema. Since each extrema in $P^*(M)$ is also an extrema in $P(M)$, the first direction of the proof is shown.

Conversely, suppose that G is not a diamond - free chordal graph. Then, we study two cases.

Case 1: G is perfect. Then, there exists a diamond or a cycle of even length as induced subgraph in G. This subgraph generates a cycle of odd length in $G \times K_3$. Let $C = \{(i_1, c_1), \ldots, (i_{2\ell+1}, c_{2\ell+1})\}$ be such a cycle in $G \times K_3$ (see Figure 1). The set of variables in the ILP corresponding to the vertices in C is $\bar{X} = \{x_{i_1,c_1}, \ldots, x_{i_{2\ell+1},c_{2\ell+1}}\}$. Let M_C be the submatrix of M containing the columns corresponding to the variables in $\bar{X}$. This matrix M_C is not perfect, and the corresponding polyhedron $P(M_C)$ contains a non-integral extremum with components $\bar{x}_{i_j,c_j} = 1/2$.

For each $1 \leq i \leq n$, there are at most two vertices in C with first index equal to i (otherwise we have a triangle). If we have two vertices $(i,c), (i,c') \in C$ with $c \neq c'$, then $\bar{x}_{i,c} + \bar{x}_{i,c'} = 1$. We define

$$\begin{aligned} A_1 &= \{i|1 \leq i \leq n, (i,c_i) \in C \text{ for exactly one } c_i \in \{1,\ldots,m\}\} \\ A_2 &= \{i|1 \leq i \leq n, (i,c) \notin C \text{ for all } c \in \{1,\ldots,m\}\}. \end{aligned}$$

Let us analyse the odd cycles in Figure 1. We may assume that A_1 is a path in G with $c_i = 1$ for each $i \in A_1$. We define the cost matrix $(k_{i,c})$ with

$$k_{i,c} = \begin{cases} -1 & \text{if } (i,c) \in C \\ 0 & \text{otherwise.} \end{cases}$$

Then, we get a feasible solution $x = (\bar{x}_{i,c})$ with respect to $P^*(M)$ as follows: We define $\bar{x}_{i,c} = \frac{1}{2}$ if $(i,c) \in C$ or if $i \in A_1$ and $c = 4$. The cardinality of A_2 is at most $n-4$. Therefore, we can distribute the value 1 to $\bar{x}_{i,5}, \ldots, \bar{x}_{i,n}$ for each $i \in A_2$ such that $\sum_{j=5}^{n} \bar{x}_{i,j} = 1$ and $\bar{x}_{i,5}, \ldots, \bar{x}_{i,n} \leq \frac{1}{n-4}$ for each $i \in A_2$. Let $x^{(int)} \in P_I^*(M)$ be an optimum solution with respect to the cost matrix $(k_{i,c})$. The objective value of $x^{(int)}$ is equal to $-\ell$, and the objective value of x is equal to $-(\ell + 1/2)$. This implies that $P^*(M)$ contains non-integral extrema.

Case 2: G is not perfect. Again, using the theorem of Chvatal [4], the clique matrix M_G corresponding to G is not perfect. Using similar arguments, we can show that $P^*(M)$ contains non-integral extrema. □

We notice that the GOCCP problem for diamond - free chordal graphs can be solved also using the tree decomposition and minimum weighted matchings.

4 Bipartite graphs

In this section we prove that the OCCP problem is NP-complete for bipartite graphs. We use the 1 - precoloring extension problem that is NP-complete for bipartite graphs and $m = 3$ proved by Bodlaender, Jansen and Woeginger [3]. Given a bipartite graph $G = (V, E)$ with vertex set $V = A \cup B$ and edge set $E \subset \{\{v,w\}|v \in A, w \in B\}$ and three specified vertices $a_1, a_2, a_3 \in A$, the problem is to decide whether there exist a 3-coloring of G with $f(a_1) = 1$, $f(a_2) = 2$ and $f(a_3) = 3$.

Theorem 13. *The OCCP problem for bipartite graphs is NP-complete if there are at least four different cost values.*

Proof. The theorem is proved by a reduction from 1-PrExt restricted to bipartite graphs. We may assume that $G = (A \cup B, E)$ contains three further vertices $b_1, b_2, b_3 \in B$ with $\{a_i, b_j\} \in E$ for $1 \leq i \neq j \leq 3$. Let n be the number of vertices in G.

Let I be an instance of 1-PrExt containing the bipartite graph $G = (A \cup B, E)$ with $a_1, a_2, a_3 \in A$ and $b_1, b_2, b_3 \in B$ as described above. An instance I' of the OCCP problem is constructed as follows. First, we define a bipartite graph $G' = (V', E')$ with vertex set $V' = \{v_{1,j}, v_{2,j} | 1 \leq j \leq 2000n\} \cup \{v_{3,j'}, v_{4,j'} | 1 \leq j' \leq 100n\} \cup \{v_5, v_6\}$ and edge set

$$E' = \{\{v_{1,j}, v_{3,j'}\}, \{v_{2,j}, v_{4,j'}\} | 1 \leq j \leq 2000n, 1 \leq j' \leq 100n\} \cup \{\{v_5, v_{3,j'}\}, \{v_6, v_{4,j'}\} | 1 \leq j' \leq 100n\} \cup \{v_5, v_6\}.$$

Then, we connect G and G' using the following edges:

$$\bar{E} = \{\{a_1, v_{3,j'}\}, \{b_1, v_{4,j'}\}, \{b_1, v_5\}, \{a_2, v_{2,j}\}, \{\{b_2, v_{1,j}\}, \{b_2, v_5\}, \{a_3, v_{3,j'}\}, \{a_3, v_{2,j}\}, \{b_3, v_{1,j}\}, \{b_3, v_{4,j'}\} \mid 1 \leq j \leq 2000n, 1 \leq j' \leq 100n\}\}.$$

In total, the bipartite graph $\bar{G}$ for I' is given by $\bar{G} = (A \cup B \cup V', E \cup E' \cup \bar{E})$. The cost values are $k_1 = 1$, $k_2 = 10$, $k_3 = 100$ and $k_4 = 15000n$.

We can prove the following statement (see our full paper): I is a yes instance of 1-PrExt if and only if the minimum total costs of coloring all vertices in I' don't exceed $6100n + 101$. □

5 Permutation graphs

In this section, we show that the OCCP and the precoloring extension problem are NP-complete for permutation graphs. The first result solves an open question in [9] and the second an open question in [5]. The proof uses a technique of Wagner [11] who proved that the cochromatic number problem is NP-complete for permutation graphs.

Let $\Pi = (i_1, \ldots, i_n)$ be a permutation of $\{1, \ldots, n\}$. We denote by $\Pi^{-1}(i)$ the position of i in Π. A graph $G = (V, E)$ with $V = \{1, \ldots, n\}$ is a permutation graph if there is a permutation Π such that $\{i, j\} \in E$ if and only if $i > j$ and $\Pi^{-1}(i) < \Pi^{-1}(j)$. In our proof, we use the point model, where the permutation $\Pi = (i_1, \ldots, i_n)$ is interpreted as a set of points $\{p_j = (j, \Pi^{-1}(j)) | 1 \leq j \leq n\}$.

Theorem 14. *1-PrExt is NP-complete for permutation graphs.*

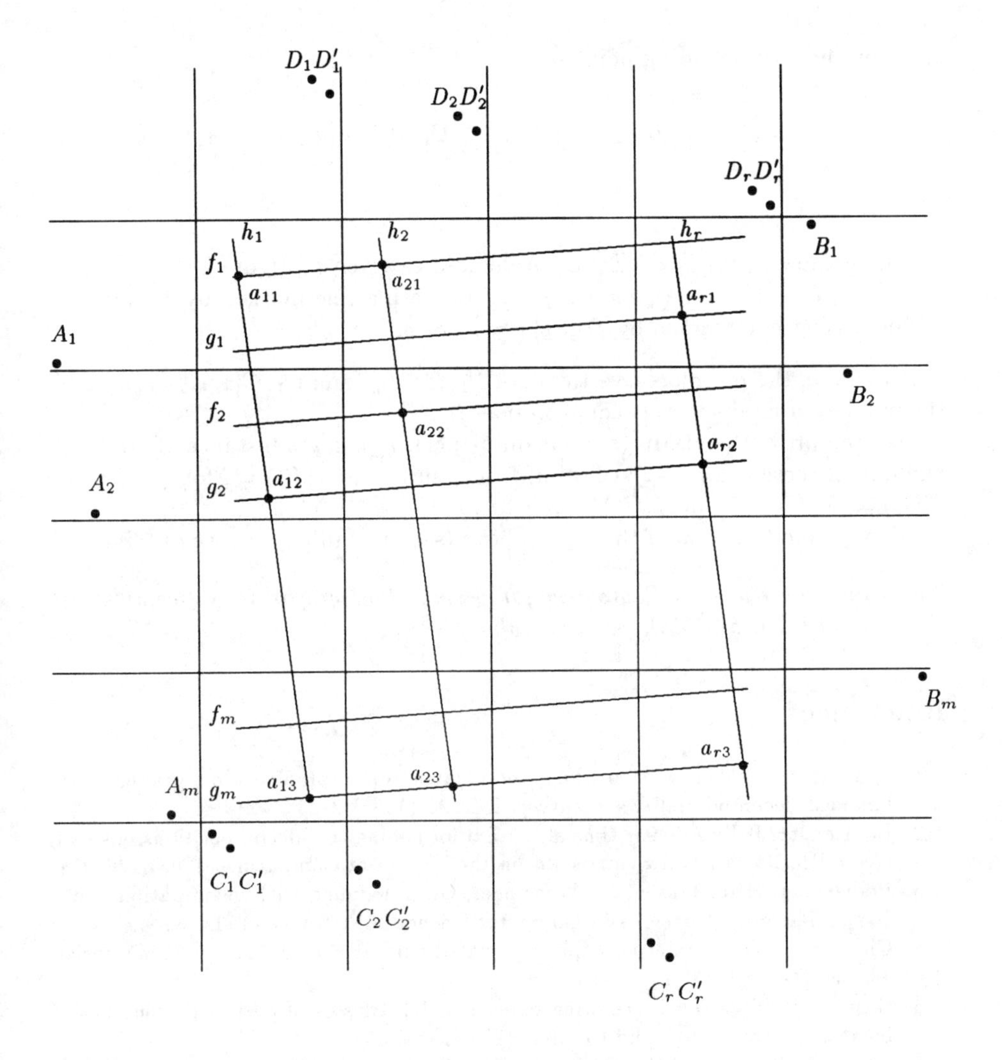

Fig. 2. The point set P constructed for the 3-SAT instance $(\bar{x}_1 \vee x_2 \vee x_m) \wedge (\bar{x}_1 \vee \bar{x}_2 \vee x_m) \wedge \ldots \wedge (x_1 \vee x_2 \vee x_m)$

Proof. The theorem is proved by a reduction from 3-SAT. Let I_1 be an instance of 3-SAT containing a collection of n clauses $c_1, \ldots, c_n$ over a set of unnegated and negated variables $\{x_1, \bar{x}_1, \ldots, x_m, \bar{x}_m\}$. In the first step, we construct a new instance I'_1 with $r = (m+1)n$ clauses such that $c'_{in+j} = c_j$ for $0 \leq i \leq m$ and $1 \leq j \leq n$. The literals of clause c'_i are denoted by $y'_{i,1}, y'_{i,2}$ and $y'_{i,3}$. Clearly, I_1 is a yes-instance if and only if I'_1 is a yes-instance of 3-SAT. An instance I_2 of the precoloring problem is constructed as follows. The permutation graph G is given by the corresponding point set

$$P = \bigcup_{i=1}^{m}(\{A_i, B_i\}) \cup \bigcup_{i=1}^{r}(\{C_i, C_i', D_i, D_i'\} \cup \{a_{i1}, a_{i2}, a_{i3}\})$$

where

- the points $A_i, B_i, C_i, C_i', D_i, D_i'$ are placed as given in Figure 2.
- a_{ik} is situated in the crossing of the decreasing line marked by h_i and the increasing line marked by f_ℓ if $y'_{i,k} = \bar{x}_\ell$ or g_ℓ if $y'_{i,k} = x_\ell$.

Furthermore, the precolored vertices are $B_1, \ldots, B_m$ and $C_1, C_1', \ldots, C_r, C_r'$ and the number of used colors is equal to $m + 2r$.

We can prove the **claim** (see our full paper): I_1' is a yes-instance of 3-SAT if and only if there is a $m + 2r$-coloring f of G such that $f(B_i) = i$ for $1 \leq i \leq m$, $f(C_j) = m + 2j - 1$ and $f(C_j') = m + 2j$ for $1 \leq j \leq r$. □

Using a modification of the proof above (see our full paper), we obtain:

Theorem 15. *The OCCP problem for permutation graphs is NP-complete if there are at least three different cost values.*

References

1. Arnborg, S.: Efficient algorithms for combinatorial problems on graphs with bounded decomposability – a survey. BIT **25** (1985) 2–23
2. Bodlaender, H.L.: A linear time algorithm for finding tree-decompositions of small treewidth. **25**.th ACM Symposium on the Theory of Computing (1993) 226–234
3. Bodlaender, H.L., Jansen, K., Woeginger, G.: Scheduling with incompatible jobs. Graph Theoretic Concepts in Computer Science, LNCS **657** (1992) 37–49
4. Chvatal, V.: On certain polytopes associated with graphs. Journal Combinatorial Theory **B-18** (1975) 138–154
5. Hujter, M., Tuza, Z.: Precoloring extension. III. Classes of perfect graphs. Combinatorics, Probability and Computing **5** (1996) 35–56
6. Kloks, T.: Treewidth. Ph. D. Thesis, Utrecht University, The Netherlands (1991)
7. Kroon, L.G., Sen, A., Deng, H., Roy, A.: The optimal cost chromatic partition problem for trees and interval graphs. Graph Theoretical Concepts in Computer Science, LNCS (1996)
8. Robertson, N., Seymour, P.D.: Graph minors. II. Algorithmic aspects of tree-width. Journal on Algorithms **7** (1986) 309–322
9. Sen, A., Deng, H., Guha, S.: On a graph partition problem with an application to VLSI layout. Information Processing Letters **43** (1992) 87–94
10. Supowit, K.J.: Finding a maximum planar subset of a set of nets in a channel. IEEE Transactions on Computer Aided Design CAD **6, 1** (1987) 93–94
11. Wagner, K.: Monotonic coverings of finite sets. Journal of Information Processing and Cybernetics - EIK **20** (1984) 633 – 639

Fault Tolerant K-Center Problems

(Extended Abstract)

Samir Khuller[1] * and Robert Pless[2] ** and Yoram J. Sussmann[2] ***

[1] Dept. of Computer Science and UMIACS
University of Maryland, College Park, MD 20742
[2] Dept. of Computer Science, University of Maryland, College Park, MD 20742

Abstract. The basic K-center problem is a fundamental facility location problem, where we are asked to locate K facilities in a graph, and to assign vertices to facilities, so as to minimize the maximum distance from a vertex to the facility to which it is assigned. This problem is known to be NP-hard, and several optimal approximation algorithms that achieve a factor of 2 have been developed for it.
We focus our attention on a generalization of this problem, where each vertex is required to have a set of α ($\alpha \leq K$) centers close to it. In particular, we study two different versions of this problem. In the first version, each vertex is required to have at least α centers close to it. In the second version, each vertex that *does not have a center placed on it* is required to have at least α centers close to it. For both these versions we are able to provide polynomial time approximation algorithms that achieve constant approximation factors for *any* α. For the first version we give an algorithm that achieves an approximation factor of 3 for any α, and achieves an approximation factor of 2 for $\alpha < 4$. For the second version, we provide algorithms with approximation factors of 2 for any α. The best possible approximation factor for even the basic K-center problem is 2. In addition, we give a polynomial time approximation algorithm for a generalization of the K-supplier problem where a subset of at most K supplier nodes must be selected as centers so that every demand node has at least α centers close to it. We also provide polynomial time approximation algorithms for all the above problems for generalizations when cost and weight functions are defined on the set of vertices.

1 Introduction

The basic K-center problem is a fundamental facility location problem and is defined as follows: given an edge-weighted graph $G = (V, E)$ find a subset $\mathcal{S} \subseteq V$

---

* Research supported by NSF Research Initiation Award CCR-9307462, and NSF CAREER Award CCR-9501355. Email addr: samir@cs.umd.edu

** Research supported by NSF Grant IRI-90-57934. Email addr: pless@cs.umd.edu

*** Research supported by NSF CAREER Award CCR-9501355. Email addr: yoram@cs.umd.edu

of size at most K such that each vertex in V is "close" to some vertex in $\mathcal{S}$. More formally, it is defined as follows:

$$\min_{\mathcal{S}\subseteq V}\max_{u\in V}\min_{v\in\mathcal{S}} d(u,v)$$

where d is the distance function. For example, one may wish to install K fire stations and minimize the maximum distance (response time) from a location to its closest fire station. The problem is known to be NP-hard [5].

An approximation algorithm with a factor of ρ, for a minimization problem, is a polynomial time algorithm that guarantees a solution with cost at most ρ times the optimal solution. Approximation algorithms for the basic K-center problem have been very well studied and are known to be optimal [6, 7, 8, 9]. These schemes present natural methods for obtaining an approximation factor of 2. Several approximation algorithms are known for interesting generalizations of the basic K-center problem as well [3, 8, 13, 15, 1, 11], including costs [8, 13, 15], weights [3, 13, 15], and capacities [1, 11]. A related problem of placing placing as few centers as possible so that each vertex without a center has at least k vertex-disjoint paths to centers is also studied in [1].

The α-neighbor K-center problem is discussed in a recent paper by Krumke [12]. The problem is formally defined as follows: given an edge-weighted graph $G = (V, E)$ find a subset $\mathcal{S} \subseteq V$ of size at most K such that each vertex in $V - \mathcal{S}$ is "close" to a set of α vertices in $\mathcal{S}$. Formally,

$$\min_{\mathcal{S}\subseteq V}\max_{u\in V-\mathcal{S}} \delta^{(\alpha)}(u,\mathcal{S})$$

where

$$\delta^{(\alpha)}(u,\mathcal{S}) = \min_{A\subseteq\mathcal{S},|A|=\alpha}\max_{a\in A} d(u,a)$$

where d is the distance function. Krumke [12] gives an algorithm with an approximation factor of 4, by generalizing the notion of an independent set of vertices.

The main motivation to study this problem is to provide some notion of fault-tolerance. Namely, if we are concerned with the placement of emergency facilities, then providing "backup" centers when one center fails to respond is useful [14].

We consider a variation of this problem as well, called the α-all-neighbor K-center problem that is formally defined as follows: given an edge-weighted graph $G = (V, E)$ find a subset $\mathcal{S} \subseteq V$ of size at most K such that each vertex in V is "close" to a set of α vertices in $\mathcal{S}$. Formally,

$$\min_{\mathcal{S}\subseteq V}\max_{u\in V} \delta^{(\alpha)}(u,\mathcal{S}).$$

We also consider a variant of this problem called the α-neighbor K-suppliers problem that is formally defined as follows: given an edge-weighted bipartite graph $G = (U, V, E)$, find a subset $\mathcal{S} \subseteq U$ of size at most K such that each vertex in V is "close" to a set of α vertices in $\mathcal{S}$. Formally,

$$\min_{\mathcal{S} \subseteq U} \max_{u \in V} \delta^{(\alpha)}(u, \mathcal{S}).$$

For all the problems considered in the paper, we also address the generalizations when vertices have *costs* and *weights*. The cost generalization is formally defined as follows: given a graph G and a cost function $c(v)$ defined on V, find a subset $\mathcal{S}$ of vertices of total cost at most K such that each vertex that needs to be covered by a center is "close" to a set of α vertices in $\mathcal{S}$. Formally, we have to pick a set $\mathcal{S}$ satisfying

$$\sum_{s \in \mathcal{S}} c(s) \leq K.$$

The weight generalization is formally defined as follows: given a graph G and a weight function $w(v)$ defined on V, find a subset $\mathcal{S}$ of vertices of size at most K such that each vertex that needs to be covered by a center is "close" to a set of α vertices in $\mathcal{S}$, where the distance from vertex u to vertex v depends on the weight of v. Formally, we change the definition of the distance measure to

$$\delta^{(\alpha)}(u, \mathcal{S}) = \min_{A \subseteq \mathcal{S}, |A| = \alpha} \max_{a \in A} d(u, a) \cdot w(u)$$

where d is the distance function.

Finally, we study the most general case, when the vertices have weights and costs. The results are summarized in the table given below.

1.1 Our Results

We improve Krumke's result, and show that we can obtain an approximation factor of 2 for the problem considered in his paper. This matches the bound for the basic K-center problem, which is the best possible [9]. The algorithm is a very natural extension of the method given by Hochbaum and Shmoys [8] for the basic K-center problem.

We also show that for the α-all-neighbor K-center problem, we can obtain an approximation factor of 3 for any α, and a similar algorithm gives an approximation factor of 2 for $\alpha < 4$ (perhaps the practically interesting case).

Recently, Chauduri, Garg and Ravi [2] independently came up with different algorithms with matching approximation factors for the α-neighbor K-center problem and the α-neighbor K-suppliers problem. Their algorithm modifies Krumke's approach. The versions for which they provide algorithms are marked in the table below by [CGR]. We do not see any way of generalizing their technique to include weights.

In addition, we provide constant approximation bounds for generalizations of the problem involving costs and weights, as well as for the α-neighbor K-suppliers problem. For the K-suppliers problem, Hochbaum and Shmoys [8] give a proof (originally due to Howard Karloff) showing that the factor of 3 is the best possible unless $P = NP$. Thus 3 is a lower bound for all the K-suppliers generalizations that we consider. Since the basic K-center problem with weights and costs is a generalization of the K-suppliers problem, a factor of 3 is also the best possible. The results are summarized in the table below. A † indicates that the bound is the best possible unless $P = NP$, while a ‡ indicates that this matches the best known bound for the basic K-center problem. Let β denote the ratio of the maximum and minimum weights.

Fault-Tolerant K-Center Approximation Factors				
	Basic	Weights	Costs	Weights + Costs
α-All-Neighbor K-Center	$3\ (2^{\dagger})^{*}$	3	$3^{\ddagger}$	$3^{\dagger}$
α-Neighbor K-Center	$2^{\dagger}$ [CGR]	3	4	$4\beta + 1$
α-Neighbor K-Suppliers	$3^{\dagger}$ [CGR]	$3^{\dagger}$	$3^{\dagger}$	$3^{\dagger}$

* if $\alpha = 2$ or 3

2 α-All-Neighbor K-Center Problems

We may assume for simplicity that G is a complete graph, where the edge weights satisfy the triangle inequality. (We can always replace any edge by the shortest path between the corresponding pair of vertices.)

The algorithm uses the threshold method introduced by Edmonds and Fulkerson in [4] and used for the K-center problem by Hochbaum and Shmoys in [8]. Sort all edge weights in non-decreasing order. Let the (sorted) list of edges be $e_1, e_2, \ldots e_m$ (where $m = \binom{n}{2}$). For each i, let the threshold graph G_i be the subgraph obtained from G by including edges of weight at most $w(e_i)$. Run the algorithm below for each i from 1 to m, until a solution is obtained. (One can also use binary search to speed up the computation as suggested by Hochbaum and Shmoys [8].) In each iteration, we work with the subgraph G_i and view it as an unweighted graph. Since G_i is an unweighted graph, when we refer to the distance between two nodes, we refer to the number of edges on a shortest path between them. In iteration i, we find a solution using some number of centers. If the number of centers exceeds K, we prove that there is no solution with cost at most $w(e_i)$. If the number of centers is at most K, we return an approximate solution.

Let G_i^2 denote the graph obtained by adding edges to G_i between nodes that have a common neighbor.

2.1 Any α

We give an algorithm that obtains an approximation factor of 3 for any value of α.

The following technique was introduced by Hochbaum and Shmoys [7, 8] and has been used extensively to solve K-center problems. Find a maximal independent set in G_i^2. Note that if the independent set has size I, then any solution with radius $w(e_i)$ must use at least αI centers, because nodes in the independent set cannot be assigned a common center. We therefore place α centers at each node in the independent set. At this point, every node in the graph is at distance at most 2 (in G_i) from α centers.

We now have to distribute the centers so that no two centers are placed on a common node. Note that if there is a solution with radius $w(e_i)$, then every node has degree at least $\alpha - 1$ in G_i. We can therefore move $\alpha - 1$ centers from each node in the independent set to a subset of its neighbors in G_i. Since every node in the graph is at distance at most 2 from a node in the independent set, we must have that every node in the graph is at distance at most 3 from α centers, which implies that this approach gives an approximation factor of 3.

Any α with weights and costs. The α-all-neighbor K-center problem with weights and costs is a generalization of the α-all-neighbor K-center problem where weight and cost functions are defined on the vertices and the objective is to pick a set of centers whose total cost is at most K, such that the radius is minimized, where the distance from u to v is now defined by $d(u, v) \cdot w(v)$, the weight of edge $e(u, v)$ multiplied by the weight of node v.

For weighted versions of the problems, consider the list of weighted distances $w_{uv} = d(u, v) \cdot w(v)$. List them in non-decreasing order as $w_1 \leq w_2 \leq \cdots \leq w_{2m}$. For each i, define the directed graph G_i as follows. Edge $e(u, v)$ is included in G_i if $d(u, v) \cdot w(v)$, the weighted distance from u to v, is at most w_i.

A small modification of the above algorithm yields a 3-approximation algorithm for the problem with weights and costs. When choosing the initial set, we always select the highest weight available node v. We then mark all nodes whose weighted distance from v is at most $2w_i$. The directed graph ensures that, if a node u has a directed edge to node v in G_i, and v has a higher weight than u, then v also has a directed edge to u in G_i, and therefore can cover in two steps any node that u can cover in one step. Once we have placed α centers at each node v in the independent set, we simply move all α centers to the α cheapest neighbors of the node (including itself), where a neighbor of v is a node with a directed edge to v in G_i. A simple extension of the proof given in [13] shows that the vertices so obtained cover all vertices in the graph with radius at most $3w_i$. Since the node must have at least α neighbors in any solution, the cost of the solution obtained is a lower bound on the cost of any solution with radius w_i.

2.2 $\alpha = 2, 3$

If α is 2 or 3, we can obtain an approximation factor of 2. The algorithm gives an approximation ratio of 3 for any α. We can prove that for $\alpha < 4$ the obtained ratio is actually 2.

The algorithm consists of α iterations. Consider the graph G_i^2. Every node is assigned a "covering number" $C(v)$ (initially 0). The multiset of centers is $\mathcal{S} = \emptyset$. In each iteration we pick an independent set of nodes. At the end of each iteration $j = 1, 2, \ldots, \alpha$, we guarantee that each node is covered by at least j centers within two steps.

In each iteration, when there is a choice for a vertex to be chosen as a center, we *prefer* picking vertices that do not have a center already assigned to them. We assign a center at the chosen vertex, and increase the covering number for all vertices within distance two in G_i.

α-ALL-NEIGHBOR K-CENTER ALGORITHM(G_i^2).

1 **for** all v
2 $\quad C(v) = 0$.
3 $\quad extra(v) = 0$.
4 $\quad helps(v) = \emptyset$.
5 **for** $j = 1$ to α **do**
$\quad$ // Phase I
6 $\quad$ **while** $\exists v \notin \mathcal{S}$ with $C(v) < j$ **do**
7 $\quad\quad$ create new center at v and set $\mathcal{S} = \mathcal{S} \cup \{v\}$.
8 $\quad\quad$ $C(v) = C(v) + 1$.
9 $\quad\quad$ $C(u) = C(u) + 1$ if $(u, v) \in E(G_i^2)$.
$\quad$ // Phase II
10 $\quad$ **while** $\exists v$ with $C(v) < j$ **do**
11 $\quad\quad$ create new center at v and set $\mathcal{S} = \mathcal{S} \cup \{v\}$.
12 $\quad\quad$ $C(v) = C(v) + 1$.
13 $\quad\quad$ $extra(v) = extra(v) + 1$.
14 $\quad\quad$ **if** $\exists u$ with $C(u) < j$ and $(u, v) \in E(G_i^2)$
15 $\quad\quad\quad$ $C(u) = C(u) + 1$.
16 $\quad\quad\quad$ Set $helps(v) = u$.
17 **for** all $v \in \mathcal{S}$ with $extra(v) \geq 1$ **do**
18 $\quad$ **if** $helps(v) = \emptyset$
19 $\quad\quad$ Shift $extra(v)$ centers to neighbors of v in G_i^2 that are not in $\mathcal{S}$.
20 $\quad$ **else**
21 $\quad\quad$ Shift one center to a common neighbor of v and $helps(v)$ in G_i^2 not in $\mathcal{S}$.
22 $\quad\quad$ Shift $extra(v) - 1$ centers to neighbors of v in G_i^2 that are not in $\mathcal{S}$.
23 **end-proc**

Lemma 1. *The above algorithm uses no more centers than the optimal solution.*

Proof. In each iteration we select an independent set in G_i^2. Let I^* be the size of the largest independent set picked in any iteration. Any solution with radius $w(e_i)$ must use at least αI^* centers, and we must have that $|\mathcal{S}| \leq \alpha I^*$. □

Notice that this algorithm produces a multiset of centers. We now show how to make the centers distinct.

Theorem 2. *The above algorithm returns a solution to the α-all-neighbor K-center problem with an approximation ratio of two if $\alpha = 2$ or 3.*

Proof. Call a node v *satisfied* in iteration j if $C(v) \geq j$. Although in each iteration we prefer to pick nodes not previously chosen as centers, after the first phase all nodes remaining with $C(v) < j$ are in $\mathcal{S}$. Define H_j to be subgraph of G_i^2 induced by these (unsatisfied) nodes in round j. Now consider the structure of H_j. The graph H_2 is a collection of singleton nodes, disconnected in G_i^2 (because they were all picked in the independent set in the first iteration). Therefore all nodes in H_2 will be added to $\mathcal{S}$.

First suppose $\alpha = 2$. Since the nodes in H_2 form an independent set, $helps(v) = \emptyset$ for all nodes in H_2. Therefore we can shift all but one center to unassigned neighbors of v in G_i. Such neighbors must exist because v must have at least one neighbor in G_i and at most two centers total are placed in the neighborhood of v in G_i^2.

Now let $\alpha = 3$. Consider a node v that was assigned as a center multiple times. If $helps(v) = \emptyset$, then we can shift all but one center to unassigned neighbors of v in G_i, by the above argument.

If $helps(v) \neq \emptyset$ then we must have $|helps(v)| = 1$, because H_3 is a graph with maximum vertex degree of 1 (since any node in H_3 with degree 2 must be satisfied). Assume $helps(v) = \{u\}$. Note that only 2 centers are assigned to v. This follows from the fact that the center on u covers v. We must shift the extra center so that it covers both u and v within distance 2. If u and v are adjacent in G_i, then we can shift the extra center on v to any neighbor of v in G_i. Otherwise, there must exist a node w adjacent to both u and v in G_i. Node w does not have any centers assigned to it because it already has 3 centers adjacent to it. Therefore we can shift the extra center to w.

Any node which does not have a center placed on it has at least α centers adjacent to it in G_i^2. As shown above, a node which has a center placed on it also has at least α centers adjacent to it in G_i^2. Therefore all nodes have at least α centers within radius $2w(e_i)$. □

We have constructed an example that causes the algorithm to fail when $\alpha = 4$. It is worthwhile to note, however, that although the algorithm does not achieve a factor of 3 on this example, it uses significantly fewer centers than the optimal solution, leaving open the possibility of adding enough centers to cover every node within distance at most twice the optimal.

3 α-Neighbor K-Center Problems

In this section, we describe an algorithm which gives an approximation factor of 2 for the α-neighbor K-center problem.

We assume that G is a complete graph with edges satisfying the triangle inequality. Iterate for each i from 1 to m until a solution is obtained.

Consider the graph G_i^2. Every node is assigned a "covering number" $C(v)$ (initially 0). The set of centers is $\mathcal{S} = \emptyset$. At the end of each iteration $j = 1, 2, \ldots, \alpha$, we guarantee that each node not chosen as a center is covered by at least j centers within distance two. In each iteration, we pick a center that is not covered by at least j centers. We assign a center at the chosen vertex, and increase the covering number for all vertices within distance two in G_i.

α-NEIGHBOR K-CENTER ALGORITHM(G_i^2).
1 **for** all v
2 $\quad C(v) = 0$.
3 **for** $j = 1$ to α **do**
4 $\quad$ **while** $\exists v$ with $C(v) < j$ **do**
5 $\quad\quad$ create center at v and and set $\mathcal{S} = \mathcal{S} \cup \{v\}$.
6 $\quad\quad$ $C(v) = \alpha$.
7 $\quad\quad$ $C(u) = C(u) + 1$ if $(u, v) \in E(G_i^2)$.
8 **end-proc**

We find at most α independent sets in α iterations.

Theorem 3. *The above algorithm finds a solution to the α-neighbor K-center problem with an approximation ratio of two.*

Proof. When the algorithm terminates, each vertex has a covering number equal to α. This guarantees that each vertex was either chosen as a center, or is covered by at least α centers within distance two. We now prove that if there is a feasible solution with K centers in some G_i, then our algorithm will not assign more than K centers in G_i.

Assume that this does not hold. In other words, there is a graph for which there is a solution that uses at most K centers, and our algorithm assigns more than K centers. Consider the smallest value of K for which the algorithm fails, and consider the smallest graph G that is a counter-example for that value of K. Assume that the centers assigned in iteration j have label j. Let S_{OPT} be the set of K vertices in graph G that have centers placed on them by the optimal solution. Note that each vertex in $V - S_{OPT}$ has at least α neighbors in S_{OPT}.

If our algorithm places centers only on vertices in S_{OPT} then we certainly do not place more than K centers. Assume that j is the highest labeled center placed at $v \in V - S_{OPT}$ by the algorithm. Let $N_{OPT}(v)$ be the neighbors of v in S_{OPT}. Clearly $|N_{OPT}(v)| \geq \alpha$. Let $V_{OPT}(v)$ be all the vertices that are adjacent to some vertex in $N_{OPT}(v)$.

We claim that there are at most α centers placed by the algorithm in $v \cup N_{OPT}(v) \cup V_{OPT}(v)$ from G. If v had a center placed on it in iteration j, then at the instant it was placed it had at most $j-1$ centers within distance 2 in G_i. Hence, there were at most $j-1$ centers with label $< j$ in this region. Since all centers with label $> j$ are placed only at nodes in S_{OPT} this implies that we cannot place two nodes with the same label in $N_{OPT}(v)$ (since the nodes placed in a single iteration form an independent set in G_i^2). Thus there can be at most $\alpha - j$ nodes of label $> j$ in $v \cup N_{OPT}(v) \cup V_{OPT}(v)$ from G. Adding gives at most α nodes in this region.

We now claim that if we delete $v \cup N_{OPT}(v) \cup V_{OPT}(v)$ from G, this gives us a smaller counter-example (unless the deleted nodes are exactly G, which is not a valid counter-example as we use only α nodes). □

3.1 Any α with weights

Using a different algorithm, we can obtain an approximation factor of three for the α-neighbor K-center problem with weights. The algorithm repeatedly selects a node which is not at least α-covered as a center and increments the covering number of all nodes within distance 3 of the center.

Consider the list of weighted distances $w_{uv} = d(u,v) \cdot w(v)$. List them in non-decreasing order as $w_1 \leq w_2 \leq \cdots \leq w_{2m}$. For each i, define the directed graph G_i as follows. Edge $e(u,v)$ is included in G_i if $d(u,v) \cdot w(v)$, the weighted distance from u to v, is at most w_i.

α-NEIGHBOR WEIGHTED-K-CENTER ALGORITHM(G_i).
1 **for** all v
2 $\quad C(v) = 0$.
3 $\mathcal{S} = \emptyset$.
4 **while** $U = \{u \mid C(u) < \alpha\} \neq \emptyset$
5 $\quad$ let $v =$ max weight vertex in U.
6 $\quad$ create center at v and and set $\mathcal{S} = \mathcal{S} \cup \{v\}$.
7 $\quad C(v) = \alpha$.
8 $\quad C(u) = C(u) + 1$ if $d(v,u) \cdot w(u) \leq 3w_i$.
9 **end-proc**

Theorem 4. *The above algorithm finds a solution to the α-neighbor K-center problem with weights with an approximation ratio of 3.*

Proof. Clearly this algorithm satisfies every vertex within a factor three of the optimal radius (because the algorithm loops until no vertex is left uncovered);

we have to argue that it does not use too many centers. Assume on the contrary that there is a graph for which there is a solution that uses at most K centers, and our algorithm assigns more than K centers. Consider the smallest value of K for which the algorithm fails, and consider the smallest graph G that is a counter-example for that value of K. Note that each vertex in $V - S_{OPT}$ has at least α neighbors in S_{OPT}.

Define the j-neighborhood of a vertex $N^j(x) = \{v \in V \mid d(v,x) \cdot w(x) \leq j \cdot w_i\}$; intuitively, the set of all nodes which could cover x within radius $j \cdot w_i$. Define the neighborhood of a vertex $N(x) = N^1(x)$. Let $N_{OPT}(x) = N(x) \cap S_{OPT}$, the nodes in the optimal solution that cover x, and let $V_{OPT}(x) = \{v \in V \mid \exists w \in N_{OPT}(x)$ such that $d(w,v) \cdot w(v) \leq w_i\}$, or equivalently, $V_{OPT}(x) = \{v \in V \mid N_{OPT}(v) \cap N_{OPT}(x) \neq \emptyset\}$, the nodes covered in the optimal solution by the nodes in $N_{OPT}(v)$. Since more than K centers were chosen by our algorithm, at least one center must have been chosen from $V - S_{OPT}$. Let v be the last such center.

Let u be the last center chosen from $N_{OPT}(v)$, after v was chosen. (If no such center exists, then set $u = v$.) We claim that in the set $v \cup N_{OPT}(v) \cup V_{OPT}(v)$ there are at most α centers placed by the algorithm. To see this observe that when u was placed, every center placed in the set is in $N^3(u)$ and thus at most $\alpha - 1$ centers were placed when we placed u, the last center in the set. Let x be a center placed in $V - S_{OPT}$, and let y be a vertex in $N_{OPT}(v) \cap N_{OPT}(x)$. (If $x \in S_{OPT}$ the proof is even easier.) To show that $x \in N^3(u)$, observe that $w(u) \cdot d(u,x) \leq w(u)(d(u,v)+d(v,y)+d(y,x)) \leq w(v) \cdot d(u,v)+w(v) \cdot d(v,y)+w(x) \cdot d(y,x)$. This is at most $w_i + w_i + w_i \leq 3w_i$. Therefore we can delete $v \cup N_{OPT}(v) \cup V_{OPT}(v)$ from G and obtain a smaller counter-example. □

3.2 Any α with costs

We describe an algorithm that gives an approximation factor of 4 for the α-neighbor K-center problem with costs. We first run the α-neighbor K-center Algorithm, to obtain an initial set of centers $\mathcal{S}$. We then shift these centers to low cost neighbors as follows.

We create a bipartite graph $H = (\mathcal{S}, V, E')$, where an edge $(s,v) \in E'$ if $v = s$ or if the edge (s,v) is in G_i and the degree of s in G_i is at least α. We define the cost of an edge $e = (s,v)$ to be $c(e)$, the cost of v. We then find a min-cost perfect matching M in H. Let $\mathcal{S}'$ be the set of nodes in V which are matched to a node in $\mathcal{S}$. Return the set $\mathcal{S}'$.

Theorem 5. *The above algorithm finds a solution to the α-neighbor K-center problem with costs with an approximation ratio of 4.*

Proof. We first show that the cost of $\mathcal{S}'$ is at most the cost of any solution with radius $w(e_i)$. Clearly a perfect matching between $\mathcal{S}$ and V exists, since each node in $\mathcal{S}$ can be matched to itself. We prove there exists a matching from $\mathcal{S}$ to the nodes in the optimal solution, which implies that the min-cost matching has cost at most that of the optimal solution. Let S_{OPT} be the set of K vertices in graph G that have centers placed on them by the optimal solution.

Consider the nodes in $\mathcal{S}$ which are also in S_{OPT}. These nodes get matched to themselves. Notice that all nodes with degree less than α must be in S_{OPT} and these nodes are all matched to themselves in the above algorithm. Now consider the last node s added to $\mathcal{S}$ that is not in S_{OPT}. Remove s from H and recursively find a matching in the remaining subgraph. There are at least α centers of S_{OPT} in the neighborhood of s in G_i and there are at most $\alpha - 1$ nodes in $\mathcal{S}$ in the neighborhood of s in G_i^2. Therefore at least one of the nodes in S_{OPT} that are in the neighborhood of s is not matched. Match s to this node.

We now prove the approximation bound. Consider a node v. If $v \notin \mathcal{S}$, then it has α neighbors in $\mathcal{S}$ in G_i^2. These neighbors are shifted by distance at most $w(e_i)$, implying that v has α neighbors in $\mathcal{S}'$ within distance $3w(e_i)$. If $v \in \mathcal{S}$, and v is matched to itself, then v is covered by itself. Otherwise, there must be a node u in the neighborhood of v in G_i which is not in $\mathcal{S}$. This node has α centers within distance $3w(e_i)$, which implies that v has α centers within distance $4w(e_i)$. □

3.3 Any α with weights and costs

A modification of the above algorithm for weights gives an approximation algorithm for the α-neighbor K-center problem with weights and costs. We first run the α-neighbor weighted-K-center Algorithm, to obtain an initial set of centers $\mathcal{S}$. We then shift these centers to low cost neighbors as in the α-neighbor K-center problem with costs.

Let β denote the ratio of the weight of the maximum weight node in G to the weight of the minimum weight node in G. A vertex that has a center placed on it may not have a center placed on it after the shifting of centers. Two points need to be noted here: if a vertex has less than α incoming edges in the directed graph G_i then we do not need to move its center. If it has at least α incoming edges, then since all those vertices either have centers, or are covered by centers, we can argue that the algorithm provides an approximation factor of $4\beta + 1$. (The problem is that these vertices may have a low weight and thus the centers that cover them may be far away.)

4 α-Neighbor K-Suppliers Problems

In [10], we prove the following:

Theorem 6. *We can obtain a polynomial time algorithm to solve the K-suppliers problem with weights and costs with an approximation factor of 3.*

References

1. J. Bar-Ilan, G. Kortsarz, and D. Peleg. How to allocate network centers. *Journal of Algorithms*, 15:385–415, 1993.
2. S. Chaudhuri, N. Garg, and R. Ravi. Best possible approximation algorithms for generalized k-Center problems. Technical Report MPI-I-96-1-021, Max-Planck-Institut für Informatik, 66123 Saarbrücken, Germany, 1996.
3. M. Dyer and A. M. Frieze. A simple heuristic for the p-center problem. *Operations Research Letters*, 3:285–288, 1985.
4. J. Edmonds and D. R. Fulkerson. Bottleneck extrema. *Journal of Combinatorial Theory*, 8:299–306, 1970.
5. M. R. Garey and D. S. Johnson. *Computers and Intractibility: A guide to the theory of NP-completeness.* Freeman, San Francisco, 1978.
6. T. Gonzalez. Clustering to minimize the maximum inter-cluster distance. *Theoretical Computer Science*, 38:293–306, 1985.
7. D. Hochbaum and D. B. Shmoys. A best possible heuristic for the k-center problem. *Mathematics of Operations Research*, 10:180–184, 1985.
8. D. Hochbaum and D. B. Shmoys. A unified approach to approximation algorithms for bottleneck problems. *Journal of the ACM*, 33(3):533–550, 1986.
9. W. L. Hsu and G. L. Nemhauser. Easy and hard bottleneck location problems. *Discrete Applied Mathematics*, 1:209–216, 1979.
10. S. Khuller, R. Pless, and Y. J. Sussmann. Fault tolerant K-Center problems. Technical Report CS-TR-3652, University of Maryland, College Park, 1996. Available by ftp at `ftp.cs.umd.edu/pub/papers/papers/3652/3652.ps.Z`.
11. S. Khuller and Y. J. Sussmann. The capacitated K-Center problem. In *Proc. of the 4th Annual European Symposium on Algorithms*, volume 1136 of *LNCS*, pages 152–166, 1996.
12. S. O. Krumke. On a generalization of the p-center problem. *Information Processing Letters*, 56:67–71, 1995.
13. J. Plesnik. A heuristic for the p-center problem in graphs. *Discrete Applied Mathematics*, 17:263–268, 1987.
14. L. Smith. Volunteers' rescue response rates worsen in Pr. William. *The Washington Post*, April 17, 1996.
15. Q. Wang and K. H. Cheng. A heuristic algorithm for the k-center problem with cost and usage weights. Technical Report UH-CS-90-15, University of Houston, 1990.

$R^{SN}_{1\text{-}tt}$(NP) Distinguishes Robust Many-One and Turing Completeness*

Edith Hemaspaandra[1]** and Lane A. Hemaspaandra[2]*** and Harald Hempel[3]†

[1] Department of Mathematics, Le Moyne College, Syracuse, NY 13214, USA.
[2] Department of Computer Science, University of Rochester, Rochester, NY 14627, USA.
[3] Inst. für Informatik, Friedrich-Schiller-Universität Jena, 07743 Jena, Germany.

Abstract. Do complexity classes have many-one complete sets if and only if they have Turing-complete sets? We prove that there is a relativized world in which a relatively natural complexity class—namely a downward closure of NP, $R^{SN}_{1\text{-}tt}$(NP)—has Turing-complete sets but has no many-one complete sets. In fact, we show that in the same relativized world this class has 2-truth-table complete sets but lacks 1-truth-table complete sets. As part of the groundwork for our result, we prove that $R^{SN}_{1\text{-}tt}$(NP) has many equivalent forms having to do with ordered and parallel access to NP and NP ∩ coNP.

1 Introduction

In this paper, we ask whether there are natural complexity classes for which the *existence* of many-one and Turing-complete sets can be distinguished. Many standard complexity classes—e.g., R, BPP, UP, FewP, NP ∩ coNP—are known that in some relativized worlds lack many-one complete (m-complete) sets, and that in some relativized worlds lack Turing-complete (T-complete) sets. However, for none of the classes just mentioned is there known any relativized world in which the class (simultaneously) has T-complete sets but lacks m-complete sets. In fact, for NP ∩ coNP and BPP, Gurevich [Gur83] and Ambos-Spies [Amb86] respectively have shown that no such world can exist. In this paper, we will show that there is a downward closure of NP, $R^{SN}_{1\text{-}tt}$(NP), that in some relativized worlds simultaneously has T-complete sets and lacks m-complete sets.

---

* A full version, containing full proofs of all results, can be found as UR-CS-TR-635 at http://www.cs.rochester.edu/trs/.

** Email: `edith@bamboo.lemoyne.edu`. Supported in part by grant NSF-INT-9513368/DAAD-315-PRO-fo-ab. Work done in part while visiting Friedrich-Schiller-Universität Jena.

*** Email: `lane@cs.rochester.edu`. Supported in part by grants NSF-CCR-9322513 and NSF-INT-9513368/DAAD-315-PRO-fo-ab. Work done in part while visiting Friedrich-Schiller-Universität Jena.

† Email: `hempel@mipool.uni-jena.de`. Supported in part by grant NSF-INT-9513368/DAAD-315-PRO-fo-ab.

In fact, $\mathrm{R}^{SN}_{1\text{-}tt}(\mathrm{NP})$ has even stronger properties. We will see that it *robustly*—i.e., in all relativized worlds, including the real world—has 2-truth-table complete (2-tt-complete) sets. Yet we will see that in our relativized world it lacks 1-tt-complete sets. Thus, this class displays a very crisp borderline between those reduction types under which it robustly has complete sets, and those reduction types under which it does not robustly have complete sets.

We now turn in more detail to describing what is currently known in the literature regarding robust completeness. Sipser [Sip82] first studied this notion, and showed that NP ∩ coNP and random polynomial time (R) do not robustly have m-complete sets. However, as alluded to in the first paragraph, Gurevich [Gur83] proved that, in each relativized world, NP ∩ coNP has m-complete sets if and only if NP ∩ coNP has T-complete sets. Thus, NP ∩ coNP cannot distinguish robust m-completeness from robust T-completeness. Ambos-Spies [Amb86] extended this by showing that no class closed downwards under Turing reductions can distinguish robust m-completeness from robust T-completeness.

Thus, the only candidates for distinguishing robust m-completeness from robust T-completeness within PSPACE are those classes in PSPACE that may lack m-complete sets yet that seem not to be closed downwards under Turing reductions. The classes R, UP, and FewP have been shown to potentially be of this form (see, respectively, [Sip82], [HH88], and [HJV93] for proofs that these classes do not robustly have m-complete sets[4]). Unfortunately, these classes are also known to not robustly have T-complete sets [HJV93], and so these classes fail to distinguish robust m-completeness from robust T-completeness.

In fact, to the best of our knowledge, the literature contains only one type of class that distinguishes robust m-completeness from robust T-completeness—and that type is deeply unsatisfying. The type is certain "union" classes—namely, certain classes that either union incomparable classes or that union certain infinite hierarchies of bounded-access classes. Both exploit the fact that if such classes have some m-complete set it must fall into some particular element of the union. An example of the "incomparable" case is that if NP ∪ coNP has m-complete sets then NP = coNP (and NP = coNP is not robustly true [BGS75]). An example (from [HJV93]) of the "infinite union of bounded-access classes"

---

[4] The study of robust completeness has been pursued in many papers. Of particular interest is the elegant work of Bovet, Crescenzi, and Silvestri [BCS92], which abstracts the issue of m-completeness away from particular classes via general conditions. Also, the other method of proving such results has been reasserted, in a very abstract and algebraic form, in the recent thesis of Borchert [Bor94], which re-poses abstractly the proof approach that was pioneered by Sipser ([Sip82], see also [Reg89]). Like the Bovet/Crescenzi/Silvestri approach, this method abstracts away from directly addressing completeness, in the case of this approach via characterizing completeness in terms of the issue of the existence of certain index sets (in the Borchert version, the discussion is abstracted one level further than this). In Section 4 we follow the Sipser/Regan/Borchert "index sets/enumeration" approach, in its non-algebraic formulation.

case is the boolean hierarchy [CGH+88], i.e.,

$$\mathrm{BH} = \{L \mid L \leq_{btt}^{p} \mathrm{SAT}\}.$$

From its definition, it is clear that SAT is T-complete (indeed, even bounded-truth-table complete) for BH. However, if BH had an m-complete set then that set (since it would be in BH) would have to be computable via some k-truth-table reduction to SAT, so there would be a $\hat{k}$ such that $\mathrm{BH} = \{L \mid L \leq_{\hat{k}\text{-}tt}^{p} \mathrm{SAT}\}$, but this is known to not be robustly true [CGH+88].

We at this point mention an interesting related topic that this paper is not about, and with which our work should not be confused. That topic, in contrast to our attempt to distinguish *the existence of* m-complete and T-complete sets for a class, is the study of whether one can merely distinguish *the set of* m-complete and T-complete sets for a class. For example, various conditions (most strikingly, NP does not "have p-measure 0" [LM96]) are known such that their truth would imply that the class of NP-m-complete sets differs from the class of NP-T-complete sets. However, this does not answer our question, as NP robustly has m-complete sets and robustly has T-complete sets. The exact same comment applies to the work of Watanabe and Tang [WT92] that shows certain conditions under which the class of PSPACE-m-complete sets differs from the class of PSPACE-T-complete sets. Also of interest, but not directly related to our interest in the *existence* of complete sets, is the work of Longpré and Young [LY90] showing that within NP Turing reductions can be polynomially "faster" than many-one reductions.

As mentioned at the start of this section, in this paper we prove that $\mathrm{R}_{1\text{-}tt}^{SN}(\mathrm{NP})$ robustly has T-complete sets but does not robustly have m-complete sets. We actually prove the stronger result that $\mathrm{R}_{1\text{-}tt}^{SN}(\mathrm{NP})$ distinguishes robust 1-tt-completeness from robust 2-tt-completeness. This of course implies that there is a relativized world in which $\mathrm{R}_{1\text{-}tt}^{SN}(\mathrm{NP})$ has T-complete (even 2-tt-complete) sets but lacks m-complete (even 1-tt-complete) sets. It is important to note that this is not analogous to the "union" examples given two paragraphs ago. $\mathrm{R}_{1\text{-}tt}^{SN}(\mathrm{NP})$ is not a "union" class. Also, the mere fact that a class is defined in terms of some type of access to NP is not, in and of itself, enough to preclude robust m-completeness, as should be clear from the fact that $\mathrm{R}_{1\text{-}tt}^{p}(\mathrm{NP})$ and $\mathrm{R}_{2\text{-}tt}^{p}(\mathrm{NP})$ robustly have m-complete sets (note: $\mathrm{R}_{1\text{-}tt}^{p}(\mathrm{NP}) \subseteq \mathrm{R}_{1\text{-}tt}^{SN}(\mathrm{NP}) \subseteq \mathrm{R}_{2\text{-}tt}^{p}(\mathrm{NP})$).

Regarding the background of the reducibility $\leq_{1\text{-}tt}^{SN}$, we mention that Homer and Longpré ([HL94, Corollary 5], see also [OW91]) have recently proven that if any set that is $\leq_{m}^{p}$-hard for NP is $\leq_{1\text{-}tt}^{SN}$-reducible (or even $\leq_{btt}^{SN}$-reducible) to a sparse set then the polynomial hierarchy equals NP. Regarding the class $\mathrm{R}_{1\text{-}tt}^{SN}(\mathrm{NP})$, we consider $\mathrm{R}_{1\text{-}tt}^{SN}(\mathrm{NP})$ to be its most natural form. However, Section 3 proves that this class has many equivalent characterizations (for example, it is exactly the class $\mathrm{P}^{(\mathrm{NP}\cap\mathrm{coNP},\mathrm{NP})}$—what a P machine can compute via one $\mathrm{NP} \cap \mathrm{coNP}$ query made in parallel with one NP query). Section 3 also gives a candidate language for $\mathrm{R}_{1\text{-}tt}^{SN}(\mathrm{NP})$ (namely $\mathrm{PrimeSAT} = \{\langle i, F\rangle \mid i \in$

PRIMES $\iff F \in \mathrm{SAT}\}$) and notes that though $\mathrm{R}^{SN}_{1\text{-}tt}(\mathrm{NP}) \subseteq \mathrm{DP}$,[5] the containment is strict unless the polynomial hierarchy collapses.

2 Preliminaries

For standard notions not defined here, we refer the reader to any computational complexity textbook, e.g., [BC93,Pap94,BDG95].

Unless otherwise stated or otherwise obvious from context, all strings will use the alphabet $\Sigma = \{0,1\}$ and all sets will be collections of such strings. For every set A we will denote the characteristic function of A by χ_A. $A^{\leq k}$ denotes $\{x \mid x \in A \wedge |x| \leq k\}$. The general notion of strong nondeterministic reducibility was introduced, though not under that name, by Selman [Sel78]. The literature contains two potentially different notions of strong nondeterministic *truth-table* reducibility, those of Long [Lon82] and Homer and Longpré [HL94]. (The notions differ, for example, regarding whether the query generation is single-valued or multivalued.) Throughout this paper, we use the notion of Homer and Longpré.

Definition 1. (see [Sel94b]) A function f is in $\mathrm{NPSV_t}$ if there exists a nondeterministic polynomial-time Turing machine N such that, on each input x, it holds that

1. at least one computation path of $N(x)$ is an accepting path that outputs $f(x)$, and
2. every accepting computation path of $N(x)$ computes the same value, i.e., $f(x)$. (Note: rejecting computation paths are viewed as having no output.)

Definition 2. [HL94] For any constant k we say A is k-truth-table strong nondeterministic reducible to B ($A \leq^{SN}_{k\text{-}tt} B$) if there is a function in $\mathrm{NPSV_t}$ that computes both (a) k strings $x_1, x_2, \cdots, x_k$ and (b) a predicate, α, of k boolean variables, such that $x_1, x_2, \cdots, x_k$ and α satisfy:

$$x \in A \iff \alpha(\chi_B(x_1), \chi_B(x_2), \cdots, \chi_B(x_k)).$$

Let $\mathcal{C}$ be a complexity class. We say $A \leq^{p,\mathcal{C}[1]}_m B$ if and only if there is a function $f \in \mathrm{FP}^{\mathcal{C}[1]}$ (i.e., computable via a deterministic polynomial-time Turing machine allowed one query to some oracle from $\mathcal{C}$) such that, for all x, $x \in A \iff f(x) \in B$.

As is standard in the literature, for any strings of symbols a and b for which $\leq^b_a$ is defined and any class $\mathcal{C}$, let $\mathrm{R}^b_a(\mathcal{C}) = \{L \mid (\exists C \in \mathcal{C})[L \leq^b_a C]\}$.

Let $\langle \cdot, \cdot \rangle$ be any fixed pairing function with the standard nice properties (polynomial-time computability, polynomial-time invertibility).

We use DPTM (NPTM) as shorthand for "deterministic (nondeterministic) polynomial-time oracle Turing machine," and we treat non-oracle Turing machines as oracle Turing machines that merely happen not to use their oracle tapes. Without loss of generality, we henceforward assume that DPTMs and

[5] $\mathrm{DP} = \{L \mid (\exists L_1, L_2 \in \mathrm{NP})[L = L_1 - L_2]\}$ [PY84].

NPTMs are clocked with clocks that are independent of the oracle. $M^A(x)$ denotes the computation of the DPTM M with oracle A on input x. At times, when the oracle is clear from context, we may write $M(x)$, omitting the oracle superscript(s) (such as $M^A(x)$).

Let $\{M_i\}$ and $\{N_i\}$ respectively be enumerations of deterministic and nondeterministic polynomial-time oracle Turing machines. Without loss of generality, let these enumerations be such that M_i and N_i run in (respectively, deterministic and nondeterministic) time $n^i + i$ and let them also be such that given i one can in polynomial time derive (as Turing machine code) M_i and N_i.

Definition 3. Let $\mathcal{C}$ and $\mathcal{D}$ be complexity classes.

1. [HHW] Let $M^{A:B}$ denote a DPTM M making one query to oracle A followed by one query to oracle B.[6] Let

$$\mathrm{P}^{\mathcal{C}:\mathcal{D}} = \{L \mid (\exists C \in \mathcal{C})(\exists D \in \mathcal{D})(\exists \text{ DPTM } M)[L = L\left(M^{C:D}\right)]\}$$

2. [HHH96] Let $M^{(A,B)}$ denote a DPTM M making, simultaneously, one query to oracle A and one query to oracle B. Let

$$\mathrm{P}^{(\mathcal{C},\mathcal{D})} = \{L \mid (\exists C \in \mathcal{C})(\exists D \in \mathcal{D})(\exists \text{ DPTM } M)[L = L\left(M^{(C,D)}\right)]\}.$$

Classes of the form $\mathrm{P}^{\mathcal{C}:\mathcal{D}}$ were introduced and studied by Hemaspaandra, Hempel, and Wechsung [HHW] for the case in which $\mathcal{C}$ and $\mathcal{D}$ are levels of the boolean hierarchy, and were studied by the present authors [HHH96] for the case in which $\mathcal{C}$ and $\mathcal{D}$ are both levels of the polynomial hierarchy. Both of those papers study the effect of the *order* of database access on the power of database-accessing machines. That line of research has led recently to the very counterintuitive result that, for each $k > 2$: $\Sigma_k^p = \Pi_k^p \iff \mathrm{P}^{\Sigma_k^p[1]} = \mathrm{P}^{\Sigma_k^p[2]}$ [HHH]. Part 2 of Definition 3 is somewhat related to work of Selivanov [Sel94a]. Selivanov studied refinements of the polynomial hierarchy. Among the classes he considered, those closest to the classes we study in this paper are his classes

$$\Sigma_i^p \triangle \Sigma_j^p = \{L \mid (\exists A \in \Sigma_i^p)(\exists B \in \Sigma_j^p)[L = A \triangle B]\},$$

where $A \triangle B = (A - B) \cup (B - A)$. Note, however, that his classes seem to be different from our classes. This can be immediately seen from the fact

[6] We do not describe the mechanics of having two oracles, as any natural approach will do in the contexts with which we are dealing. For example, oracle machines can all have one oracle tape, with the query to the first oracle being contained in the tape cells to the right of the origin and the query to the second oracle being contained in the tape cells to the left of the origin, and with only the appropriate half being erased after entering the distinguished state denoting a query to that half. Alternatively and perhaps more naturally, one can allow the oracle machine to have one oracle tape per oracle.

that all our classes are closed under complementation, but the main theorem of Selivanov [Sel94a] states that no class of the form $\Sigma_i^p \triangle \Sigma_j^p$, with $i > 0$ and $j > 0$, is closed under complementation unless the polynomial hierarchy collapses. Nonetheless, the class $\Sigma_i^p \triangle \Sigma_j^p$ is not too much weaker than $\mathrm{P}_{1,1\text{-tt}}^{\Sigma_i^p,\Sigma_j^p}$, as it is not hard to see (by easy manipulations if $i \neq j$, and from the work of Wagner [Wag90] and Köbler, Schöning, and Wagner [KSW87] for the $i = j$ case) that, for all i and j, it holds that $\{L \mid (\exists L' \in \Sigma_i^p \triangle \Sigma_j^p)[L \leq_{1\text{-tt}}^p L']\} = \mathrm{P}_{1,1\text{-tt}}^{\Sigma_i^p,\Sigma_j^p}$.

3 Equivalent forms of $\mathrm{R}_{1\text{-}tt}^{SN}(\mathrm{NP})$

In this section, we consider the class $\mathrm{R}_{1\text{-}tt}^{SN}(\mathrm{NP})$ and note that this class is quite oblivious to definitional variations; it has many equivalent forms.

The following lemma was proven in [HHH96] and will be useful here.[7]

Lemma 4. *[HHH96] If $\mathcal{C}_1$ and $\mathcal{C}_2$ are classes such that $\mathcal{C}_1$ is closed downwards under $\leq_m^{p,\mathcal{C}_2[1]}$ then*

$$\mathrm{P}^{\mathcal{C}_1:\mathcal{C}_2} = \mathrm{P}^{\mathcal{C}_2:\mathcal{C}_1} = \mathrm{P}^{(\mathcal{C}_1,\mathcal{C}_2)}.$$

Now we are prepared to state the main theorem of this section, Theorem 5. It will follow easily from this theorem that $\mathrm{R}_{1\text{-}tt}^{SN}(\mathrm{NP})$ is equivalent to ordered access to NP and NP∩coNP, and also to parallel access to NP and NP ∩ coNP—with one query to each of NP and NP ∩ coNP allowed in each case. Theorem 5's proof uses the following technique. The theorem deals with $\mathrm{R}_{1\text{-}tt}^{SN}(\mathcal{C})$, i.e., with a certain type of 1-truth-table reduction. A 1-truth-table has two bits of information—what to do (accept versus reject) if the answer is yes, and what to do (accept versus reject) if the answer is no—contained in the truth-table itself. Additionally, in $\mathrm{R}_{1\text{-}tt}^{SN}(\mathcal{C})$ there is information in the (yes/no) answer from the $\mathcal{C}$ query. The key trick in the proof (this occurs in the proof that $\mathrm{R}_{1\text{-}tt}^{SN}(\mathcal{C}) \subseteq \mathrm{P}^{(\mathrm{NP}\cap\mathrm{coNP},\mathcal{C})}$) is to restructure this so that the effect of the 1-truth-table reduction to a $\mathcal{C}$ query is simulated by one query each to NP ∩ coNP and $\mathcal{C}$. In effect, the NP ∩ coNP returns, in its one-bit answer, enough information about the two-bit truth-table that the base machine, working hand-in-hand with the $\mathcal{C}$ query, can make do with the one bit rather than two.

Theorem 5. *For every class $\mathcal{C}$ that is closed downwards under $\leq_m^{p,\mathrm{NP}\cap\mathrm{coNP}[1]}$ we have*

$$\mathrm{R}_{1\text{-}tt}^{SN}(\mathcal{C}) = \mathrm{P}^{\mathrm{NP}\cap\mathrm{coNP}:\mathcal{C}} = \mathrm{P}^{\mathcal{C}:\mathrm{NP}\cap\mathrm{coNP}} = \mathrm{P}^{(\mathrm{NP}\cap\mathrm{coNP},\mathcal{C})}.$$

For any classes $\mathcal{C}_1$ and $\mathcal{C}_2$, let $\mathcal{C}_1 \ominus \mathcal{C}_2 =_{\mathrm{def}} \{L \mid (\exists A \in \mathcal{C}_1)(\exists B \in \mathcal{C}_2)[L = A - B]\}$.

Theorem 6. *For every class $\mathcal{C}$ that is closed downwards under $\leq_m^{p,\mathrm{NP}\cap\mathrm{coNP}[1]}$, it holds that*

$$\mathrm{P}^{\mathcal{C}[1]} \subseteq \mathrm{R}_{1\text{-}tt}^{SN}(\mathcal{C}) \subseteq \mathcal{C} \ominus \mathcal{C}.$$

---

[7] The theorem, its asymmetry notwithstanding, is not mistyped. One does not need to additionally assume that $\mathcal{C}_2$ is closed downwards under $\leq_m^{p,\mathcal{C}_1[1]}$.

Let us apply to $\mathrm{R}^{SN}_{1\text{-}tt}(\mathrm{NP})$ the results just obtained. The following well-known fact will be helpful.

Lemma 7. $\mathrm{R}_m^{p,\mathrm{NP}\cap\mathrm{coNP}[1]}(\mathrm{NP}) = \mathrm{NP}$.

From Lemma 7 and Theorems 5 and 6 we have the following two corollaries for $\mathrm{R}^{SN}_{1\text{-}tt}(\mathrm{NP})$.

Corollary 8. $\mathrm{R}^{SN}_{1\text{-}tt}(\mathrm{NP}) = \mathrm{P}^{\mathrm{NP}\cap\mathrm{coNP}:\mathrm{NP}} = \mathrm{P}^{\mathrm{NP}:\mathrm{NP}\cap\mathrm{coNP}} = \mathrm{P}^{(\mathrm{NP}\cap\mathrm{coNP},\mathrm{NP})}$.

Corollary 9. $\mathrm{P}^{\mathrm{NP}[1]} \subseteq \mathrm{R}^{SN}_{1\text{-}tt}(\mathrm{NP}) \subseteq \mathrm{DP}$.

Since $\mathrm{R}^{SN}_{1\text{-}tt}(\mathrm{NP})$ is closed under complementation but DP is suspected not to be, the second inclusion probably is strict (we note in passing that, due to the closure under complementation of $\mathrm{R}^{SN}_{1\text{-}tt}(\mathrm{NP})$, $\mathrm{R}^{SN}_{1\text{-}tt}(\mathrm{NP}) \subseteq \mathrm{DP} \iff \mathrm{R}^{SN}_{1\text{-}tt}(\mathrm{NP}) \subseteq \mathrm{DP} \cap \mathrm{coDP}$).

Corollary 10. *If* $\mathrm{R}^{SN}_{1\text{-}tt}(\mathrm{NP}) = \mathrm{DP}$ *(equivalently, if* $\mathrm{R}^{SN}_{1\text{-}tt}(\mathrm{NP}) = \mathrm{DP} \cap \mathrm{coDP}$*) then the boolean (and thus the polynomial [Kad88]) hierarchy collapses.*

Though Corollary 10 gives strong evidence that the second inclusion of Corollary 9 is strict, we know of no class collapse that follows from the assumption that the first inclusion is not strict (though it is easy to directly construct an oracle relative to which the first inclusion is strict, and clearly the first inclusion must be strict in the relativized world we are going to construct in Section 4 in which $\mathrm{R}^{SN}_{1\text{-}tt}(\mathrm{NP})$ lacks m-complete sets). Can one prove that $\mathrm{P}^{\mathrm{NP}[1]} = \mathrm{R}^{SN}_{1\text{-}tt}(\mathrm{NP})$ implies some surprising collapse of complexity classes?

What types of sets are in $\mathrm{R}^{SN}_{1\text{-}tt}(\mathrm{NP})$? Define $\mathrm{PrimeSat} = \{\langle i, f\rangle \mid i \in \mathrm{PRIMES} \iff f \in \mathrm{SAT}\}$. Clearly $\mathrm{PrimeSat} \in \mathrm{P}^{(\mathrm{NP}\cap\mathrm{coNP},\mathrm{NP})}$ and thus, by Corollary 8, $\mathrm{PrimeSat} \in \mathrm{R}^{SN}_{1\text{-}tt}(\mathrm{NP})$. On the other hand, $\mathrm{PrimeSat} \in \mathrm{P}^{(\mathrm{ZPP}\cap\mathrm{UP}\cap\mathrm{coUP},\mathrm{NP})}$ (since $\mathrm{PRIMES} \in \mathrm{ZPP} \cap \mathrm{UP} \cap \mathrm{coUP}$ [AH87,FK92]), so it seems somewhat unlikely that PrimeSat is m-complete for $\mathrm{R}^{SN}_{1\text{-}tt}(\mathrm{NP})$. In fact, though (see the discussion in Section 4) $\mathrm{R}^{SN}_{1\text{-}tt}(\mathrm{NP})$ robustly has 2-tt-complete sets, nonetheless $\mathrm{R}^{SN}_{1\text{-}tt}(\mathrm{NP})$ may well lack 1-tt-complete sets. In fact, we will in the next section construct a relativized world in which $\mathrm{R}^{SN}_{1\text{-}tt}(\mathrm{NP})$ has no 1-tt-complete set.

Finally, we mention that $\mathrm{P}^{\mathrm{NP}\cap\mathrm{coNP}:\mathrm{NP}} = \mathrm{R}^{SN}_{1\text{-}tt}(\mathrm{NP})$ is a case where guarded database (oracle) access seems more powerful than standard access. Due to space limitations here, details can be found in the full version of this paper, which is available as UR-CS-TR-635 at http://www.cs.rochester.edu/trs/.

4 Completeness

In this section we prove that there is a relativized world in which $\mathrm{R}^{SN}_{1\text{-}tt}(\mathrm{NP})$ has no 1-tt-complete sets (and thus no m-complete sets). We will note that $\mathrm{R}^{SN}_{1\text{-}tt}(\mathrm{NP})$ robustly has 2-tt-complete sets. Thus we show even more, namely that $R^{SN}_{1\text{-}tt}(NP)$ distinguishes robust 1-tt-completeness from robust 2-tt-completeness (and thus it also distinguishes robust m-completeness from robust T-completeness).

To discuss relativized completeness we must define relativized reductions and the natural relativizations of our classes. So that our theorems are fair, we choose full relativizations (see [Rog67]), i.e., relativizations in which both the reductions and the classes may access the oracle. However, as Theorem 12 will show, many different statements regarding completeness—some involving partial relativizations—are equivalent. In fact, we will make use of some of these equivalences in proving our result.

Definition 11. 1. Let $\leq^{SN,A}_{1\text{-}tt}$ be as in Definition 2, except with $NPSV_t$ replaced by $NPSV^A_t$.

2. (Full relativization of $R^{SN}_{1\text{-}tt}(NP)$) $(R^{SN}_{1\text{-}tt}(NP))^A =_{def} R^{SN,A}_{1\text{-}tt}(NP^A)$, i.e., $\{L \mid (\exists C \in NP^A)[L \leq^{SN,A}_{1\text{-}tt} C]\}$.

3. Let $\leq^{p,A}_{1\text{-}tt}$ (respectively, $\leq^{p}_{1\text{-}tt}$ [LLS75]) be as in Part 1 of the present definition (respectively, as in Definition 2) except with $NPSV^A_t$ (respectively, $NPSV_t$) replaced by FP^A_t (respectively, FP_t), where FP_t denotes the deterministic polynomial-time computable functions.[8] Many-one and 2-truth-table reductions are relativized in the obvious analogous ways.

4. (Full relativization of $P^{(NP\cap coNP,NP)}$) $(P^{(NP\cap coNP,NP)})^A =_{def} \{L \mid L$ is recognized by a deterministic polynomial-time Turing machine that makes, in parallel, at most one query to $NP^A \cap coNP^A$ and at most one query to NP^A, and that additionally has—before the parallel round or after the parallel round or both—unlimited access to $A\}$.

Theorem 12. *Let A be any set. All of the following twelve statements are equivalent.*

1. $(R^{SN}_{1\text{-}tt}(NP))^A$ *has* $\leq^{p,A}_{1\text{-}tt}$*-complete sets.*

2. $(R^{SN}_{1\text{-}tt}(NP))^A$ *has* $\leq^{p}_{1\text{-}tt}$*-complete sets.*

3. $(R^{SN}_{1\text{-}tt}(NP))^A$ *has* $\leq^{p,A}_{m}$*-complete sets.*

4. $(R^{SN}_{1\text{-}tt}(NP))^A$ *has* $\leq^{p}_{m}$*-complete sets.*

5–8. The same as Parts 1–4, with $(R^{SN}_{1\text{-}tt}(NP))^A$ *replaced by* $(P^{(NP\cap coNP,NP)})^A$.

9–12. The same as Parts 1–4, with $(R^{SN}_{1\text{-}tt}(NP))^A$ *replaced by* $P^{(NP^A\cap coNP^A,NP^A)}$.

[8] This definition is equivalent to the more traditional "generator plus evaluator" 1-truth-table definition with, as is natural, both the generator and the evaluator relativized.

Theorem 12 follows from Lemma 13, Lemma 14, and the fact that $\left(\mathrm{R}^{SN}_{1\text{-}tt}(\mathrm{NP})\right)^A$ is clearly closed downwards under $\leq^{p,A}_{1\text{-}tt}$ reductions.

Lemma 13. *For each set A,*

$$\left(\mathrm{R}^{SN}_{1\text{-}tt}(\mathrm{NP})\right)^A = \left(\mathrm{P}^{(\mathrm{NP}\cap\mathrm{coNP},\mathrm{NP})}\right)^A = \mathrm{P}^{(\mathrm{NP}^A\cap\mathrm{coNP}^A,\mathrm{NP}^A)}.$$

Lemma 13 is essentially a relativized version of part of Corollary 8, plus the observation that the technique used in the second half of the proof of that result (that is, the second half of the proof (omitted here due to space) of Theorem 5, in the case $\mathcal{C} = \mathrm{NP}$) in fact can easily show not just $\left(\mathrm{R}^{SN}_{1\text{-}tt}(\mathrm{NP})\right)^A \subseteq \left(\mathrm{P}^{(\mathrm{NP}\cap\mathrm{coNP},\mathrm{NP})}\right)^A$, but even $\left(\mathrm{R}^{SN}_{1\text{-}tt}(\mathrm{NP})\right)^A \subseteq \mathrm{P}^{(\mathrm{NP}^A\cap\mathrm{coNP}^A,\mathrm{NP}^A)}$.

Most non-completeness proofs of the Sipser/Regan/Borchert school (i.e., proofs based on tainting enumerations) use a bridge between the existence of $\mathcal{C}^A$-$\leq^{p,A}_m$-complete sets and the existence of $\mathcal{C}^A$-$\leq^p_m$-complete sets. Here, we extend that link to also embrace 1-truth-table reductions.

Lemma 14. *Let $\mathcal{D}$ be any class (quite possibly a relativized class, such as $\left(\mathrm{R}^{SN}_{1\text{-}tt}(\mathrm{NP})\right)^A$) that is closed downwards under $\leq^{p,A}_{1\text{-}tt}$ reductions. Then the following four statements are equivalent: (a) $\mathcal{D}$ has $\leq^p_m$-complete sets, (b) $\mathcal{D}$ has $\leq^{p,A}_m$-complete sets, (c) $\mathcal{D}$ has $\leq^p_{1\text{-}tt}$-complete sets, (d) $\mathcal{D}$ has $\leq^{p,A}_{1\text{-}tt}$-complete sets.*

As is standard in the Sipser/Regan/Borchert approach to establishing non-completeness, we wish to characterize the existence of complete sets via the issue of the existence of a certain index set. Lemma 15 does this. Since it is quite similar to the analogous lemmas in previous non-completeness papers (see, e.g., [HH88, Lemmas 2.7 and 4.2]) we do not include the proof. We do, however, mention the following points. The lemma draws freely on Theorem 12. Also, the claim in Lemma 15 regarding P and P^A being equivalent (in that context) is an invocation of a trick from the literature [HH88, p. 134].

Lemma 15. *For every oracle A, $\left(\mathrm{R}^{SN}_{1\text{-}tt}(\mathrm{NP})\right)^A$ has $\leq^{p,A}_{1\text{-}tt}$-complete sets if and only if there exists a* P *set (equivalently, a P^A set) I of index quadruples such that*

1. $(\forall\langle i,j,k,l\rangle)[\langle i,j,k,l\rangle \in I \Rightarrow L(N^A_j) = \overline{L(N^A_k)}]$*, and*
2. $\mathrm{P}^{(\mathrm{NP}^A\cap\mathrm{coNP}^A,\mathrm{NP}^A)} = \{L\left(M_i^{(L(N^A_j),L(N^A_l))}\right) \mid (\exists k)[\langle i,j,k,l\rangle \in I]\}$*, and*
3. $(\forall\langle i,j,k,l\rangle)[\langle i,j,k,l\rangle \in I \Rightarrow (\forall x)[$*In the run of* $M_i^{(L(N^A_j),L(N^A_l))}(x)$*, M_i makes at most one round of truth-table queries and that round consists of at most one query (simultaneously) to each part of its "$(L(N^A_j),L(N^A_l))$" oracle*$]]$.

We now state our non-completeness theorem.

Theorem 16. *There is a recursive oracle A such that $\left(\mathrm{R}^{SN}_{1\text{-}tt}(\mathrm{NP})\right)^A$ lacks $\leq^{p,A}_{1\text{-}tt}$-complete sets.*

For each A, since $\left(\mathrm{R}^{SN}_{1\text{-}tt}(\mathrm{NP})\right)^A \subseteq (\mathrm{R}^{p}_{2\text{-}tt}(\mathrm{NP}))^A$ and $\mathrm{NP}^A \subseteq \left(\mathrm{R}^{SN}_{1\text{-}tt}(\mathrm{NP})\right)^A$, it follows that $\left(\mathrm{R}^{SN}_{1\text{-}tt}(\mathrm{NP})\right)^A$ clearly has $\leq^{p,A}_{2\text{-}tt}$-complete sets. In particular, all sets $\leq^{p,A}_{m}$-complete (or even $\leq^{p,A}_{1\text{-}tt}$-complete) for NP^A are $\leq^{p,A}_{2\text{-}tt}$-complete for $\left(\mathrm{R}^{SN}_{1\text{-}tt}(\mathrm{NP})\right)^A$. (We note in passing that it is also immediately clear that, for each A, $\left(\mathrm{R}^{SN}_{1\text{-}tt}(\mathrm{NP})\right)^A$ has $\leq^{SN,A}_{1\text{-}tt}$-complete sets, e.g., all sets $\leq^{p,A}_{m}$-complete for NP^A.) We may now state the following from Theorem 16.

Corollary 17. *There is a recursive oracle A so that $\left(\mathrm{R}^{SN}_{1\text{-}tt}(\mathrm{NP})\right)^A$ has $\leq^{p,A}_{2\text{-}tt}$-complete sets but has no $\leq^{p,A}_{1\text{-}tt}$-complete sets.*

Weakening this, we have the following.

Corollary 18. *There is a recursive oracle A so that $\left(\mathrm{R}^{SN}_{1\text{-}tt}(\mathrm{NP})\right)^A$ has $\leq^{p,A}_{T}$-complete sets but has no $\leq^{p,A}_{m}$-complete sets.*

In summary, we showed that $\mathrm{R}^{SN}_{1\text{-}tt}(\mathrm{NP})$ distinguishes robust m-completeness from robust T-completeness. Indeed it distinguishes robust 1-tt-completeness from robust 2-tt-completeness. We conjecture that $\mathrm{DP} \cap \mathrm{coDP}$ will also distinguish robust 1-tt-completeness from 2-tt-completeness. However, note that this does not generalize to a claim that $\mathrm{BH}_k \cap \mathrm{coBH}_k$ (where BH_k is the kth level of the boolean hierarchy—see [CGH+88] for the definition of BH_k) distinguishes robust $k-1$-tt-completeness from robust k-tt-completeness. In fact, it is clear that, for each $k \geq 2$, $\mathrm{BH}_k \cap \mathrm{coBH}_k$ robustly has 2-tt-complete sets (in fact, for $k \geq 2$, it follows from the structure of the boolean hierarchy that all $\leq^{p,A}_{m}$-complete sets for BH_{k-1} are $\leq^{p,A}_{2\text{-}tt}$-complete for $\mathrm{BH}_k \cap \mathrm{coBH}_k$).

Acknowledgments

We are deeply indebted to Gerd Wechsung for both his kind encouragement and his valuable guidance. We thank Bernd Borchert for a helpful literature reference and Maren Hinrichs for enjoyable conversations.

References

[AH87] L. Adleman and M. Huang. Recognizing primes in random polynomial time. In *Proceedings of the 19th ACM Symposium on Theory of Computing*, pages 462–469. ACM Press, May 1987.

[Amb86] K. Ambos-Spies. A note on complete problems for complexity classes. *Information Processing Letters*, 23:227–230, 1986.

[BC93] D. Bovet and P. Crescenzi. *Introduction to the Theory of Complexity.* Prentice Hall, 1993.

[BCS92] D. Bovet, P. Crescenzi, and R. Silvestri. A uniform approach to define complexity classes. *Theoretical Computer Science*, 104(2):263–283, 1992.

[BDG95] J. Balcázar, J. Díaz, and J. Gabarró. *Structural Complexity I.* EATCS Monographs in Theoretical Computer Science. Springer-Verlag, 2nd edition, 1995.

[BGS75] T. Baker, J. Gill, and R. Solovay. Relativizations of the P=?NP question. *SIAM Journal on Computing*, 4(4):431–442, 1975.

[Bor94] B. Borchert. *Predicate Classes, Promise Classes, and the Acceptance Power of Regular Languages.* PhD thesis, Universität Heidelberg, Mathematisches Institut, Heidelberg, Germany, 1994.

[CGH$^+$88] J. Cai, T. Gundermann, J. Hartmanis, L. Hemachandra, V. Sewelson, K. Wagner, and G. Wechsung. The boolean hierarchy I: Structural properties. *SIAM Journal on Computing*, 17(6):1232–1252, 1988.

[FK92] M. Fellows and N. Koblitz. Self-witnessing polynomial-time complexity and prime factorization. In *Proceedings of the 7th Structure in Complexity Theory Conference*, pages 107–110. IEEE Computer Society Press, June 1992.

[Gur83] Y. Gurevich. Algebras of feasible functions. In *Proceedings of the 24th IEEE Symposium on Foundations of Computer Science*, pages 210–214. IEEE Computer Society Press, November 1983.

[HH88] J. Hartmanis and L. Hemachandra. Complexity classes without machines: On complete languages for UP. *Theoretical Computer Science*, 58:129–142, 1988.

[HHH] E. Hemaspaandra, L. Hemaspaandra, and H. Hempel. A downward translation in the polynomial hierarchy. In *Proceedings of the 14th Annual Symposium on Theoretical Aspects of Computer Science.* Springer-Verlag *Lecture Notes in Computer Science.*

[HHH96] E. Hemaspaandra, L. Hemaspaandra, and H. Hempel. Query order in the polynomial hierarchy. Technical Report TR-634, University of Rochester, Department of Computer Science, Rochester, NY, September 1996.

[HHW] L. Hemaspaandra, H. Hempel, and G. Wechsung. Query order. *SIAM Journal on Computing.* To appear.

[HJV93] L. Hemaspaandra, S. Jain, and N. Vereshchagin. Banishing robust Turing completeness. *International Journal of Foundations of Computer Science*, 4(3):245–265, 1993.

[HL94] S. Homer and L. Longpré. On reductions of NP sets to sparse sets. *Journal of Computer and System Sciences*, 48(2):324–336, 1994.

[Kad88] J. Kadin. The polynomial time hierarchy collapses if the boolean hierarchy collapses. *SIAM Journal on Computing*, 17(6):1263–1282, 1988. Erratum appears in the same journal, 20(2):404.

[KSW87] J. Köbler, U. Schöning, and K. Wagner. The difference and truth-table hierarchies for NP. *RAIRO Theoretical Informatics and Applications*, 21:419–435, 1987.

[LLS75] R. Ladner, N. Lynch, and A. Selman. A comparison of polynomial time reducibilities. *Theoretical Computer Science*, 1(2):103–124, 1975.

[LM96] J. Lutz and E. Mayordomo. Cook versus Karp-Levin: Separating completeness notions if NP is not small. *Theoretical Computer Science*, 164(1–2):123–140, 1996.

[Lon82] T. Long. Strong nondeterministic polynomial-time reducibilities. *Theoretical Computer Science*, 21:1–25, 1982.

[LY90] L. Longpré and P. Young. Cook Reducibility Is Faster than Karp reducibility in NP. *Journal of Computer and System Sciences*, 41(3):389–401, 1990.

[OW91] M. Ogiwara and O. Watanabe. On polynomial-time bounded truth-table reducibility of NP sets to sparse sets. *SIAM Journal on Computing*, 20(3):471–483, June 1991.

[Pap94] C. Papadimitriou. *Computational Complexity.* Addison-Wesley, 1994.

[PY84] C. Papadimitriou and M. Yannakakis. The complexity of facets (and some facets of complexity). *Journal of Computer and System Sciences*, 28(2):244–259, 1984.

[Reg89] K. Regan. Provable complexity properties and constructive reasoning. Manuscript, April 1989.

[Rog67] H. Rogers, Jr. *The Theory of Recursive Functions and Effective Computability.* McGraw-Hill, 1967.

[Sel78] A. Selman. Polynomial time enumeration reducibility. *SIAM Journal on Computing*, 7(4):440–457, 1978.

[Sel94a] V. Selivanov. Two refinements of the polynomial hierarchy. In *Proceedings of the 11th Annual Symposium on Theoretical Aspects of Computer Science*, pages 439–448. Springer-Verlag *Lecture Notes in Computer Science #775*, February 1994.

[Sel94b] A. Selman. A taxonomy of complexity classes of functions. *Journal of Computer and System Sciences*, 48(2):357–381, 1994.

[Sip82] M. Sipser. On relativization and the existence of complete sets. In *Proceedings of the 9th International Colloquium on Automata, Languages, and Programming*, pages 523–531. Springer-Verlag *Lecture Notes in Computer Science #140*, 1982.

[Wag90] K. Wagner. Bounded query classes. *SIAM Journal on Computing*, 19(5):833–846, 1990.

[WT92] O. Watanabe and S. Tang. On polynomial-time Turing and many-one completeness in PSPACE. *Theoretical Computer Science*, 97(2):199–215, 1992.

Syntactic Characterization in Lisp of the Polynomial Complexity Classes and Hierarchy

Salvatore Caporaso[1], Michele Zito[2], Nicola Galesi[3], and Emanuele Covino[4]

[1] Univ di Bari, Dip. di Informatica, v.Amendola 173, I-70126, logica@gauss.uniba.it
[2] University of Warwick Coventry, CV4 7AL, UK, M.Zito@dcs.warwick.ac.uk
[3] Universitat Politecnica de Catalunya, Barcelona, Galesi@goliat.up.es
[4] Abt. th. Informatik, Universität Ulm, Covino@theorie.informatik.uni-ulm.de

Abstract. The definition of a class $\mathcal{C}$ of functions is *syntactic* if membership to $\mathcal{C}$ can be decided from the construction of its elements. Syntactic characterizations of PTIMEF, of PSPACEF, of the polynomial hierarchy PH, and of its subclasses Δ_n^p are presented. They are obtained by progressive restrictions of recursion in Lisp, and may be regarded as *predicative* according to a foundational point raised by Leivant.

1 Introduction

At least since 1965 [6] people think to complexity in terms of *TM's plus clock or meter*. However, understanding a complexity class may be easier if we define it by means of *operators* instead of *resources*. Different forms of *limited recursion* have been used to this purpose. After the well-known characterizations of LINSPACEF [15] and PTIMEF [5], further work in this direction has been produced (see, for example, [11], [8], [4]).

Both approaches (resources and limited operators) are not *syntactic*, in the sense that membership to a given class cannot be decided from the construction of its elements (for example, if f is primitive recursive (PR) in g and h, we cannot decide whether f is actually bounded above by a third function k). And both approaches may be criticized on foundational grounds. The definition of an entity E is *impredicative* (see Poincaré [14], p. 307) if it uses a variable defined on a domain including E. Examples of impredicative definitions are $\sqrt{2} =_{df} \max z(z^2 \leq 2)$ and $\mathrm{Pow}(x) =_{df} \{y | y \subseteq x\}$. The definition of, say, PTIMEF, by means of the (predicative) class of all T-computable functions, might be regarded as impredicative too. For a better position of the problem, and for a remarkable solution in proof-theoretic terms, see Leivant [9].

The first purely syntactic definition of PTIMEF, based on a form of unlimited PR on binary numerals is in [1]. Further characterizations of the same class are in [12] and [10], using finite automata and, respectively, λ-calculus. A syntactic definition of PTIMEF and LINTIMEF, by a tortuous variant of TM's, is in [2]. PSPACEF has been studied less. We are not aware of any recursive characterization (even impredicative) of the polynomial hierarchy PH.

In this paper we define a number of fragments of Lisp, by means of a progressive sequence of restrictions to (unlimited) recursion; and we show the equivalence between these fragments and the polynomial classes. Lisp has been chosen, instead of other models of computation, because it offers the obvious advantages of a richer data type and of a higher-level language, and because it fits traditional mathematical methods of investigation, like induction on the construction of functions and arguments. A preliminary validation of this choice is discussed in the last section of this paper, together with perspectives and other aspects of our work.

We now outline the adopted recursion schemes. A function $f[\mathbf{x}; y]$ is defined by *course-of-values recursion* if its value depends on a pre-assigned number n of values $f[\mathbf{x}; y_i]$ for n previous values of y. What makes the difference is the meaning of *previous.* For PSPACEF and PH we mean any z such that $|z| < |y|$ (that is we may choose n values among $O(2^{c|y|})$ previous values). For PTIMEF we mean any subexpression of y (n among $O(|y|)$ values). The restriction of PSPACE to PH is obtained by asking that the *invariant* function of the recursion be in the form $f[\mathbf{x}; y_1]$ *or* ... *or* $f[\mathbf{x}; y_n]$. Classes Δ_k are defined by counting in the most obvious way the levels of nesting of this form of recursion.

A rather extreme formulation of an aspect of the work presented here is that it allows a position of some celebrated problems in terms of comparison between similar operators, of an apparently increasing strength, instead than in terms of contrast between heterogeneous resources.

2 Recursion Free Lisp

An *atom* is a sequence of capital letters and decimal digits. A special role is assigned to atoms $T(F)$, associated with the truth-values *true (false)*, and NIL. An (S-)*expression* is an atom, or a *dotted couple* $(x \cdot y)$, where x and y are expressions. $\omega, \omega_1, \ldots$ are (variables defined on the) atoms; $s, \ldots, z, s_1, \ldots$ are S-expressions. $\mathbf{s}, \ldots, \mathbf{z}$ are *tuples of expressions* of the form $x_1; \ldots; x_n$ $(n \geq 0)$. An *(S-)function* f takes a tuple of arguments $\mathbf{x}$ into an expression $f[\mathbf{x}]$; d, e, f, g, h are functions, and $\mathbf{d}, \mathbf{d}^{\,1} \ldots$ are tuples of functions. If a tuple of syntactical entities has been introduced by means of a notation of the form $\mathbf{E}$, we denote by E_i its i-th member (for example x_i, z_i^j are the i-th expression of $\mathbf{x}, \mathbf{z}^{\,j}$; and f_i^j is the i-th function of $\mathbf{f}^{\,j}$).

A *list* is an expression of the particular form we now describe: atom NIL is the *empty list*, also denoted by (); all other lists x are in the form $(x_n \cdot (\ldots \cdot (x_1 \cdot \mathrm{NIL}) \ldots)$, and are shown as $(x_n, \ldots, x_1)$; $(x)_i = x_i$ is the *i-th component of* x, and $\#(x) = n \geq 0$ is its *number of components*.

Sometimes, along a computation, we mark an (occurrence of an) expression x by a superscript $\tau = A, B, AB$, and we say that x^τ is of type τ; when x has not been marked, we say that it is of type 0, and we write x^0. (Thus marked S-expressions are the actual constants of our language.) The type of all atoms is 0. The type of all non-atomic sub-expressions of x^τ is τ. A relation of compatibility is established by stating that:

1. all expressions are compatible with those of type 0;
2. all expressions of type $\tau \neq 0$ are incompatible with those of their same type τ and with those of type AB.

$\mathbf{x}^{\tau}$ ($\mathbf{x}^{\neq\tau}$) is a tuple of variables of the same type τ (of type $\neq \tau$). Types are not specified in the definition of a function, when they don't change (cf. 2.2.2) or when they don't affect the result (cf. 2.2.1).

2.1 Basic functions

The class $\mathcal{B}$ of the *basic functions* consists of:

1. *predicates at* and *eq*, such that $at[x] = T(F)$ if x is (not) an atom, and $eq[x; y] = T(F)$ if x and y are (not) the same atom;
2. the *conditional* $cond[x; y; z] = y$ if $x = T$, and $= z$ if $x \neq T$; $cond[x_1; y_1; \ldots; cond[x_n; y_n; z] \ldots]$ is usually displayed as $[x_1 \to y_1; \ldots; x_n \to y_n; T \to z]$.
3. the *selectors* $sel^n_j[\mathbf{x}] = x_j$, and, for every atom ω, the *constant functions* $\omega[\mathbf{x}] = \omega$; we often let these functions be denoted by their results; *id* is the *identity* sel^1_1;
4. the *predecessors car* and *cdr*, such that: $car[\omega] = cdr[\omega] = \omega$; $car[(y \cdot z)] = y$ and $cdr[(y \cdot z)] = z$; sometimes we write x' for $car[x]$ and x'' for $cdr[x]$;
5. the *constructor*
$$cons[x^{\tau_1}; y^{\tau_2}] = \begin{cases} (\mathrm{NIL})^{AB} & \text{if the arguments are incompatible} \\ (x \cdot y)^0 & \text{if both } \tau_i \text{ are } 0 \\ (x \cdot y)^{\tau} & \text{if one of the } \tau_i \text{ is } \tau \text{ and the other is } 0 \\ (x \cdot y)^{AB} & \text{if one of the } \tau_i \text{ is A and the other is B;} \end{cases}$$
6. functions α, β, ζ, which leave un-changed the atoms, and such, otherwise, that $\alpha[x^0] = x^A, \beta[x^0] = x^B$, $\alpha[x^{\neq 0}] = \beta[x^{\neq 0}] =$NIL; $\zeta[x^{\tau}] = x^0$;
7. function *unite*, which leaves its argument x un-changed if x' or x'' are not lists; and takes $((x_1, \ldots, x_m), x_{m+1}, \ldots, x_{m+n})$ into $(x_1, \ldots, x_{m+n})$ otherwise; for example $unite[((A, B, C), D, E)] = (A, B, C, D, E)$.

These basic functions differ from those of pure Lisp for a few changes, adopted to handle the types and to exclude marginal cases of undefined functions.

2.2 Substitutions

By a composite notation like $f[\mathbf{x}]$, we mean that all arguments of f occur (not necessarily once) in $\mathbf{x}$, but we don't imply that every x_i is an actual argument of f.

A main difference with pure Lisp is that we renounce to its λ's to show substitutions (SBST) explicitly, by replacing the substituted variable with the substituend function. This rudimental way allows simpler definitions and space-complexity evaluations, at the price of a systematic ambiguity between functions and values. Thus, deciding for example whether $car[x]$ and $car[y]$ are *the same thing* is left to context. A SBST to an absent variable has no effect; all occurrences of the substituted variable are replaced by the substituend function. No kind of disjunction between original and new variables is assumed.

We write $\mathbf{f}[\mathbf{x}]$ for $f_1[\mathbf{x}]; \ldots; f_n[\mathbf{x}]$. Given n functions $\mathbf{h}[\mathbf{u}]$, and given $g[\mathbf{x}; \mathbf{z}]$, we write $g[\mathbf{x}; \mathbf{h}[\mathbf{u}]]$ for the *simultaneous SBST* of $\mathbf{h}[\mathbf{u}]$ to the n variables $\mathbf{z}$ in g. The special form of substitution we now introduce allows to by-pass the type-restrictions on the *cons*'s one should otherwise handle, in order to re-assemble the parts of the argument, after processing them separately.

Definition 1. The unary function f is defined by *internal substitution* (INSBST) in $g_1, \ldots, g_k$ if we have

$$f[x] = \begin{cases} \text{NIL} & \text{if } \#(x) < k \\ (g_1[(x)_1], \ldots, g_k[(x)_k], (x)_{k+1}, \ldots, (x)_{\#(x)}) & \text{otherwise,} \end{cases}$$

or

$$\begin{cases} f[\omega] & = \text{NIL} \\ f[(u \cdot w)] & = (g_1[u] \cdot g_2[w]); \end{cases}$$

Notation: $f = \pi(\mathbf{g})$. Functions $\mathbf{g}$ are the *scope* of the *INSBST*.

Given a class $\mathcal{C}$ of functions, we denote by $\mathcal{C}^*$ its closure under SBST and INSBST. For example, the class of all *recursion-free* functions is $\mathcal{B}^*$.

2.3 Lengths

The *length* $|z|$ of z is the number of atoms and dots occurring in (the value assigned to) z. $|\mathbf{x}|$ and $\max(\mathbf{x})$ are respectively $\sum_i |x_i|$ and $\max_i(|x_i|)$.

$|f[\mathbf{x}]|$ is the length of the value of $f[\mathbf{x}]$ when a system of values is assigned to $\mathbf{x}$; $|\mathbf{f}[\mathbf{x}]|$ is $\sum_i |f_i[\mathbf{x}]|$. For example $|cons[x; x]| = 2|x| + 1$; $|y''| \leq \max(1, |y| - 2)$. We say that $f[\mathbf{x}]$ is *limited* by the numerical function ϕ (possibly a constant) if for all $\mathbf{x}$ we have $|f[\mathbf{x}]| \leq \phi(|\mathbf{x}|)$.

Define $lh_c(f)$ to be $2n + 1$, where n is the number of *cons* occurring in the construction of f.

The idea of next lemma is rather simple: types allow cobbling together, without any limitation, the arguments of type 0; but at most one A and/or one B may contribute to the function being computed.

Lemma 2. *For all recursion-free function f in which ζ doesn't occur, we have*

$$|f[\mathbf{x}^{\,0}; \mathbf{s}^{\,1A}; \mathbf{s}^{\,2B}]| \leq lh_c(f) \max(1, |\mathbf{x}|) + \max(\mathbf{s}^{\,1}) + \max(\mathbf{s}^{\,2}).$$

Proof. Let us write m for $lh_c(f)\max(1, |\mathbf{x}|)$, and M_i for $\max(\mathbf{s}^{\,i})$. We show that $z^{\tau} = f[\mathbf{x}; \mathbf{s}^{\,1}; \mathbf{s}^{\,2}]$ implies $|z| \leq m + n$, where:

$\tau = 0$ (case 1) implies $n = 0$;

$\tau = A$ ($\tau = B$) (case 2) implies $n \leq M_1$ ($n \leq M_2$); and

$\tau = AB$ (case 3) implies $n \leq M_1 + M_2$.

Induction on the construction of f. Base. Assume that f is *cons*, since else the result is trivial. We have, for example, $|cons[t^A; t^B]| = 2|t^{AB}| + 1 \leq 3 + 2|t|$. Etc.

Step. (1) Let us first assume that f begins by a basic function. Then we may assume further that the form of f is $cons[g_1[\mathbf{x}; \mathbf{s}^{1A}; \mathbf{s}^{2B}]; g_2[\mathbf{x}; \mathbf{s}^{1A}; \mathbf{s}^{2B}]]$, since the lemma is an immediate consequence of the ind. hyp. for all other basic functions. Let $g_i[\mathbf{x}; \mathbf{s}^1; \mathbf{s}^2] = z_i^{\tau_i}$, $i = 1, 2$; thus $z^\tau = cons[z_1; z_2]$. Let us write m_i for $lh_c(g_i)\max(1, |\mathbf{x}|)$ Cases 1-3 as above.
Case 1. We have $\tau_1 = \tau_2 = 0$ The ind. hyp. gives $|z_i| \leq m_i$. The result follows, since $lh_c(f) = lh_c(g_1) + lh_c(g_2) + 1$.
Case 2. We have $\tau = A$ or $\tau = B$, one of the τ_i is τ, and the other is 0; let for example $\tau_1 = B$. The ind. hyp. gives $|g_1| \leq m_1 + M_2$, $|g_2| \leq m_2$. The result follows by immediate computations.
Subcase 3.1. One of the τ_i, say τ_1, is A, and the other is B. The ind. hyp. gives $|z_i| \leq m_i + M_i$ and the result follows immediately, since $m_1 + m_2 \leq m$.
Subcase 3.2. One of the τ_i is AB, and the other is 0. Similarly.
(2) The possibility remains that the form of f is $\pi(g_1, \ldots, g_k)[h[\mathbf{x}; \mathbf{s}^1; \mathbf{s}^2]]$. Let $h[\ldots] = y^\tau$. Assume $\#(y) = k$. Case 1. $\tau \neq 0$. Then, since all components of y are of the same type τ, the ind. hyp. gives $|f| \leq \sum_i(|(y)_i| + lh_c(g_i)) + k - 1 \leq |y| + \sum_i(lh_c(g_i))$, and the result follows by the ind. hyp. applied to h, since $lh_c(f) \geq lh_c(h) + \sum_i lh_c(g_i)$. Case 2. $\tau = 0$. Immediately from the ind. hyp., applied to h and to the g's.

2.4 Some classes of recursion-free functions

(1) A *proper cut of order n* is a composition of $n \geq 0$ predecessors. We regard the identity as an *improper cut* of order 0. Two cuts g_1, g_2 are *disjoint* if they don't return two overlapping sub-expressions of their argument. In syntactic terms, the g's are disjoint if for no g_i there is a cut h, such that $g_i[x] = h[g_{3-i}[x]]$. A *fully disjoint tuple (of cuts) C* is a tuple e, such that: (a) every e_i is either a cut g_i, or is in the form $\pi(\mathbf{h}^i)$, where every h_j^i is a cut; and (b) all couples g_i, h_j^i are disjoint.

Define the cuts $1st, 2d, 3d, \ldots, i\text{-}th$, such that $i \leq \#(x)$ implies $i\text{-}th[x] = (x)_i$; any tuple of cuts of this form is an example of fully disjoint tuple.
(2) For every list $y = (\omega_1, \ldots, \omega_n)$, we call *unary append (of order n)*, and we denote by $app(y)$, function $cons[\omega_1; [\ldots; cons[\omega_n; x] \ldots]]$; if x is a list, it appends its components to those of y. For example, for $y = (A, B)$ and for every list $x = (x_1, \ldots, x_n)$, we have $app(A, B)[x] = (A, B, x_1, \ldots, x_n)$.
Define $list[x] = cons[x; \mathrm{NIL}[x]]$; for all n define $list[x_{n+1}; \ldots; x_1] = cons[x_{n+1}; list[x_n; \ldots; x_1]]$.

For example, we have $list[()] = (())$; $list[A, (A), ((A))] = (A, (A), ((A)))$.
(3) Define the *sentential connectives not, or, and* from

$$not[x] = [x \to F; T \to T], \quad x \text{ or } y = [x \to T; y \to T; T \to F].$$

A *simple boolean* function is built-up from eq, at and the connectives. A *boolean* function is obtained by substitution of some cuts to some variables in a simple boolean function.
(4) For all list of atoms q, s, t define functions $g(q, s, t)$ by (see proof of Lemma 3 for their use) $g(q, s, t)[x] = \pi(app(s), q, t, cdr)$; we have

$$g(q, s, t)[((x_1, \ldots, x_n), u, w, (y_1, \ldots, y_m))] = ((s, x_1, \ldots, x_n), q, t, (y_2, \ldots, y_m)).$$

(5) A function is *trivially decreasing* if is a proper cut; or if it is in the form $\pi(g_1, \ldots, g_m)$, and: (a) every g_i is a cut, or a unary *app*; and (b) the sum of the orders of all cuts is higher than the sum of the orders of all unary *app*'s. For example, $\pi(app(T), 3d, id)$ is trivially decreasing. If $f = \pi(g_1, \ldots, g_m)$ is trivially decreasing, and if no $g_i[y]$ is an atom, then $|f[y]| < |y|$.

3 Recursion schemes

An obvious condition to ensure that a recursion scheme defines total functions is that its recursive calls refer to values of the recursion variable, which preceed, according to some (partial) order, its current value. In the Conclusion, doubts are expressed about closure of the polynomial classes under recursion schemes based on an order isomorphic to the natural numbers. Hence our first restriction is to the order $x < y$ iff $|x| < |y|$.

Definition 3. Given (1) m *parameters* $\mathbf{x}$, a *principal variable* y, and n *auxiliary variables* $\mathbf{s}$;
(2) an n-ple $\mathbf{d}$ of trivially decreasing functions, together with a *terminating* boolean function $g^*[y]$, depending on the form of the d's in some trival way that we don't specify here;
(3) an *initial* function $g[\mathbf{x}; y]$ and an *invariant* function $h[\mathbf{x}; y; \mathbf{s}]$;
function f is defined by *course-of-values recursion* (CVR) *in* g, h if we have

$$f[\mathbf{x}; y] \begin{cases} g[\mathbf{x}; y] & \text{if } g^*[y] = T \\ h[\mathbf{x}; y; f[\mathbf{x}; d_1[y]]; \ldots; f[\mathbf{x}; d_n[y]]]] & \text{otherwise.} \end{cases}$$

The following example shows that an exponential space complexity may easily be reached with very poor means: no nesting, and a single recursive call to the most obvious sub-expression of the recursion variable. Thus restrictions to the invariant h have to be adopted. We have two alternatives: either we drastically impose that h be boolean, or we use the types machinery to rule its growth.

$$\begin{cases} ex[x; \omega] = cons[x; x] \\ ex[x; y] = cons[ex[x; y'']; ex[x; y'']]. \end{cases}$$

Definition 4. 1. Function f is *(recursively) boolean* if is boolean and recursion-free, or if is defined by CVR with boolean invariant function.

2. Function $f[\mathbf{x}; y]$ is defined by *short CVR* (SCV) if it is defined by CVR, if the initial function g is in the class $\mathcal{PL}$ defined below, and if the invariant is boolean.
3. Function f is defined by *or-SCV* (OR-CV) if it is defined by SCV, and the form of its invariant is

$$h[\mathbf{x}; \mathbf{y}; \mathbf{s}] = s_1 \ or \ s_2 \ or \ \ldots \ or \ s_n.$$

4. Function f is defined by *fast CVR* (FCV) if: is defined by SCV; the decreasing functions form a fully disjoint tuple of cuts; and the invariant h is
   (a) either boolean, or
   (b) is recursion-free, and there is a function h^*, in which ζ doesn't occur, and a tuple $\mathbf{e}$ of α's and β's, such that

$$h[\mathbf{x}; y; \mathbf{s}] = \zeta[h^*[\mathbf{x}; y; e_1[s_1]; \ldots; e_n[s_n]]].$$

The sense of clause (b) above is that, if $z_1, \ldots, z_n$ are the previous values of f, then h is not allowed to *cons* any z_i with itself, though it may *cons* at most one of the z's in the scope of an α with at most one of those in the scope of a β.

Examples of FCV. Define the *numeral* $num(m)$ for m to be the list whose $m+1$ components are all 0. A function *mult*, such that $mult[num(h); num(k)] = num(hk)$ may be obtained from function $mult_0$ below, by some trivial changes

$$mult_0[x; y] = \begin{cases} x & \text{if } y \text{ is an atom} \\ \zeta[unite[cons[x; \alpha[mult_0[x; cdr[y'']]]]]] & \text{otherwise.} \end{cases}$$

Thus FCV, with *cdr* as decreasing function, may be regarded as an analogue of number-theoretic PR. Next example shows that, with *car*, *cdr* as decreasing functions, FCV is the analogue of the form of recursion known in Literature as *tree PR*. In the concluding section the advantages of taking less trivial cuts as decreasing functions are discussed. The following function lh computes $num(|y|)$

$$\begin{cases} lh[\omega] = (0) \\ lh[y] \;= \zeta[cons[0; unite[list[\alpha[lh[y']]; \beta[lh[y'']]]]]]. \end{cases}$$

Define the equality by $x = y := eqc[cons[x; y]]$, where eqc is defined by FCV, with $d_1 = \pi(car, car)$ and $d_2 = \pi(cdr, cdr)$, by

$$eqc[y] = \begin{cases} eq[y'; y''] & \text{if } at[y'] \text{ or } at[y''] = T \\ eqc[d_1[y]] \text{ and } eqc[d_2[y]] & \text{otherwise.} \end{cases}$$

Example of OR-CV: SAT. Assume defined function $true[(v, u, w, z)]$, which, if v is a list of atoms and z is (the code for) a sentential formula: (a) assigns true (false) to the i-th literal of z if the i-th component of v is (not) T; (b) returns $T(F)$ if z is true (false) under this truth-assignment. Define by OR-CV, with decreasing tuples

$$d_1 = \pi(app(T), cdr, cdr, id) \quad d_2 = \pi(app(F), cdr, cdr, id)$$

$$st[y] = \begin{cases} true[y] & \text{if } at[(y)_2] \\ st[d_1[y]] \text{ or } st[d_2[y]] & \text{otherwise.} \end{cases}$$

Satisfiability is decided by $sat[x] = st[list[(); lh[x]; lh[x]; x]]$.

Example of SCV : QBF. We show that thePSPACE-complete language QBF is accepted by a function qbf definable in $\mathcal{PSL}$. Let $b, b_1, \ldots$ be *(boolean) literals*, and let ϕ, χ be *quantified boolean formulas.* Let $num_2(i)$ be the binary numeral for i, and define the code ϕ^* for ϕ by

$0^* = T$; $1^* = F$; $b_i^* = (VAR, num_2(i))$; $(\neg\phi)^* = (NOT, \phi^*); (\forall b\phi)^* = (ALL, b^*, \phi^*)$;

$(\exists b\phi)^* = (EX, b^*, \phi^*)$; $(\chi \wedge \psi)^* = (AND, \chi^*, \psi^*)$; $(\chi \vee \psi)^* = (OR, \chi^*, \psi^*)$.

We associate each occurrence $\hat{b}$ of literal b in formula ϕ with a list $AV(\hat{b}, \phi)$, to be used as *address and truth-assignment*, and defined by

1. let ϕ be $\chi \vee \psi$ or $\phi = \chi \wedge \psi$; if $\hat{b}$ is in χ (is in ψ) then $AV(\hat{b}, \phi)$ is $(L, AV(\hat{b}, \chi))$ (is $(R, AV(\hat{b}, \psi))$); it says that $\hat{b}$ is in the left (right) principal sub-formula of ϕ;
2. if ϕ is $\forall(\exists)b_i\chi$, and we wish to assign T, F to the occurrences of b_i in the scope of the indicated quantifier, then $AV(\hat{b}, \phi) = (T, AV(\hat{b}, \chi))$ or, respectively, $(F, AV(\hat{b}, \chi))$.

A function $val[(x, u, z)]$ can be defined in $\mathcal{PL}$, which, by an input of the form $(AV(\hat{b}, \phi), u, \phi^*)$ returns $T(F)$ if $AV(\hat{b}, \phi)$ is the address and truth-assignment of an occurrence in ϕ of a true (false) literal. Define

$$qb[y] = \begin{cases} val[y] & \text{if } at[(y)_2'] \\ [(y)_2' = AND \rightarrow qb[d_{11}[y]] \text{ and } qb[d_{12}[y]]; & \\ (y)_2' = OR \quad \rightarrow qb[d_{11}[y]] \text{ or } qb[d_{12}[y]]; & \\ (y)_2' = ALL \quad \rightarrow qb[d_{21}[y]] \text{ and } qb[d_{22}[y]]; & \\ (y)_2' = EX \quad \rightarrow qb[d_{21}[y]] \text{ or } qb[d_{22}[y]]; & \\ (y)_2' = NOT \quad \rightarrow not[qb[d_3[y]]]; & \\ (y)_2' = VAR \quad \rightarrow qb[d_3[y]]] & \text{otherwise;} \end{cases}$$

function qb is defined by SCV, with the following trivially decreasing tuples

$d_{11} = \pi(app(L), 2d, id)$ $d_{12} = \pi(app(R), 3d, id)$

$d_{21} = \pi(app(T), 3d, id)$ $d_{22} = \pi(app(F), 3d, id)$ $d_3 = \pi(id, 2d, id)$

We can now define $qbf[x] = qb[list[(); x; x]]$.

4 Characterization

Given an operator O taking functions to functions, and a class $\mathcal{C}$ of functions, we write $O(\mathcal{C})$, for the class of all functions obtained by at most one application of O to the elements of $\mathcal{C}$; $O^*(\mathcal{C})$ is the closure of $\mathcal{C}$ under O. Thus, $O(\mathcal{C})^*$ and $O^*(\mathcal{C})^*$ are the closures of $O(\mathcal{C})$ and $O^*(\mathcal{C})$ under substitution.

Definition 5. Define

POLYTIMEF LISP($\mathcal{PL}$, also $\Delta_1^p\mathcal{L}$) $= \text{FCV}^*(\mathcal{B}^*)^*$;

$\Delta_{n+2}^p\mathcal{L}$ $= \text{OR-SCV}(\Delta_{n+1}^p\mathcal{L})^*$;

POLYNOMIAL HIERARCHY LISP($\mathcal{PHL}$) $= \text{OR-SCV}^*(\mathcal{PL})^*$.

POLYSPACEF LISP($\mathcal{PSL}$) $= \text{SCV}^*(\mathcal{PL})^*$.

Theorem 6. *All Lisp classes above are equivalent to the complexity classes their names suggest.*

Proof. We have POLYTIMEF$\subseteq \mathcal{PL}$ by lemma 8. By lemma 7, all functions in $\mathcal{PL}$ are limited by a polynomial; hence, by lemma 9, $\mathcal{PL} \subseteq$POLYTIMEF. By the same lemma, since the invariant in definitions by SCV is boolean, we have $\mathcal{PSL} \subseteq$PSPACEF. We have PSPACEF$\subseteq \mathcal{PSL}$, since, by the example above, the PSPACE-complete set QBF can be decided in $\mathcal{PSL}$, and since $\mathcal{PL} \subseteq \mathcal{PSL}$. Lemma 10 shows the equivalence of the two hierarchies.

5 Equivalence

Lemma 7. 1 *If $f[\mathbf{x}; y]$ is FCV in g and h, with recursion variable y, then there is a constant m such that*

$$|f[\mathbf{x}; y]| \leq m|\mathbf{x}; y| \times |y|.$$

2 *Every function definable in $\mathcal{PL}$ is limited by a polynomial.*

Proof. 1 Notations like under definition 4(4). Assume that h is not boolean, and define $M = \max(lh_c(g), lh_c(h))$. Induction on $|y|$. Base. Immediately by lemma 1 (with s absent). Step. Assume $N := |\mathbf{x}| \geq 1$. Let $\mathbf{s}^{1A}$ denote the tuple of expressions such that $e_j = \alpha$ and $s_j^{1A} = e_j[f[\mathbf{x}; d_j[y]]]$ for some j; similarly for $\mathbf{s}^{2B}$. By lemma 1, since $lh_c(h) \leq M$, we have

$$|f[\mathbf{x}; y]| \leq M(N + |y|) + \max(\mathbf{s}^{1A}) + \max(\mathbf{s}^{2B}).$$

Since $\mathbf{d}$ is fully disjoint, there exist two sub-expressions u and w, such that $\max(\mathbf{s}^{1A}) = |f[\mathbf{x}; u]|$, $\max(\mathbf{s}^{2B}) = |f[\mathbf{x}; w]|$ and $|u| + |w| < |y|$. By the ind. hyp. we then have

$$|f[\mathbf{x}; y]| \leq M(N+|y|)+M(N+|u|)|u|+M(N+|w|)|w| \leq m(n+|y|)(1+|u|+|w|).$$

2 Induction on the construction of f. Step. If f is defined by FCV, part 1 applies. If $f[\mathbf{x}]$ is defined by SBST in $g_1[\mathbf{x}; u]$ of $g_2[\mathbf{x}]$ to u, by the ind. hyp. there are k_1, k_2, such that g_i is limited by $\lambda n.m_i n^{m_i} + m_i$, with $m_i = 2^{k_i}$; f is then limited by $\lambda n.mn^m + m$, with $m = 2^{k_1(k_2+1)}$. If $f[y] = \pi(\mathbf{g})[y]$, the result follows immediately by the ind. hyp. applied to the g's.

5.1 Simulation of TM's

Lemma 8. *All functions computable in polynomial time are definable in $\mathcal{PL}$.*

Proof. We restrict ourselves to TM's with a single semitape, that conclude their operations by entering an endless loop. *Productions* are in the form $(q_i S_j \Rightarrow q_{ij} I_{ij})$ $(i \leq s,\ j \leq t)$ where q_i, q_{ij} are states, S_j is a tape symbol, and *instruction* I_{ij} is a new symbol or $\in \{$*right,left*$\}$. We use the same notations for states (tape symbols) and for their codes, which are lists of s (t) atoms. *Instantaneous*

descriptions (i.d.) are coded by a list of the form (l, q, o, r), where: q and o are the state and the observed symbol; the j-th component of list r (list l) is the list of t atoms coding the j-th symbol at the right (left) of the observed symbol. A recursion-free function $next_M$ can be defined that takes an i.d. of a given TM M into the next one. Its form is
$[eql(q_1)[2d[x]] \to [eql(S_1)[3d[x]] \to exec_{11}; \ldots; eql(S_t)[3d[x]] \to exec_{1t}[x]];$
$\ldots;$
$[eql(q_s)[2d[x]] \to [eql(S_1)[3d[x]] \to exec_{s1}; \ldots; eql(S_t)[3d[x]] \to exec_{st}[x]];,$
where, for all lists of atoms p, predicate $eql(p)[x]$ is true iff $x = p$, and where $exec_{ij}$ is the function that executes instruction I_{ij}. For example, if q_{ij} is q, and I_{ij} is right, then $exec_{ij}$ is
$[eql(S_1)[car[4th[x]] \to ex_{ij1}[x]; \ldots; eql(S_t)[car[4th[x]] \to ex_{ijt}[x]],$
where ex_{ijk} is obtained from functions $g(q_i, S_j, S_k)$ in 2.5(4), by replacing (in order to add a *blank symbol* BL when M moves right to visit for the first time a new cell) the indicated *cdr* by
$[eq[\mathrm{NIL}; cdr[u]] \to (BL); T \to cdr[u]].$
Let a TM M be given, together with an input (coded by) x, and with a polynomial bound of the form $\lambda n.(h+n)^k$. From functions *mult* and *lh* of Section 3, a function p_{hk} can be defined which takes x into $num((h+|x|)^k)$; a function *start* can be defined, which takes x into the initial i.d. $(x, q_1, BL, (BL))$, where BL is the code for M's blank symbol. The following function s_M, by input x and $y = num(h)$, simulates the behaviour of M for h steps

$$\begin{cases} s_M[x;\omega] = x \\ s_M[x;y] \;= next_M[s_M[x;y'']]; \end{cases}$$

the required simulation is performed by $sim_M[x] = s_M[start[x]; p_{hk}[x]]$.

5.2 Simulation of CVR by TM's

Lemma 9. *If f is defined by CVR (FCV) and is limited by a polynomial, if its initial function is in* POLYTIMEF, *then f is in* POLYSPACEF (POLYTIMEF).

Proof. Outline of the simulation. Let f be defined by CVR with notations of Definition 2. Let $g, g^*, h, \mathbf{d}$ be simulated by the TM's G, G^*, H, D_i. Assume that f is limited by a polynomial p. A TM F simulating f can be defined, which behaves in the following way.

Let θ be a n-ary tree of height $\leq |y|$, whose root is (labelled by) y, and such that: every internal node z has n children $d_1[z], \ldots, d_n[z]$; and every leaf satisfies the terminating condition decided by G^*. F visits θ in the mode known as *post-order*. It records in a stack Σ_1 the sequence of recursive calls; and it stores in a second stack Σ_2 the values $f[\mathbf{x}; d_j[z]]$ which are needed to compute $h[\mathbf{x}; z; f[\mathbf{x}; d_1[z]]; \ldots; f[\mathbf{x}; d_n[z]]]$.
Space complexity. In addition to space used by G and H, F needs space for the stacks; the amount for Σ_1 is linear in $|y|$, since we have to store $\leq |y|$ objects, each $\leq n$. When in Σ_1 there are r numbers j_q, in Σ_2 there are $\sum_{q=1}^{r}(n - j_q) \leq n|y|$ values of f; thus space for $|\Sigma_2|$ is linear in $p(|\mathbf{x}; y|) \cdot |y|$.

Time complexity. Let f be defined by FCV in g, h. Since $\mathbf{d}$ is fully disjoint, the tree θ has $\leq |y|$ nodes, and, therefore, G, H are applied less than $|y|$ times. The result follows, since their input is bounded above by p.

5.3 Equivalence of PH and PHL

Lemma 10. *For all n we have $\Delta_n^p = \Delta_n^p \mathcal{L}$.*

Proof. (Outline) $\subseteq$. Induction on n. Step. Let language $L \in \Delta_n^p$ over alphabet $\Gamma = \{S_1, \ldots, S_q\}$ be given. Let atom ω_i code S_i, and let word $w = S_{i(1)}, \ldots, S_{i(n)} \in \Gamma^*$ be coded by the list of atoms $X = (\omega_{i(1)}, \ldots, \omega_{i(n)})$. Let $g[\mathbf{x}; u] \in \Delta_n^p \mathcal{L}$ be the characteristic function of L, which is granted by the ind. hyp. We show that the characteristic function f of

$$L' = \{(X_1, \ldots, X_m, Y) : \exists U(|U| \leq |Y| \wedge (X_1, \ldots, X_m, U) \in L)\}$$

is in $\Delta_{n+1}^p \mathcal{L}$. With decreasing tuples $\pi(app(\omega_i), cdr, cdr)$, define by OR-SCV

$$f^*[\mathbf{x}; y] = \begin{cases} g[\mathbf{x}; y] & \text{if } \mathrm{at}[3\mathrm{d}[\mathrm{y}]] \\ f^*[\mathbf{x}; d_1[y]] \ or \ \ldots \ or \ f^*[\mathbf{x}; d_q[y]] & \text{otherwise.} \end{cases}$$

Language L' is accepted by $f[\mathbf{x}; y] = f^*[\mathbf{x}; (); y; y]$.
$\supseteq$. Induction on n and on the construction of function $f \in \mathcal{PHL}$ to be simulated. Assume that $f[\mathbf{x}; y]$ is defined by OR-SCV in $g \in \Delta_n^p \mathcal{L}$ and h, with decreasing functions d_j (since else the result is an immediate consequence of the induction on f or of the fact that PTIMEF$= \Delta_1^p \mathcal{L}$). Let g decide language L. A nondeterministic TM M_f with oracle L can be defined, which: (1) at each call to h, iterates an invariant cycle, including, at each *or* of h, the choice of a j and the simulation of d_j; (2) at each g, queries the oracle. The time complexity of M_f is quadratic ($\leq |\mathbf{y}|$ applications of the TM's simulating functions d_j).

6 Conclusions

Normal form From proof of Lemma 8 (from the example on QBF), we see that only one level of nesting of FCV (SCV) is actually needed to compute POLYTIMEF (POLYSPACEF). This may be used to give an analogue for these classes of Kleene's *normal form theorem* for PR functions.

Classes DTIMEF(n^k). The fact above implies that to characterize these classes we have to rule the number and quality of the SBST's. For example, let $\mathcal{PL}_3$ be the Lisp class which is obtained from FCV$(\mathcal{B}^*)^*$ by excluding substitutions of the arguments of a recursive function by other recursive functions; and let $\mathcal{PL}_2$ be the further retriction of $\mathcal{PL}_3$ to recusively boolean functions; it can be proved that $\mathcal{PL}_3 \subseteq$DTIMEF(n^3), that $\mathcal{PL}_2 \subseteq$DTIME(n^2); and that if f is recursively boolean in functions in DTIMEF(n^k), then it is in DTIME(n^{k+1}). A classification of all classes DTIMESPACEF(n^k, n^m) can be easily obtained by following this approach.

Validation By scanning [7], §51,57 we see that all algorithms for the first Gödel theorem and for predicate T (a universal function) are written in a language quite close to our $\mathcal{PL}$ (besides notations, we have just to replace all bounded quantifiers by a program for search of sub-expressions). This is not surprising, since Kleene's arithmetization methods are based on his *generalized arithmetic* ([7], §50) which, in turn, may be regarded as a form of *primitive recursive Lisp.* This might point out a certain adequacy of our dialects of Lisp to represent algorithms. It might then be sensible to show the time/space complexity of an algorithm by just describing it in the language of $\mathcal{PL}$, and then checking to which element of the classification above does it belong.

Improvements to the language Types are only an apparent burden for concrete use of $\mathcal{PL}$, since we may forget them, and just watch that the previous values of the function under definition by FCV be not *cons-ed* together by the invariant, if not boolean. A more serious obstacle is that we are free to nest any number of boolean FCV's above at most one not-boolean FCV. We plan to remove this limitation by means of a re-definition of the types.

A point dividing these authors We have defined only the Δ-subclasses of PH, and not the Σ's and Π's, like NP, co-NP, etc. Some among us believe that class OR-SCV($\mathcal{PL}$) characterizes NP, while others maintain that it is *too large*. To discuss this point, let us say that language L is *accepted* by f when we have $x \in L$ iff $f[x] = T$. Indeed $\overline{SAT}$ is accepted by function $not[sat[x]]$, and this function is in OR-SCV($\mathcal{PL}$)*, and not in OR-SCV($\mathcal{PL}$), since is defined by substitution of $sat[x]$ in function $not[x]$. Thus, from a strictly syntactic point-of-view, we might pretend that classes OR-SCV($\Delta^p_k\mathcal{L}$) are characterizations of Σ^p_{k+1}, and define $\Pi^p_k\mathcal{L}$ to be the class of all functions of the form $not[f[x]]$, $f \in \Sigma^p_k\mathcal{L}$. But perhaps we should look at more substantial facts than mere syntax: it is undeniable that, so to say, *sat knows* $\overline{SAT}$; thus one is entitled to say that OR-SCV($\mathcal{PL}$) is not well-defined with respect to resources, and is not an acceptable characterization of NP.

Stronger forms of recursion. Let $<_S$ be a total order of the S-expressions. Let us say that f is defined by *n-strong* CVR if $f[y]$ depends on n values $f[y_i]$, such that, for all i, we have $y_i <_S y$. It can be easily proved that POLYSPACEF is closed under 1-strong CVR. Apparently ([3]), it can be proved that POLYTIMEF is not closed under 2-strong CVR; and that if POLYSPACEF is closed under 2-strong CVR, then POLYSPACE = EXPTIME. The proof of this result fails when relativized to oracle-TM's.

References

1. S. Bellantoni and S. Cook, A new recursive characterization of the polytime functions, in 24th Annual ACM STOC (1992) 283-293.

2. S. Caporaso, Safe Turing machines, Grzegorczyk classes and Polytime (to appear on Intern. J. Found. Comp. Sc.).
3. S. Caporaso and M. Zito, On a relation between uniform coding and problems of the form DTIME($\mathcal{F}$) =? DSPACE($\mathcal{F}$), submitted to Acta Informatica.
4. P. Clote, A time-space hierarchy between polynomial time and polynomial space, Math. Systems Theory, 25(1992) 77-92.
5. A. Cobham, The intrinsic computational complexity of functions, in Y. Bar Hillel, ed., Proc. Int. Conf. Logic, Methodology and Philosophy Sci. (North-Holland, Amsterdam, 1965) 24-30.
6. J. Hartmanis and R.E. Stearns, On the computational complexity of algorithms. Trans. A.M.S. 117(1965)285-306.
7. S.C. Kleene, Introduction to metamathematics (North-Holland, Amsterdam, 1952).
8. M. Kutylowski, A generalyzed Grzegorczyk hierarchy and low complexity classes, Information and Computation, 72.2(1987) 133-149.
9. D. Leivant, A foundational delineation of computational feasibility. 6th Annual IEEE symposium on Logic in computer science (1991).
10. D.Leivant and J.Y. Marion, Lambda calculus characterization of Poly-time (to appear on Fundamenta Informaticae).
11. M. Liskiewicz, K. Lorys, and M. Piotrow, The characterization of some complexity classes by recursion schemata, in Colloquia Mathematica Societatis János Bolyai, 44, (Pécs, 1984) 313-322.
12. Mecca and Bonner, Sequences, Datalog and transducers, PODS 1995 (reference to be revs'd in the final version of this paper).
13. C.H. Papadimitriou, A note on expressive power of PROLOG, Bull. EATCS, 26(1985) 21-23.
14. H. Poincaré Les mathématiques et la logique, Revue de Métaphisique et de Morale, 14(1906)297-317.
15. R.W. Ritchie, Classes of predictably computable functions, Trans. Am. Math. Soc. 106(1963) 139-173.
16. H.E. Rose, Subrecursion: Functions and hierarchies, (Oxford Press, Oxford, 1984).

On the Drift of Short Schedules

(Extended abstract)

Uriel Feige * and Giora Rayzman

Department of Applied Math and Computer Science
The Weizmann Institute
Rehovot 76100, Israel

Abstract. For the Job Shop Scheduling (JSS) problem, the drift of a schedule is the maximum difference between the number of operations performed by two jobs within a time interval. We show instances of the JSS problem for which any short schedule must allow for nonconstant drift.

1 Introduction

In the Job Shop Scheduling (JSS) problem there are n jobs and m machines. Each job consists of a sequence of operations, where each operation is specified in terms of the machine on which it is to be processed, and the duration of the operation (in terms of units of time). Each machine can perform at most one operation at any time unit. The operations in each job need to be performed in order, where a new operation cannot begin until the previous one finishes. The goal is to schedule the operations on the machines in a way that does not violate the above constraints, such that the time until the last operation terminates is minimized. This problem is NP-hard.

The case of *unit length no repetition* JSS, in which each operation is of unit length and no job contains two operations to be performed on the same machine, is of special interest. In particular, it can model the issue of managing the queues in a packet routing problem in which the route of each packet within the network (from source to destination) is prespecified. Each packet can be viewed as a job, and each link can be viewed as a machine, assuming that no two packets are allowed to cross the same link at the same unit of time. Scheduling the packets amounts to specifying, for each unit of time and each link, which of the packets (if any) waiting in queue is to cross the link. Let d be the maximum distance that a packet needs to cross in the network (or more generally, the maximum length of a job), and let c denote the maximum number of packets whose paths cross the same link (or more generally, the maximum number of jobs that have operations on the same machine). Clearly, and schedule has to have length at least $\max[c, d]$. A celebrated result of Leighton, Maggs and Rao [3] shows that for such problems, there always is a schedule whose length is $O(c+d)$. Moreover, this

* Incumbent of the Joseph and Celia Reskin Career Development Chair. Yigal Alon fellow. feige@wisdom.weizmann.ac.il.

schedule has the property that the length of the queue at each link is bounded by some universal constant at all time units (assuming that each node of the network is a source of at most a constant number of packets). In the more general jobs/machines interpretation, the results of [3] show that there is a universal constant that bounds the number of time steps between any two consecutive operations of a job (though the number of time steps that a job waits until its first operation is performed may not be bounded). It is not known whether results similar to those of [3] hold when operations are not of unit length, or when repetitions are allowed. This open question motivated our current study of analysing the structure of short schedules.

In this paper we investigate the *drift* of short schedules. For a schedule, the active period of a job is the time interval between the performance of the first and last operations of the job. For any time interval and two jobs that are active in it, their drift with respect to the time interval is the difference in the number of operations performed by the two jobs within the time interval. The drift of a schedule is the maximum drift of any two jobs and any time interval in which both jobs are active. We are interested in schedules that are both short ($O(c+d)$ time units) and have small drift (bounded by a constant). A schedule with a short drift can be interpreted as a schedule for which in every time interval, the rate at which different (active) jobs receive service is similar. This is a very strong notion of fairness of a schedule, which may be appropriate in situations where one does not want to be discriminated against, compared to others. (It is frustrating to be in heavy traffic where your lane is advancing at a slow pace. It may be even more frustrating if you notice that some other lane is advancing at a more rapid pace. Here a job is to drive the car for a distance of one kilometer. Each operation is to advance by ten meters. The drift you experience is the the distance that keeps growing between your car and that red car in the other lane.) Leighton, Maggs and Rao [3] made no attempt to minimize the drift. Unlike bounded queues, drift does not seem to be a natural parameter for packet routing algorithms. The drift in their schedules is not bounded by a constant, but can readily be seen to be $O(\log n)$.

Observe that for unit length JSS problems, it is always possible to design schedules with drift 1, by scheduling jobs in a "round robin" fashion. However, the resulting schedule may be very long. Our main result is that for some unit length no repetition JSS problems, any short schedule must have a nonconstant drift.

Theorem 1. *There is an infinite family of unit length no repetition JSS problems (one problem for each value of n), in which there are n jobs and n machines, such that any schedule of length $O(n)$ requires a drift of $\Omega(\log n/\log\log n)$.*

It is possible to modify the construction of Thm. 1 so that it gives a routing problem for which there is no linear schedule with constant drift. However, the resulting routing problem is rather bizarre, with packets following routes that make unexplicable detours. We do not know whether something similar to Thm. 1 is true for routing problems in which packets follow shortest paths.

The notion of drift is very natural in some contexts (such as clock synchronization in distributed computation). We are not aware of previous work relating to this notion in the context of Job Shop Scheduling (where it is perhaps less natural). Part of our interest in the drift of schedules comes from an attempt to design scheduling algorithms which are guaranteed to produce (relatively) short schedules. A common first step in such algorithms (e.g., in [3] and [5]) is to assign random $O(n)$ delays to jobs. This has the effect of spreading out randomly the time units in which there is demand for any particular machine, and with high probability, at every time unit there are at most $O(\log n / \log\log n)$ jobs that want to use this machine. Inspired by work of [2, 1] we investigated the possibility that by assigning initial delays in a more clever way, it is possible to decrease the demand per machine below $o(\log n / \log\log n)$. However, an extended version of Theorem 1 (see Theorem 13), and in particular, the first part of its proof (summarized in Theorem 9), implies that in some cases this is impossible. More generally, any algorithm for finding short schedules for (unit length no repetition) JSS problems must allow for $\Omega(\log n / \log\log n)$ drift.

The proof of Theorem 1 has two parts. In the first part (Section 3) we use the probabilistic method to show that for some JSS problem with n jobs, no matter how initial delays are chosen, there will be a linear number of "fat" time units in which there is some machine that is requested by $\Omega(\log n / \log\log n)$ jobs. In the second part (Section 4) we show that any short schedule with small drift can be modified to give a delay pattern in which there is a sublinear number of fat time units. The two parts combined imply the theorem.

An extended version of this paper is available [4].

2 The Model

The definitions that follow are specialized for the unit length no repetition version of the JSS problem.

There are:
a set of m machines $\mathcal{M} = \{M_1, \ldots, M_m\}$,
a set of n jobs $\mathcal{J} = \{J_1, \ldots, J_n\}$,
a set of operations $\mathcal{O} = \{O_{ij} || i = 1, \ldots, \mu_j; j = 1, \ldots, n\}$,
with
O_{ij} being the i-th operation of job J_j,
μ_j being the number of operations of job J_j.

Each operation O_{ij} requires one unit of processing time, and has to be performed on a machine with index k_{ij}. A *job* $J_j = \{O_{ij} | i = 1, \ldots, \mu_j\}$ is a sequence of operations that are to be performed in order. We assume that a job has at most one operation on a given machine. Each machine can perform a single operation during a unit of time. The following properties are of interest:

Input properties (in the introduction, these properties where denoted by d and c respectively):
P_{max} = the largest job length, i.e. $\max_j \mu_j$.
Π_{max} = the largest machine load, i.e. $\max_{M_k} |\{k_{ij}\}|_{k_{ij}=k}$.

Specific schedule properties:
C_{ij} = the completion time of O_{ij},
$C_{max} = \max_{i,j} C_{ij}$.

A *job shop schedule* is a collection of one-machine schedules

$$\sigma_k : \{O_{ij} | O_{ij} \text{ is to be performed on } M_k\} \overset{1-1}{\mapsto} \mathcal{Z}_0^+, k = 1, \ldots, m,$$

such that for every j and $\forall J_j \in \mathcal{J}$, and $i = 1, \ldots, \mu_j - 1$, $\sigma(O_{i+1j}) \geq \sigma(O_{ij}) + 1$ (assuming appropriate subscripts on σ).

The objective is to produce a schedule that minimizes C_{max}.

3 Arbitrary Delay Patterns

Here we consider the case of n jobs, each with n operations, to be performed on n machines. Each job has precisely one operation on each machine. Hence there are $n! < n^n$ possible sequences that make up a job, and $(n!)^n < n^{n^2}$ possible instances of our problem. For each job independently we allow an initial delay in the range $0, \ldots, d$, where $d < cn$ for some c. The range of interest for c is $1 \leq c \leq \log n$.

In this section we consider the schedule that arises by a delay pattern. This schedule is not required to be legal – several jobs might request the same machine at the same unit of time. We will not try to resolve these momentary loads on machines, just to analyse them.

Definition 2. For an instance of the problem and a choice of n delays, we call a column (time step) *fat* if some machine is requested by $\Omega(\log n / \log\log n)$ jobs at this time step.

We show that for some instance of the problem, there are $\Omega(n)$ fat columns, regardless of how the delays are chosen.

Our proof is structured as follows. We use a probabilistic argument to show that any choice of the delays handles at most a fraction of $d^{-n}/2$ of all problem instance, in the sense that there are less than $n/4$ fat columns. Since there are only d^n choices of delays, it follows that at least half of the problem instances are not handled by any choice of delays.

For a fixed choice of delays and a fixed column, the situation resembles the known experiment of throwing balls into bins. Here the n machines are the bins, and each job gets to throw one ball (operation) in one of the bins, at random. If the number of balls (active jobs in the column) is $\Omega(n)$, we expect one bin (machine) to receive $\Omega(\log n / \log\log n)$ balls. The analysis gets a bit more complicated once several columns are considered, because of dependencies between different columns – if a job J has an operation on machine M in column i, then J cannot have an operation on the same machine M in any other column.

We now proceed with our proof.

Definition 3. After selecting initial delays in the range $0, \ldots, cn$, a block of (jobs $\times$ time units) shall be called *active* or *busy* if it has length $n/2$ (i.e. the number of time units), and has $n/2c$ jobs active in it, that is each of these jobs is running during each and every column (time step) of the block.

Proposition 4. *Given an instance of n jobs each of length n and an assignment of initial delays in the range $0, \ldots, cn$ to the jobs, there is a busy block.*

Proof. There are at most $(c+1)n$ time steps in which jobs are active. Partition these time steps into $2(c+1)$ disjoint blocks, each containing $n/2$ consecutive time steps. Remove the first and last of these blocks. Each job is active for at least one of the remaining $2c$ blocks. Hence there is at least one block in which at least $n/2c$ jobs are active. □

Our analysis proceeds as follows:

- We first fix the initial delays pattern.
- Then the operations of each job belonging to the resulting busy block are chosen at random, column after column.
- Only afterwards the rest of the operations of the job (outside the block) are selected.

We now fix an arbitrary delay pattern. This gives a busy block. Number the columns (times steps) in the busy block from 1 to $n/2$. Start assigning at random operations to the active jobs in the busy block, column after column.

Definition 5. A machine is *alive* at a certain point in time, if it has not already been used by more than $\frac{n}{4c}$ of the jobs.

Definition 6. A job J_j *wants* machine M_k at a certain point in time, if none of the operations already assigned to job J_j required machine M_k.

Lemma 7. *After fixing the initial delays pattern, in the busy block, for $j < n/4c$, for any assignment of operations to columns $1, \ldots, j$, the probability that column $j+1$ is fat is $\geq 1 - n^{-c_1}$, for any constant c_1 and large enough n.*

Proof. Consider column $j+1$ of the busy block. The number of machines that are alive at this column is at least:

$$n - \frac{\frac{n}{2c} \cdot j}{\frac{n}{4c}} = n - 2 \cdot j > n/2$$

We use the following notation. The number of jobs that *want* machine i is denoted by n_i. If the machine is alive then $n_i \geq \frac{n}{4c}$. A job that may choose machine i in column j has $n-j$ possible machines to choose from. We denote this number by N_j.

$$P(N_j, n_i, i, j, \geq k) \stackrel{\text{def}}{=} \text{probability machine } \#i, \text{ which is alive, is requested by at least } k \text{ jobs in column } j.$$

$$P(N_j, n_i, i, j, = k) \stackrel{\text{def}}{=} \text{probability machine } \#i, \text{ which is alive, is requested by exactly } k \text{ jobs in column } j.$$

$$Q(N_j, n_i, j, < k) \stackrel{\text{def}}{=} \text{probability that at column } j \text{ there is no live machine requested by at least } k \text{ jobs.}$$

Estimating $P(N_j, n_i, i, j, \geq k)$ we have:

$$\begin{aligned} P(N_j, n_i, i, j, \geq k) &\geq P(N_j, n_i, i, j, = k) \\ &\geq \binom{n_i}{k} \cdot \frac{1}{(N_j)^k} \cdot (1 - \frac{1}{N_j})^{n_i - k} \\ &\geq \frac{(n_i)^k}{k^k} \cdot \frac{1}{(N_j)^k} \end{aligned}$$

where the last inequality holds for our parameters when $6 \leq k \leq n_i/2$, and n is sufficiently large.

We call a machine *light* if less then k jobs request it in column j. Thus:

$$\begin{aligned} Q(N_{ij}, n_i, j, < k) &\leq \Pr[\cap_{l=1}^{n/8ck} \text{ machine } \#l \text{ is light}] \\ &\leq \Pr[\cap_{l=1}^{n/8ck-1} \text{ machine } \#l \text{ is light}] \\ &\quad \cdot \Pr[\text{ machine } \#\frac{n}{8ck} \text{ is light} \mid \cap_{l<\frac{n}{8ck}} \text{ machine } \#l \text{ is light}] \\ &\vdots \\ &\leq \prod_{l=1}^{n/8ck} (1 - P(N_{ij} - l - 1, n_i - k(l-1), j, \geq k)) \\ &\leq \left[1 - \frac{(\frac{n}{8c})^k}{k^k} \cdot (\frac{1}{N_{ij} - \frac{n}{8ck}})^k\right]^{\frac{n}{8ck}} \\ &\leq \left[1 - \frac{(\frac{n}{8c})^k}{k^k} \cdot (\frac{1}{N_{ij}})^k\right]^{\frac{n}{8ck}} \\ &\leq \left[1 - \frac{(\frac{n}{8c})^k}{k^k} \cdot (\frac{1}{n})^k\right]^{\frac{n}{8ck}} \\ &\leq \left[1 - \frac{(\frac{1}{8c})^k}{k^k}\right]^{\frac{n}{8ck}} \end{aligned}$$

For $c < \log n$, for $k < c_0 \cdot \frac{\lg n}{\lg \lg n}$ with $c_0 \leq 1/2$, and for large enough n, we get this probability to be $< 1/n^{c_1}$. □

Lemma 8. *For* any *assignment of initial delays to the instance of the random permutations problem, for any constant $\beta < 1/2$, with probability $> 1 - n^{-c_2 n}$, there are $\geq (1/2 - \beta)n$ fat columns in the busy block, where $c_2 > 0$ is an arbitrary constant and n is sufficiently large.*

Proof. Let E the event that there exist at least βn non-fat columns in the active block.

$$\begin{aligned}
\Pr[E] &\leq \Sigma_{B=\beta n}^{n/2} \binom{n/2}{B} \Pr[\cap_{i=1}^{B} \text{ block column } \#i \text{ is not fat}] \\
&\leq \Sigma_{B=\beta n}^{n/2} \binom{n/2}{B} \Pr[\cap_{i=1}^{B-1} \text{ block column } \#i \text{ is not fat}] \\
&\quad \cdot Pr[\text{ block column } \#B \text{ is not fat} | \cap_{i=1}^{B-1} \text{ block column } \#i \text{ is not fat}] \\
&\vdots \\
&\leq \frac{n}{2} 2^{n/2} n^{-c_1 \beta n} \\
&\quad \text{(since the upper bound proved in Lemma 7 is w.r.t. } \textit{any} \text{ history)} \\
&\leq n^{-c_2 n}
\end{aligned}$$

where $c_2 \simeq \beta c_1 - o(1)$ can be made arbitrarily large by taking c_1 from Lemma 7 arbitrarily large. □

Set β above to be $\beta = 1/4$. Making c_1 (and hence c_2) large enough so that $n^{c_2 n} > 2(n \log n)^n$, any choice of delays handles at most a fraction of $d^{-n}/2$ of all problem instances – in the sense that there are less than $n/4$ fat columns for that instance. This implies:

Theorem 9. *For sufficiently large n, there is an instance of the unit-length no-repetitions JSS problem (and thus of the general JSS problem), s.t. for any assignment of the initial delays in the range up to $n \log n$, there are $n/4$ time units in each of which, there is at least one machine with a load of $\frac{\log n}{2 \log \log n}$.*

4 Small Drifts versus Good Delay Patterns

In this section we shall explore how well synchronized (fair) an optimal schedule can be and how short a well synchronized schedule can be made. In fact we shall show that one of the properties of the schedule comes at the expense of the other.

Definition 10. An *s-fair* schedule is a schedule in which for any pair of jobs J_1, J_2 and any two time steps $t_1 \leq t_2$ for which both of these jobs are *active* (i.e. have already started but have not terminated yet) $|C_1(t_1, t_2) - C_2(t_1, t_2)| \leq s$, where $C_i(t_1, t_2)$ is the number of operations job J_i has completed between time steps t_1 and t_2. We shall call the difference $|C_1(t_1, t_2) - C_2(t_1, t_2)|$ – the *drift* between J_1 and J_2 at time interval (t_1, t_2).

Definition 11. We shall say that a job *touches* a time-unit if the time unit falls within the time interval in which the job is active.

Definition 12. For a given schedule, the process of *shrinking* a job with respect to time t denotes removing recursively all the intermediate delays scheduled for the job in a way such that:

if a unit of delay was at some time $t_0 < t$ then the entire job's part preceding t_0, originally scheduled to start at some time t^{old}_{start} is rescheduled to start at time $t^{new}_{start} = t^{old}_{start} + 1$;

if a unit of delay was at some time $t_0 \geq t$ then the entire job's part following t_0, originally scheduled to start at some time t^{old}_{start} is reschedule to start at time $t^{new}_{start} = t^{old}_{start} - 1$.

The job is said to be *clustered* around t after all the intermediate delays are removed in the above fashion.

Theorem 13. *For some instances of the unit-length, no machine repetitions problem (and thus of the general JSS problem) there does not exist an s-fair schedule of length $O(P_{max} + \Pi_{max})$, for any constant s. More generally, for these instances, a schedule of length $c_l n$ cannot be s-fair for $s = \frac{1}{16c_l} \frac{\lg n}{\lg \lg n}$.*

Proof. Let the instance of the general JSS problem be a permutations jobs instance for which in Section 3 we showed that *any* initial delay assignment causes $n/4$ fat columns. Assume for contradiction that there exist a linear schedule for this instance. W.l.o.g., assume that the jobs of the instance are sorted in ascending order of their initial delays, i.e. J_1 has the shortest initial delay, J_n – the longest. The length of the schedule is $c_l n$, for some $c_l < \log n / \log \log n$. Partition the linear schedule into blocks of length n. Since the time-span of any job's activity for the instance of the problem is $\geq n$, each job must "touch" at least one of the blocks boundaries. Let $b(j)$ denote the time at the first boundary that job J_j "touches".

Now transform the linear schedule that we have into another linear schedule, which will be at most twice as long, in the following way:

- First *shrink* each job J_j with respect to its $b(j)$ by removing its intermediate delays. See Fig. 2 for the result.
- Double the distance between block boundaries. The new block boundary $b'(j)$ is located at $2b(j)$. Schedule each job J_j to be clustered around its new block boundary $b'(j)$ (i.e., shifted by $b(j)$ steps relative to the former clustered schedule). See Fig. 3

Thus, in the resulting linear schedule "clusters" of jobs are formed around each block boundary b, without any overlaps between different "clusters".

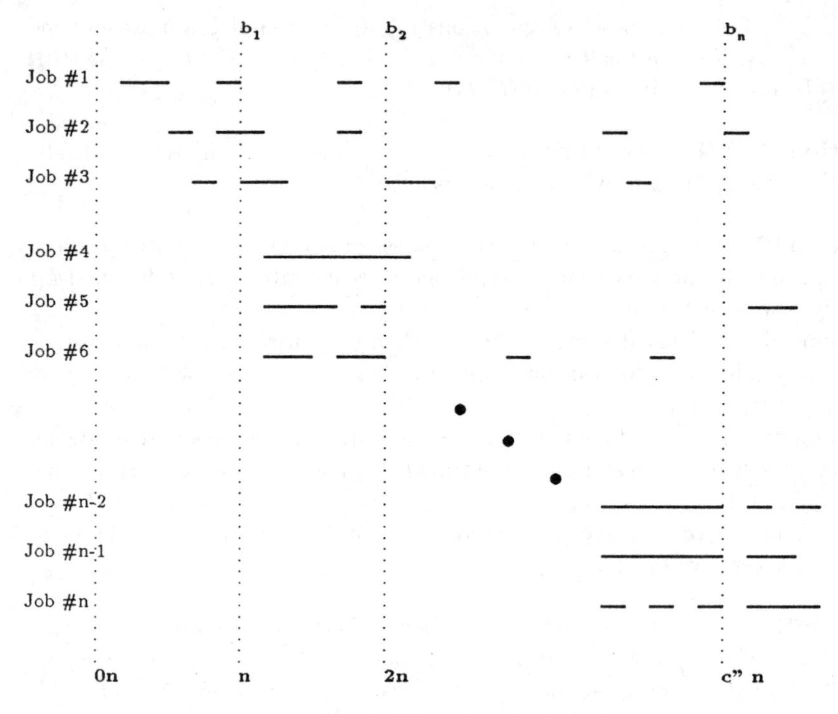

Fig. 1. The original schedule.

The new "shrunk" and "clustered" linear schedule can be viewed as an initial delays schedule in the sense of Section 3. Beforehand we have proved that this schedule must have a busy block with $n/4$ fat columns. All these fat columns must belong to the same cluster.

W.l.o.g., renumber the operations of each job involved in this cluster with respect to the cluster's boundary, i.e. the first operation of job J_j in the cluster after the boundary shall be called O_j^0, the first before the boundary shall be called O_j^{-1} etc.

Let J_1 be a job of the cluster that is active during the busy block. Mark the positions $p_0, \ldots, p_\nu$, $\nu = n/4$, corresponding to the operations of J_1 that belong to the fat columns. Locate the places of these operations of J_1 in the original schedule and consider the time intervals corresponding to the s J_1 operations before each such "fat column" operation and s J_1's operations after the "fat column" operation (recall that s is the postulated drift). Denote I^{p_i} such an interval corresponding to $O_1^{p_i}$. Now, consider the following subset of the intervals: $I^{p_0}, I^{p_{2s}}, I^{p_{4s}}, \ldots$. No two of these intervals overlap.

Lemma 14. *At least one of the time intervals* $I^{p_{2hs}}, h = 0, \ldots, \frac{\nu}{2s}$ *described above is short, i.e. has length at most* $8s \cdot c_l$.

Proof. The entire original schedule has length $\leq c_l \cdot n$. In it we identified $n/8s$ disjoint time intervals. Thus at least one of them is of length at most $8sc_l$. □

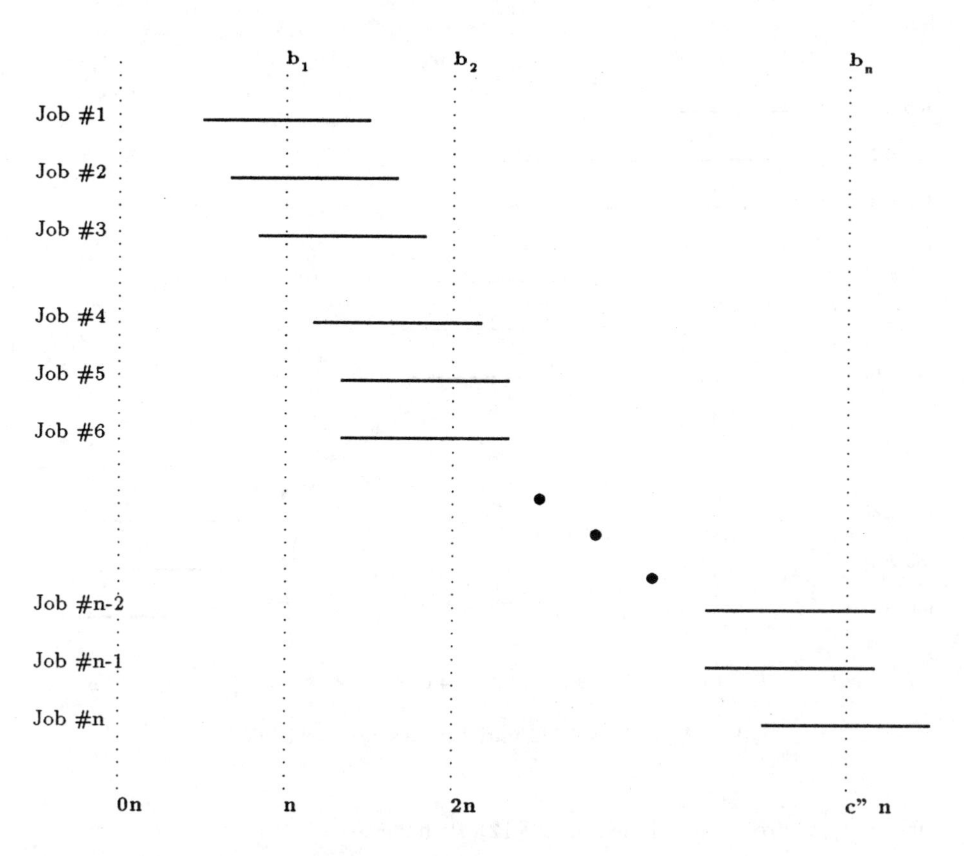

Fig. 2. The clustered schedule.

Let $t(1;k)$ denote the time at which operation k of job J_1 was originally scheduled (before clustering).

Lemma 15. *In the original* s-fair *schedule, each job J_j belonging to J_1's cluster of the "clustered" and "spread" schedule, has performed its O_j^k operation in the time interval between $t(1;k-s)$ and $t(1;k+s)$.*

Proof. Suppose otherwise. Let the b be the block boundary around which all the jobs in J_1's cluster are concentrated. Consider the job J_j that hasn't completed its O_j^k before $t(1;k+s), k \geq 0$. Recall our definition of shrinking and the renumbering of the cluster's jobs operation. Since $k \geq 0$, we may conclude that O_j^k was originally scheduled after the block boundary b. In addition, $O_1^0, \ldots, O_1^{k+s}$ were also scheduled after the block boundary b. Therefore, in the time interval between the block boundary b and $t(1;k+s)$, setting

$$t_1 = b,$$
$$t_2 = t(1;k+s).$$

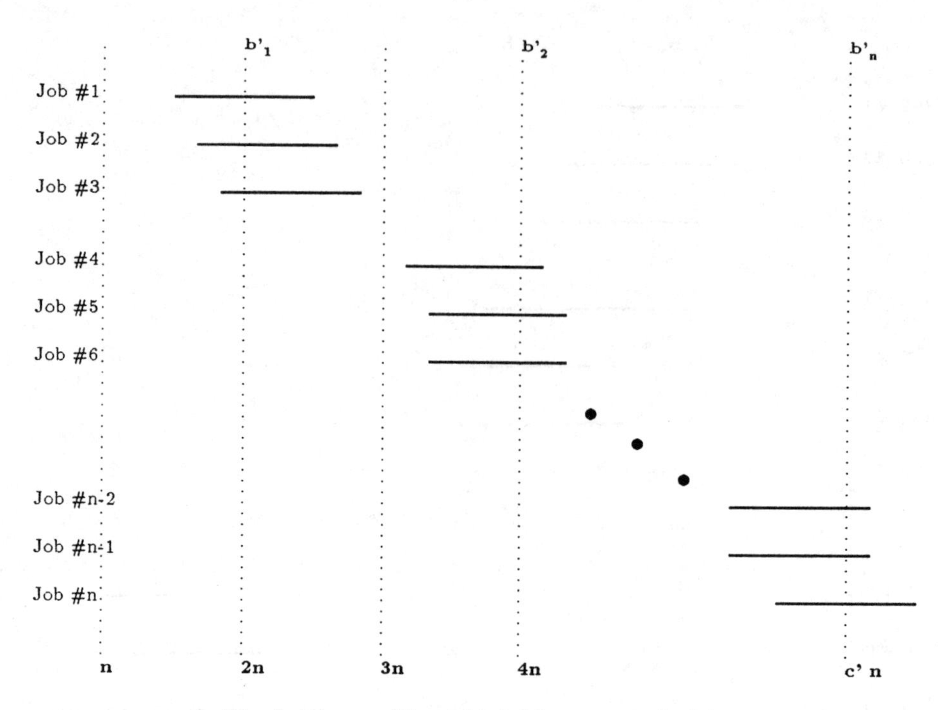

Fig. 3. The resulting initial-delays type schedule.

(where t_1, t_2 are as mentioned in Def 12) we have:

$$|C_1(t_1, t_2) - C_j(t_1, t_2)| > s,$$

contradicting the fact that the original schedule was s-fair.
The case where $k < 0$ is treated similarly.
Hence, we conclude that J_j must have completed its O_j^k in the time interval defined by $t_1(k-s)$ and $t_1(k+s)$. □

As the result of Lemma 15 we may conclude, that all of the operations on the loaded machine in the fat column were within distance s of the corresponding operation of J_1.

According to Lemma 14 there exists a short time interval in the original schedule, which corresponds to s operations of J_1 before the fat column operation of J_1 and s operations after it. By s-fairness, all the operations of the fat column on the most loaded machine M_i were scheduled within this time interval. However, the size of the time interval in which they should be scheduled is $8s \cdot c_l$, while there are $\frac{\log n}{2\log\log n}$ jobs with an operation on M_i to be scheduled in that interval. For the schedule to be legal, machine M_i performs at most one operation at each time unit. Hence $s \geq \log n / 16 c_l \log\log n$. □

References

1. Y. Azar, A.Z. Broder, A.R. Karlin, and E. Upfal. Balanced allocations. In *Proc. 26th ACM Symp. on Theory of Computing*, pages 593–602, 1994.
2. R.M. Karp, M. Luby, and F.M. auf der Heide. Efficient PRAM simulation on a distribute memory machine. *Algorithmica* (1996) 16:517–542.
3. T. Leighton, B. Maggs, and S. Rao. Packet routing and job-shop scheduling in O(congestion + dilation) steps. *Combinatorica*, 14(2):165–186, 1994.
4. G. Rayzman. Approximation techniques for job-shop scheduling problems. Master's thesis, Department of Applied Math and Computer Science, The Weizmann Institute of Science, 1996.
5. D. Shmoys, C. Stein, and J. Wein. Improved approximation algorithms for shop scheduling problems. *SIAM Journal on Computing*, 23(3):617–632, June 1994. A preliminary version of these results appeared in SODA '91.

On Removing Non-degeneracy Assumptions in Computational Geometry*

(EXTENDED ABSTRACT)

Francisco Gómez[1], Suneeta Ramaswami[2] and Godfried Toussaint[2]

[1] Dept. of Applied Mathematics, Universidad Politecnica de Madrid, Madrid, Spain
[2] School of Computer Science, McGill University, Montréal, Québec, Canada

Abstract

Existing methods for removing degeneracies in computational geometry can be classified as either *approximation* or *perturbation* methods. These methods give the implementer two rather unsatisfactory choices: find an *approximate* solution to the *original* problem given, or find an *exact* solution to an *approximation* of the original problem. We address an alternative approach that has received little attention in the computational geometry literature. Often a typical computational geometry paper will make a non-degeneracy assumption that can in fact be removed (without perturbing the input) by a global rigid transformation of the input. In these situations, by applying suitable *pre-* and *post-* processing steps to an algorithm, we obtain the *exact* solution to the *original* problem using the algorithm that assumes a non-degenerate input, even when that input is in fact degenerate.

We consider several non-degeneracy assumptions that are typically made in the literature, propose algorithms for performing the pre- and post- processing steps that remove these degeneracies, analyze their complexity and, for some of these problems, give lower bounds on their worst-case complexity. The assumptions considered here include: (1) no two points in the plane on a vertical line (2) no two points in space lie on a vertical line (3) no two points in space have the same x-coordinate (4) no three points in space lie on a vertical plane and (5) no two line segments lie on a vertical plane. We propose low-degree polynomial-time solutions for the decision, computation and optimization versions of all these problems. For the optimization version of problem (5) we give an $O(n^4)$ time algorithm, reducing the previous best running time of $O(n^6 \log n)$.

1 Introduction

Algorithms in computational geometry are usually designed for the real RAM (random access machine) assuming that the input is in *general position.* More specifically, the general position assumption implies that the input to an algorithm for solving a specific problem is free of certain degeneracies. Yap [26] has distinguished between intrinsic or *problem-induced* and *algorithm-induced* degeneracies.

* The first author's research was carried out during his visit to McGill University in 1995 and was self-supported. The second and third authors were supported by NSERC Grant no. OGP0009293 and FCAR Grant no. 93-ER-0291.

A survey of various techniques available for coping with degeneracies, and their interaction with the real RAM assumption, was written by Fortune [17]. In one class of methods to deal with degeneracies for solving a problem, the approach is to compute some *approximation* to the exact solution of the problem when a degenerate input is encountered [20]. The other well known class of methods for handling degeneracies is via *perturbation*. Here the input is perturbed in some way by an infinitesimal amount so that the degeneracies are no longer present in the perturbed input. For a review of other examples of problem-specific perturbation schemes, see [15, 25]. More general and powerful methods have recently been proposed by Edelsbrunner and Mücke [13], Yap [26] and Emiris and Canny [15]. An insightful discussion of perturbation methods and their short-comings is given by Seidel [25].

The above methods give the implementer two rather unsatisfactory choices: find an *approximate* solution to the *original* problem given, or find an *exact* solution to an *approximation* of the original problem. Sometimes it may be possible to convert the approximate solution obtained from perturbation methods to the exact solution by using some kind of *post-processing* step but this step may be difficult and complicated [7]. In fact, Seidel [25] questions the wisdom of using perturbation methods altogether.

In this paper we address an issue concerning non-degeneracy assumptions which has received little attention in the computational geometry literature. Often a typical computational geometry paper will make a non-degeneracy assumption that can in fact be removed (without perturbing the input) by a global rigid transformation of the input (such as a rotation, for example). Once the solution is obtained on the transformed non-degenerate input, the solution can be transformed back trivially (by an inverse rotation) to yield the solution to the original problem.

We consider several non-degeneracy assumptions that are typically made in the literature, propose efficient algorithms for performing the pre- and post-processing steps (suitable rotations) that remove these degeneracies, analyze their complexity in the real RAM model of computation and, in some cases, give lower bounds on their worst-case complexity. While degeneracies are a practical concern and the real RAM may be, from the practical standpoint, a poor model if numbers become long due to computation of rotations, we restrict ourselves in this work to infinite precision in order to focus on the purely geometric aspects of removing algorithm-induced degeneracies. In the practical implementation of our algorithms, where numerical robustness is important as well, the rational rotation methods of Canny, Donald and Ressler [8] may be incorporated.

The assumptions considered here include:

(1) no two planar points lie on a vertical line,
(2) no two points in space lie on a vertical line,
(3) no two points in space have the same x-coordinate,
(4) no three points in space lie on a vertical plane,
(5) no two line segments lie on a vertical plane

These problems are also intimately related to those of obtaining "nice" orthogonal projections of polygons and skeletons of spatial subdivisions in such disciplines as knot-theory, graph-drawing [5] and visualization in computer graphics [4]. Removing assumptions (1), (2) and (3) is equivalent to computing so-called *regular* and *Wirtinger* projections of points. Removing assumption (4) is equivalent to finding a plane such that the orthogonal projection of the points onto that plane has no three projected points collinear. Removing assumption (5) ensures that no two projected edges are collinear and is a well known projection method for visualizing "wire-frame" objects [21]. We would like to point out that the techniques presented in this paper may be used to find projections (of points in space) that contain no four co-circular points. However, this method would require an oracle that computes intersections between high-degree curves on a sphere. Since such a model of computation is impractical, we do not present the result here.

Looking at these degeneracy problems from the point of view of finding "nice" projections, uncovers additional desirable properties that we would like the transformed non-degenerate input to have. In order to increase the *robustness* of our solution even further, we are interested in obtaining the *most* non-degenerate input possible. Hence the five problems itemized above also have their optimization counterparts. We also give algorithms for such *optimal* removal of degeneracies.

We propose low-degree polynomial-time solutions for the decision, computation and optimization versions of all these problems. For the optimization version of problem (5) we give an $O(n^4)$ time algorithm, reducing the previous best running time of $O(n^6 \log n)$ [21].

2 No Two Planar Points on a Vertical Line

Many papers in computational geometry assume that no two points of a given planar set S have the same x-coordinate. For example, in [23] the assumption is used to obtain a random monotonic polygon with vertex set S in $O(K)$ time, where $n < K < n^2$ is the number of edges in the visibility graph of the monotonic chain with vertex set S. In this section, we consider the problem of removing this non-degeneracy assumption.

A projection of S on a line L is said to be *regular* if each point in S projects to a distinct point on L, i.e., there are no vertical line degeneracies. In other words the projection contains n distinct points. Without loss of generality, when a line L is given, we will assume L to be the x-axis.

2.1 The Decision Problem

We consider first the decision problem: given a set S, does it contain a degeneracy or does it give a regular projection on the x-axis. A very simple algorithm gives $O(n \log n)$ worst-case time. We simply check the x-coordinate values of the points for duplicates. The projection is regular if and only if there are no duplicates. Furthermore, by reducing element-uniqueness to the regular projection decision problem, a lower bound of $\Omega(n \log n)$ follows. Thus we have:

Theorem 1. *Given a set S of n distinct planar points and a line L, whether S admits a regular projection onto L can be determined in $\Theta(n \log n)$ time.*

The expected complexity of the algorithm can be reduced to $O(n)$ by allowing floor functions as primitive operations and using *bucket sorting* [11].

2.2 The Computation Problem

We now consider the computation problem: given a set S of n distinct points in the plane, find a rotation of S that removes the degeneracies, or equivalently, a line L such that S yields a regular projection on L. That a regular projection always exists follows from the fact that the only forbidden directions of projections are those determined by the lines through pairs of points of S. These forbidden directions are represented as follows: let the circle C^2, representing the set of directions, be the unit circle in the plane centered at the origin. For every pair of points in S, translate the line through them so that it intersects the origin. Intersect each such line with C^2 to yield a pair of forbidden points on C^2.

Although there are $O(n^2)$ such forbidden directions they have measure zero on the circle of directions C^2. Hence there is indeed no shortage of directions that allow regular projections. Observe that using the $O(n \log n)$ slope-selection algorithm [9] to find the k-th and $(k+1)$-st slope will not help, because the slopes are not necessarily unique. In fact, there may be several slopes (as many as $\Omega(n^2)$) with the same value.

By combining the decision problem discussed above with a simple bounding technique we obtain an efficient algorithm. Given S, and our goal to determine a regular projection, first check if the vertical projection of S on the x-axis is regular or not. This takes $O(n \log n)$ time, as described above. Assume that the projection of S onto the x-axis is *not* regular. This implies that the maximum slope, denoted by M_1, is equal to infinity. As mentioned above, looking for the next-largest non-duplicate slope, i.e, the maximum slope with the *constraint* that it not be infinite, could be costly. Let us denote this slope by M_2. The key realization for obtaining an efficient algorithm is that we do not need this constrained maximum slope. All we need is the value M of some slope such that $M_2 \leq M < M_1$, which may be found as follows.

Let Δx and Δy denote, respectively, the difference in x and y coordinates of two distinct points in S. The slope of the line going through this pair of points in S is given by $\Delta y / \Delta x$. Let $\max_S\{\Delta y\}$ denote the maximum value that Δy can take over all pairs of points in S, i.e., the width of S in the y-direction. Similarly, let $\min_S\{\Delta x\}$ denote the minimum value that Δx can take over all pairs of points in S, i.e., the *smallest gap* in the x-coordinate values of points in S. Of course, since the projection on the x-axis is not regular, we have that $\min_S\{\Delta x\} = 0$. To find a non-infinite upper bound on the second highest slope, delete all duplicates from the projected set of points in $O(n \log n)$ time: call this reduced set of points S^*. It follows that the desired direction (slope) for rotating S is then given by any finite value greater than $M = \max_S\{\Delta y\}/\min_{S^*}\{\Delta x\}$. Note that $M_2 \leq M < M_1 (= \infty)$. Computing $\max_S\{\Delta y\}$ takes linear time and computing $\min_{S^*}\{\Delta x\}$ takes $O(n \log n)$ time, giving us the following result.

Theorem 2. *Given a set S of n distinct points in the plane, finding a regular projection onto the x-axis takes $O(n \log n)$ time.*

As an example of the application of our algorithm to remove degeneracies, let us return to the method of [23] for generating random monotonic polygons with vertex set S in $O(K)$ time where $n < K < n^2$. We remark that the non-degeneracy assumption made in [23] is not a convenience but crucial for their algorithm. Our results imply that with a little extra work, their algorithm will run without having to invoke the assumption that no two points of S have the same x-coordinate. In particular, combining our results implies that we can solve the problem in $O(n \log n + K)$ time.

2.3 The Optimization Problem

We consider here the problem of removing the degeneracies in the "best way possible", i.e., of finding the projection that is most robust or tolerant in a way that will be made precise below. One natural definition of tolerance is the idea, which comes from computer graphics, that the projected points are the result of viewing the points from some directional angle, perhaps by a viewer or a camera. Once the regular projection has been obtained it is desirable that the projection remain regular during subsequent perturbations of the camera position. In fact we would like to find, among all the regular projections, the one that has maximum tolerance in this sense, namely, the projection that allows the greatest angular deviation of the viewpoint without ever creating degeneracies. We call this the regular projection with *maximum projective tolerance.* The previous discussion implies that the solution to this problem is determined by the mid-point of the *largest gap* among consecutive forbidden points on C^2, which can be found in $O(n^2 \log n)$ time.

A more general *weighted* version of this problem was first solved by Ramos in his thesis [24] on the tolerance of geometric structures. He uses much more complicated techniques involving lower envelopes of trigonometric functions, but obtains the same time complexity as the above algorithm. It is interesting to note that such a simple algorithm exists for the unweighted case. Thus we have:

Theorem 3. *Given a set S of n distinct points in the plane, a regular projection with the maximum projective tolerance can be computed in $O(n^2 \log n)$ time and $O(n^2)$ space.*

If the model of computation allows the use of the floor function as a primitive operation then there is an and elegant linear time algorithm based on the "pigeon-hole" principle, due to Gonzalez [19], for computing the largest gap of a set of numbers on the real line. This algorithm can be extended to work on the circle C^2, giving us an $O(n^2)$ time algorithm for the above problem.

3 No Two Points in Space on a Vertical Line

The remaining sections of this paper are concerned with removing non-degeneracy assumptions in 3-D (three-dimensional space), where there exists a greater variety of possible degeneracies than in the plane. In this section we consider the type of degeneracy that occurs when two points in space differ in only one of their coordinates, say the z-coordinate, i.e., when two (or more) points lie on a vertical line. This is one natural generalization of the planar degeneracy considered in the previous section.

Let S be a set of n distinct points in Euclidean space and let H be a plane. A projection of S on H is said to be regular if each point in S projects to a distinct point on H, i.e., if no vertical line degeneracies occur when H is the xy-plane. Without loss of generality, assume that the given plane H is the xy-plane. It is convenient in 3-D to represent the directions of projection by points on the surface of the unit sphere centered at the origin, which we denote as C^3.

3.1 The Decision Problem

In the decision problem we wish to decide, given a set of points S in 3-D, if it contains any vertical-line degeneracies, i.e., if it gives a regular projection on the xy-plane. Let S_{xy} denote the set of points obtained by projecting S onto the xy-plane. Sort the points in S_{xy} lexicographically by x and y coordinates to check if two (or more) points in S_{xy} have the same x *and* y coordinates. If there are no such points, we conclude that there are no vertical line degeneracies in S. Once again, element uniqueness can be reduced to this decision problem, giving us the following.

Theorem 4. *Given a set S of n distinct points in space and a plane H, determining whether S admits a regular projection onto H is $\Theta(n \log n)$.*

3.2 The Computation Problem

In this section, we want to compute, given a set S of n distinct points in space, a rotation of S that removes vertical-line degeneracies or equivalently, a plane H such that S yields a regular projection onto H. The only forbidden directions of projections are given by lines in space going through pairs of points in S, since S does not admit a regular projection onto planes perpendicular to these lines. By translating every such line to the origin and intersecting it with the sphere of directions, we obtain $O(n^2)$ (not necessarily distinct) forbidden points on C^3.

We obtain an efficient algorithm for the computation problem in 3-D by applying a strategy similar to the one in 2-D. Assume that S does *not* project regularly onto the xy-plane (this can be checked in $O(n \log n)$ time). This implies that there is at least one forbidden point at the "north-pole" of the sphere of directions C^3. Pairs of points in S that lead to a fixed slope in space correspond to points on "latitude" circles on C^3. The proper second highest slope corresponds to the smallest such circle containing the north pole. To find a regular projection, it suffices to find a point, other than the "north-pole", that lies inside this smallest latitude circle.

Let S_{xy}, S_x and S_y denote the sets of planar points obtained by projecting S onto the xy-plane, x-axis and y-axis, respectively. Sort each of S_{xy}, S_x and S_y lexicographically. Scan through each sorted list to remove multiple occurrences of a point (keeping one copy). Call these new sorted lists S^*_{xy}, S^*_x and S^*_y, respectively. As before, let Δx, Δy and Δz denote the difference in x, y and z-coordinate values, respectively, of any two distinct points in 3-D. Consider a line going through two points in S. Its slope is given by $\Delta z / \sqrt{(\Delta x)^2 + (\Delta y)^2}$.

Since the projection onto the xy-plane is not regular, it follows that there is a pair of points in S such that the line through them has maximum slope $M_1 = \infty$, which leads to a forbidden point at the north-pole. Our objective is to find a non-infinite upper bound on the second-highest slope. In other words, we find a pair of points from S that maximizes Δz and a pair of points from S that gives the smallest possible non-zero value for $(\Delta x)^2 + (\Delta y)^2$. The former is given by a pair of points in S that define the width of S in the z-direction i.e. $\max_S\{\Delta z\}$ and can be found in linear time. The latter is given by the closest-pair in S^*_{xy}.

Computing the closest pair in S^*_{xy} in $O(n \log n)$ time requires a non-trivial method, such as Voronoi diagram construction. However, to keep the algorithm as *simple* as possible, we avoid the use of such a method. Observe that it is not necessary to actually find the closest pair; a positive value less than the closest-pair distance suffices. Let $d = \min\{\min_{S^*_x}\{\Delta x\}, \min_{S^*_y}\{\Delta y\}\}$. A simple argument shows that $d \leq$ the closest-pair distance and can be computed in $O(n \log n)$ time. Therefore, $M = \max_S\{\Delta z\}/d$ (clearly $d > 0$) gives us the required value lying between the largest slope M_1 and the second-largest slope M_2, i.e., $M_2 \leq M < M_1 = \infty$. S admits a regular projection onto a plane perpendicular to any line with finite slope greater than M.

Theorem 5. *Given a set S of n distinct points in space, finding a regular projection of S onto the xy-plane takes $O(n \log n)$ time.*

3.3 The Optimization Problem

As in the two-dimensional case, it is desirable to remove vertical-line degeneracies in an optimal way, i.e., a direction of projection from which we can deviate the most without introducing degeneracies. As before, we call this a regular projection with the maximum projective tolerance.

The above discussion implies that the solution to this problem is determined by the center of the largest empty circle among the $O(n^2)$ forbidden (and not necessarily distinct) points on C^3. We first discard all duplicate points on C^3. The largest empty circle can be found by computing the Voronoi diagram of the remaining points on the sphere. Augenbaum and Peskin [3] give an $O(m^2)$ time insertion method for computing the Voronoi diagram of a set of m points on a sphere. There is a faster algorithm given by Brown [6], in which the Voronoi diagram is obtained in linear time from the convex hull of the set of points (treating them as a set of points in three space). The latter can be constructed in $O(m \log m)$ time (even in the presence of degeneracies such as four or more coplanar points; see [12]). We then have the following result.

Theorem 6. *Given a set S of n distinct points in space, a regular projection plane with the maximum projective tolerance can be computed in $O(n^2 \log n)$ time and $O(n^2)$ space.*

4 No Two Points in Space with the Same x-coordinate

Many geometric algorithms dealing with point sets in 3-D assume that no two points have the same coordinate along one or more axes (see [10], for example). Note that in 2-D the assumptions that (1) no two points lie on a vertical line and (2) no two points have the same x-coordinate, are equivalent. In 3-D on the other hand, the two problems are different. In this section we first consider the problem of rotating a set of points in 3-D so that no two points have the same x-coordinate. We say then, as before, that the projection of S onto the x-axis is regular. Then we show how to generalize this result so that no two points have the same coordinate on any of the three axes.

4.1 The Decision Problem

Given a set S of n distinct points and a line L in 3-D, the decision problem is to decide if S contains a degeneracy or if it admits a regular projection onto L. Assume, without loss of generality, that L is the x-axis. The decision problem then is to determine if any two points of S have the same x-coordinate. Once again, we can reduce element-uniqueness to this problem to show an $\Omega(n \log n)$ lower bound.

Theorem 7. *Given a set S of n distinct points in space, whether S admits a regular projection onto the x-axis can be determined in $\Theta(n \log n)$ time.*

4.2 The Computation Problem

Given a set S of n distinct points in 3-D, the problem is to find a rotation of S that removes the degeneracies or, equivalently, find a line L such that S yields a regular projection onto L. This can be done as follows: first find a plane of regular projection for S. From Theorem 5 we know that this can be done in $O(n \log n)$ time. Let H be such a plane and let S_H be the planar set of points obtained by projecting S onto H. In the plane H, find a line of regular projection for S_H. From Theorem 2 we know that this can be done in $O(n \log n)$ time as well. This line is a line of regular projection for S and so we have the following.

Theorem 8. *Given a set S of n distinct points in space, a rotation of S so that no two points have the same x-coordinate can be computed $O(n \log n)$ time.*

In [10], it is actually assumed that the points have distinct coordinates in all the axes i.e. no two points have the same x, no two points have the same y and no two points have the same z-coordinate. We can extend our method to handle this more general case as well, within the same time bound.

4.3 The Optimization Problem

In this section, we give an algorithm for the optimization problem, i.e., for removing degeneracies in such a way that the resulting line L of regular projection is the most robust. The projection of S onto L will then be said to have the maximum projective tolerance. We make this precise in the following.

Consider a pair of points in S. Now construct a plane through them and consider the direction orthogonal to this plane. In this direction the two points have the same coordinate. Therefore this is a forbidden direction for the final x-axis. This situation is true for every plane through the pair of points. Therefore, a pair of points in S yields a great circle on the sphere C^3 as the set of forbidden directions for the rotation of S. Thus the entire set S yields an arrangement of $O(n^2)$ great circles on C^3. To find a regular projection we must find a point in the interior of any face of this arrangement.

To find a projection of maximum projective tolerance, we need to find a point on the sphere C^3 which is the center of the largest spherical disc contained in a face of the arrangement of great circles. There are $O(n^2)$ great circles and hence the arrangement can be computed in $O(n^4)$ time. It follows therefore that a line L in space such that S yields the regular projection onto L with the maximum projective tolerance can be found in $O(n^4)$ time and space.

Theorem 9. *Given a set S of n distinct points in 3-D, a rotation of S such that S yields a regular projection of maximum projective tolerance onto the x-axis, can be found in $O(n^4)$ time and space.*

It is possible, with a slight increase in run-time, to reduce the above space complexity to $O(n^2)$ as follows: Let G be the set of great circles on C^3. For each great circle α in G, there are two anti-podal points on C^3 that represent the normal to the plane through α. Let P be the set of all such normal points. It can be shown that computing the largest spherical disc contained in a face of the arrangement of G is equivalent to computing a line through the origin that minimizes the maximum distance of the points in P to that line. This is the problem of computing the *thinnest anchored cylinder* containing P. Follert [16] gives an $O(m\lambda_6(m)\log m)$ time and $O(m)$ space algorithm for this problem, where $\lambda_6(m)$ is the maximal length of any Davenport-Schinzel sequence of order six over an alphabet of size m and is known to be slightly superlinear [1]. In our case, $m = |P| = O(n^2)$.

5 No Three Points in Space on a Vertical Plane

Another non-degeneracy assumption that is commonly made for a set of points in three dimensions is that no three of them lie on a vertical plane. For example, Guibas et al. [2], give a linear-time algorithm for constructing the convex hull of a set of points in three dimensions with certain special properties. In their algorithm, one of the assumptions that is made is that no three of the input set of points lie on a vertical plane. Some geometric algorithms use the projected points in some of their steps on which they apply other algorithms. Some of these intermediate algorithms make the non-degeneracy assumption that no three of the projected points are collinear, which implies that the input set of points in space has no three on a vertical plane. We consider here the problem of removing this non-degeneracy assumption i.e., of finding a direction of projection in which this degeneracy does not hold.

Let S be a set of distinct points in Euclidean space and let H be a plane. The projection of S onto H is said to be a *non-collinear projection* if no three of the projected points are collinear, i.e., there are no vertical-plane degeneracies when H is the xy-plane. Note that each of the projected points must also be distinct, so that a non-collinear projection is also a regular projection.

5.1 The Decision Problem

Assume, without loss of generality, that H is the xy-plane. We would like to decide if the input set S of n points in space has vertical-plane degeneracies or if the projection onto the xy-plane is a non-collinear projection. To solve the decision problem, first determine if there are any vertical-line degeneracies using Theorem 4. If there are, then S has vertical-plane degeneracies. If there are none, project S onto the xy-plane to obtain S_{xy} and find an orientation for the x-axis such that no two points in S_{xy} have the same x-coordinate (see Theorem 2). A simple $O(n^2 \log n)$ algorithm for detecting collinearities is obtained by sorting, for each point in S_{xy}, the remaining points by polar angle. A faster method is obtained by using the dual map $(a, b) \mapsto y = ax + b$ of S_{xy}. The dual is an arrangement of n lines. In the arrangement, a vertex of degree six corresponds to three collinear points. Constructing the whole arrangement and checking for vertices of degree six or more can be done in $O(n^2)$ time and space (see [14]).

The problem of deciding if three points from a set of points in space lie on a vertical plane admits the obvious reduction from the problem of deciding if there are three collinear points in a set of points in the plane (the 3-collinear problem). This reduction shows that the decision problem belongs to the class of 3SUM-hard problems [18]. Problems in the class are at least as hard as the problem of determining, given a set of integers, if there are three that sum to zero.

Theorem 10. *Given a set S of n distinct points in space and a plane H, determining whether S admits a non-collinear projection onto H takes $O(n^2)$ time and space.*

5.2 The Computation Problem

In this section, we give an efficient algorithm to compute, for a given set S of n distinct points in space, a rotation of S that removes vertical-plane degeneracies; in other words, an algorithm to find a plane H such that S yields a non-collinear projection onto H. We will assume for the remainder of the section that the input is a set of points in three dimensions such that no three of the points in *space* are collinear, since a non-collinear projection exists if and only if this is the case. It follows therefore that computing a direction of non-collinear projection is 3SUM-hard. It is possible to determine whether three points in space are collinear in $O(n^2)$ time and space. Assume also that no two points in S have the same x coordinate. As shown in Section 4.2, we can find in $O(n \log n)$ time an orientation of the axes such that this assumption is true.

We now give an algorithm for the computation problem. To see which directions are forbidden (for the plane normals), consider a triple of points in S. Each such triple defines a plane and every direction contained in that plane gives a forbidden direction. By intersecting those planes with the sphere of directions C^3 we obtain a set of $\binom{n}{3}$ great circles as forbidden directions. To compute a non-collinear direction we must find a point that does not lie on any of the great circles. If S_{xy} is a non-collinear projection we are done. Otherwise, a certain number of great circles pass through the north pole of C^3. We denote by A the set of great circles containing the north pole (vertical great circles). For each great circle *not* in A we compute its distance to the north pole. The smallest such distance, which can be found in $O(n^3)$ time, determines a circle γ whose center is at the north pole and which does not contain any non-vertical great circles.

Among all great circles in A we find the two distinct a, b that are adjacent when viewed from above. These can be found in $O(n^3)$ time. If c is a great circle through the north pole whose slope is between a and b, its intersection with γ leads to an open arc of non-collinear directions of projection. Thus a non-collinear projection can be computed in $O(n^3)$ time and $O(n)$ space.

We show in the remainder of this section that this straightforward $O(n^3)$ time algorithm can in fact be improved to an $O(n^2)$ time algorithm, which now requires $O(n^2)$ space. This improvement is obtained by making two observations that give us faster run-times for each step in the algorithm described above: (a) A pair of distinct great circles from the set A that are adjacent when viewed from above can, in fact, be found in $O(n^2)$ time, and (b) Instead of finding the largest circle centered at the north pole that does not contain any non-vertical great circles (the circle γ that we computed earlier), it is enough, in fact, to compute some circle *smaller* than γ, and this can indeed be done in $O(n^2)$ time. The result is stated below, with a sketch of the proof.

Theorem 11. *Given a set S of n distinct points in space, a non-collinear projection, if it exists, can be computed in $O(n^2)$ time and space.*

Proof sketch: Observe that step (a) can be done in $O(n^2)$ time by constructing the arrangement of lines given by the dual of S_{xy} (Note: no two lines in the dual are parallel) and by finding, in the set of vertices of degree 6 or more, the two with the smallest and second-smallest x-coordinates.

Now let us proceed to (b), i.e., the question of computing a circle (call it δ) containing the north pole so that no non-vertical circles go through it. Let S be the input set of n points in 3-space. Consider any point of S and call it p. Of all the planes defined by triples of points in S, point p has $O(n^2)$ planes going through it. These planes define a set of $O(n^2)$ great circles on C^3: call this set G_p and let T_p be the *non-vertical* great circles from G_p. We find a circle δ_p centered at the north pole so that no great circle from T_p goes through it, and we do this for every p. The required circle δ will then simply be the smallest circle from the set $\{\delta_p \mid p \in S\}$.

We now briefly sketch a linear-time algorithm to find δ_p. For any $\alpha \in G_p$, let N_α be the point on C^3 that represents the normal to (the plane through) α. There are two points on C^3 that represent the normal to α; it suffices, for our purposes, to consider only those that lie on or above the equator. If α is a *vertical* great circle, N_α will lie on the equator of C^3 and the z-coordinate of N_α, $z(N_\alpha)$, will be zero. However, for every $\alpha \in T_p$, $z(N_\alpha)$ will be non-zero. To find a non-vertical plane that is closest to the north-pole, we need to find an N_α with the smallest z-coordinate greater than zero. In order to find a δ_p, however, it is sufficient to find *some* non-zero value that is smaller than $z(N_\alpha)$ for all $\alpha \in T_p$.

Consider now a great circle $\beta \in T_p$. β is given by a plane that goes through p and two other distinct points, say u and v. We denote by S' the set of points obtained from the input set S by identifying p with the origin and every other point with its *central projection* on the unit sphere centered at the (new) origin. By a slight abuse of notation, we will continue to use the same symbol to refer to a point in S' as the corresponding point in S.

Let $u \equiv (x_u, y_u, z_u)$ and $v \equiv (x_v, y_v, z_v)$. The cross-product of the vectors $\overrightarrow{pu}$ and $\overrightarrow{pv}$ gives us a vector N'_β normal to β (assume it lies above the equator). It can be shown that $z(N_\beta) \geq z(N'_\beta)$. Therefore, when we find a non-zero value less than $z(N'_\beta)$ for all such pairs of vectors, we have the value we are seeking; namely, a non-zero value less than $z(N_\alpha)$ for all $\alpha \in T_p$.

It can be shown that a non-zero value less than $z(\overrightarrow{pu} \times \overrightarrow{pv})$, for all distinct $u, v \in S'$ (not collinear with p), can be found in linear time. We omit this part of the proof for lack of space. Consequently, we obtain a δ_p in linear time, giving us the desired result.

■

5.3 The Optimization Problem

The problem of finding a direction of projection that allows the most deviation without introducing vertical-plane degeneracies, i.e., a projection with maximum projective tolerance, is reduced to the problem of finding the largest empty circle contained in a face of the arrangement of the forbidden great circles. Since there are $O(n^3)$ great circles in the arrangement, we have the following:

Theorem 12. *Given a set S of n distinct points in space (such that no three are collinear), a direction such that S yields a non-collinear projection with the maximum projective tolerance can be computed in $O(n^6)$ time and space.*

As before, the technique outlined in Section 4.3 can be used to reduce the space complexity with a slight increase in run-time.

6 Non-degeneracy Assumptions for Line Segments

In this section, we consider the problem of removing certain non-degeneracy assumptions for line segments in three-dimensional space. We focus on the optimization version of the problem; the decision and computation versions are straightforward. In 3-D computer graphics, the selection of a position of the eye, i.e., a direction of orthogonal projection, is an important problem. For instance, in [21], Kamada and Kawai assert that the resulting image of the 3-D object, which is typically a "wire-frame" object made up of a collection of line segments, should be "easy to comprehend". This means that there should be as little loss of information as possible: information is lost, for example, when two line segments in space project to collinear line segments, or when line segments project to a single point. Therefore, when the line segments are projected onto the xy-plane, this condition is equivalent to the non-degeneracy requirement that no two line segments lie in a vertical plane. In [21], the authors call such a direction of projection (eye-position), in which degeneracies are removed, a "general eye-position".

In order to compute a "most general eye-position", i.e., a direction of projection in which such degeneracies are removed in the best way possible, Kamada and Kawai [21] give an algorithm that takes $O(n^6 \log n)$ worst-case time, where n is the size of the input set of line segments in space. This is the direction of projection that allows the most deviation without introducing any degeneracies of the type described above.

We give here a more efficient algorithm for this problem. Let S be the input set of line segments in 3-D (as in [21], assume no two line segments in space are collinear). Consider the plane, if it exists, containing a pair of line segments in S. Every line contained in this plane gives a forbidden direction of projection (an eye-position which is not a "general eye-position"). Therefore, this pair of line segments describes a great circle of forbidden directions on C^3. By doing this for every pair of line segments, we obtain an arrangement of $O(n^2)$ great circles on C^3. The "most-general eye position", a direction giving the projection of maximum projective tolerance, is given by the center of the largest spherical disc contained in a face of the arrangement of great circles. Therefore for the case of wire frame objects we have the following theorem:

Theorem 13. *Given a set S of n line segments in space, a direction such that S yields a projection with no two line segments collinear and with the maximum projective tolerance can be computed in $O(n^4)$ time and space.*

As in Section 4.3, let G be the set of great circles on C^3 and let P be the set of pairs of points on C^3 normal to each great circle in G. As stated there, the above problem is equivalent to computing the thinnest anchored cylinder of P and can be solved using Follert's algorithm [16]. This is equivalent to computing the two smallest anti-podal spherical caps containing P. There are applications, such as GIS and computer-aided architecture, in which it is desirable that the "eye-position" lies "above" every forbidden plane, as given by the great circles in G. In such cases, the problem reduces to finding the smallest enclosing spherical cap containing the points $P' \subset P$ that lie in the upper hemisphere. We can show that this can be found by computing the smallest enclosing sphere of P' and intersecting it with C^3 in constant time. The former can be computed in $O(|P'|)$ time by using Megiddo's technique [22], giving us an $O(n^2)$ time and space algorithm for this special case.

Acknowledgments: The authors would like to thank Jean-Daniel Boissonnat, Hervé Brönnimann, Olivier Devillers, Fady Habra, Pedro Ramos, Monique Teillaud and Mariette Yvinec for useful discussions and references.

References

1. P. K. Agarwal, M. Sharir, and P. Shor. Sharp upper and lower bounds for the length of general davenport-schinzel sequences. *J. Combin. Theory, Ser. A*, 52:228–274, 1989.
2. A. Aggarwal, L. J. Guibas, J. Saxe, and P. W. Shor. A Linear-Time Algorithm for Computing the Voronoi Diagram of a Convex Polygon. *Disc. and Comp. Geo.*, 4:591–604, 1989.
3. J. M. Augenbaum and C. S. Peskin. On the construction of the Voronoi mesh on a sphere. *J. of Comp. Physics*, 59:177–192, 1985.

4. P. Bhattacharya and A. Rosenfeld. Polygons in three dimensions. *J. Visual Comm. and Image Representation*, 5(2):139–147, June 1994.
5. P. Bose, F. Gómez, P. Ramos, and G. Toussaint. Drawing nice projections of objects in space. In *Proc. of Graph Drawing*, Passau, Germany, September 1995.
6. K. Q. Brown. *Geometric transforms for fast geometric algorithms.* Ph.D. thesis, Department of CS, CMU, Pittsburgh, PA, 1980. Report CMU-CS-80-101.
7. C. Burnikel, K. Mehlhorn, and S. Schirra. On degeneracy in geometric computations. In *Proc. 5th SODA*, pages 16–23, 1994.
8. J. Canny, B. Donald, and E. K. Ressler. A rational rotation method for robust geometric computations. In *Proc. 8th ACM Symp. on Comp. Geo.*, 1992.
9. R. Cole, J. S. Salowe, W. Steiger, and E. Szemeredi. An optimal time algorithm for slope selection. *SIAM J. on Comp.*, 18:792–810, 1989.
10. A. M. Day. The Implementation of an Algorithm to Find the Convex Hull of a Set of Three-Dimensional Points. *ACM Trans. on Graphics*, 9(1):105–132, 1990.
11. L. Devroye and T. Klincsek. Average time behavior of distributive sorting algorithms. *Computing*, 26:1–7, 1981.
12. H. Edelsbrunner. *Algorithms in Combinatorial Geometry.* Springer-Verlag, 1987.
13. H. Edelsbrunner and E. P. Mücke. Simulation of simplicity: a technique to cope with degenerate cases in geometric algorithms. *ACM Trans. Graph.*, 9:67–104, 1990.
14. H. Edelsbrunner, J. O'Rourke, and R. Seidel. Constructing arrangements of lines and hyperplanes with applications. *SIAM J. Comp.*, 15(2):341–363, 1986.
15. I. Z. Emiris and J. F. Canny. A general approach to removing degeneracies. *SIAM J. of Comp.*, 24(3):650–664, 1995.
16. F. Follert. Lageoptimierung nach dem maximin-kriterium. Master's thesis, Universität des Saarlandes, Saarbrücken, 1994.
17. S. Fortune. *Directions in Geometric Computing*, chapter Progress in computational geometry, pages 82–127. Antony Rowe Ltd., 1993.
18. A. Gajentaan and M. H. Overmars. On a class of $O(n^2)$ problems in computational geometry. *Comp. Geo.: Th. and App.*, 5(3):165–185, 1995.
19. T. Gonzales. Algorithms on sets and related problems. Technical report, Department of Computer Science, University of Oklahoma, 1975.
20. M. Iri and K. Sugihara. Construction of the Voronoi diagram for "one million" generators in single-precision arithmetic. *Proc. of the IEEE*, 80(9):1471–1484, 1992.
21. T. Kamada and S. Kawai. A Simple Method for Computing General Position in Displaying Three-Dimensional Objects. *Comp. Vision, Graphics and Image Proc.*, 41:43–56, 1988.
22. N. Megiddo. Linear-time Algorithms for Linear Programming in R^3 and Related Problems. *SIAM J. on Comp.*, 12(4):759–776, 1983.
23. J. S. B. Mitchell, J. Snoeyink, G. Sundaram, and C. Zhu. Generating random x-monotone polygons with given vertices. In *Proc. 6th CCCG*, pages 189–194, 1994.
24. P. Ramos. *Tolerancia de estructuras geometricas y combinatorias.* PhD thesis, Univ. Politec. de Madrid, 1995.
25. R. Seidel. The nature and meaning of perturbations in geometric computing. *Disc. and Comp. Geo.*, 1996. to appear.
26. C. K. Yap. Symbolic treatment of geometric degeneracies. *J. of Symbolic Comp.*, 10:349–370, 1990.

Maintaining Maxima under Boundary Updates*
(Extended abstract)

Fabrizio d'Amore[1] Paolo Giulio Franciosa[1] Roberto Giaccio[1] Maurizio Talamo[1]

Dipartimento di Informatica e Sistemistica,
Università di Roma "La Sapienza",
via Salaria 113, I–00198 Roma, Italy.
E-mail: {damore,pgf,giaccio,talamo}@dis.uniroma1.it

Abstract. Given a set of point $P \in \mathbb{R}^2$, we consider the well-known *maxima problem,* consisting of reporting the maxima (not dominated) points of P, in the dynamic setting of boundary updates. In this setting we allow insertions and deletions of points at the extremities of P: this permits to move a resizable vertical window on the point set.
We present a data structure capable of answering maxima queries in optimal $O(k)$ worst case time, where k is the size of the answer. Moreover we show how to maintain the data structure under boundary updates operation in $O(\log n)$ time per update.
This is the first technique in a dynamic setting capable of both performing update operations and answering maxima queries in optimal worst case time.

1 Introduction

Dominance and maxima problems are fundamental ones in combinatorics and in computational geometry. Among the former typical examples are "given a set of point S and a query point p report all elements of S dominating (or dominated by) p," or "given a set of point S report all pairs $\langle p, q\rangle$ in S such that p dominates (or is dominated by) q." Typical examples of the latter have the form "given a set of points S report its maxima." Equivalent formulations in terms of posets are also possible.

In this paper we present a data structure capable of answering maxima queries in two dimensions, and show how to maintain the data structure under "boundary update" operations [dG96], consisting of arbitrary (feasible) sequences of point insertion/deletion operations occurring at the extremities of the point set S; in other words, S can be considered as a point set delimited at left and at right by two vertical sweep lines, which can independently move rightward and leftward: each movement of one sweep line gives rise to a boundary update operation. This can be also seen as a moving and resizable vertical window on the point set.

* Work partially supported by EU ESPRIT LTR Project ALCOM-IT under contract no. 20244, and by the Italian MURST National Project "Efficienza di Algoritmi e Progetto di Strutture Informative."

Several papers appear in the literature studying the maxima problem in a non-static setting; for the fully dynamic case there are several solutions in which maxima queries can be answered in $O(k)$, where k is the size of the answer. A first breakthrough is introduced in [OvL81] where updates require $O(\log^2 n)$ worst case time; faster solutions are proposed in [FR90,Jan91], allowing insertions to be performed in $O(\log n)$ and deletions in $O(\log^2 n)$ worst case time. Up to now no solution is known supporting both updates in $O(\log n)$ while retaining the optimal query time. In [Kap94] Kapoor presents a technique that supports both kinds of updates in logarithmic time, but the resulting query time is optimal only in amortized sense. More specifically, he proves that the set of maxima of a set of points in the plane can be maintained in $O((n+m)\log n + r)$ operations when there are n insertions, m deletions and r is the number of points reported.

The data structure we propose answers maxima queries by returning in $O(1)$ a balanced structure containing all maxima; this is similar to what done in [OvL81,FR90,Jan91], while [Kap94] only returns the sequential list of maxima. In addition, our solution supports boundary updates in $O(\log n)$ time per update. Even if our setting is a restriction of the fully dynamic one, it is the first non-static setting supporting insertions and deletions in logarithmic time, while retaining the optimal query time.

The paper is organized as follows. After giving in Sect. 2 some basic definitions and notations, in Sect. 3 we describe our data structure, and in Sect. 4 we describe how it can be maintained under boundary updates. Finally, in Sect. 5 we give some final remarks and address future research on the topic.

2 Basic definitions

We introduce some basic definitions concerning dominance in two dimensions. Given a point $p \in \mathbb{R}^2$, we denote its abscissa by $x(p)$ and its ordinate by $y(p)$. Moreover, let left(P) (right(P)) be the leftmost (rightmost) element of a set P. Given a set of points $P = \{p_1, p_2, \ldots, p_n\} \subset \mathbb{R}^2$, we say p_i *dominates* p_j ($p_j \prec p_i$) if and only if $x(p_j) < x(p_i)$ and $y(p_j) < y(p_i)$. For the sake of simplicity we assume $x_i \neq x_j$ and $y_i \neq y_j$ for any $i \neq j$. We call *covering relation* the transitive reduction of the dominance relation: in this case, if p_i covers p_j we write $p_j \prec: p_i$; in other words $p_j \prec: p_i$ if and only if there is no point p_k such that $p_j \prec: p_k \prec: p_i$. Such relation has also been referred to as *direct dominance* [GNO89]; a strictly related relation is the so-called *rectangular visibility*, which corresponds to the cover relation for any orientation of the axis [OW88]. Two points p and q are said *incomparable* ($p \parallel q$) if neither $p \prec q$ nor $q \prec p$.

A set of points P is a *chain* if the dominance relation is a total order on P, i.e. for any pair $\{p,q\} \subseteq P$ either $p \prec q$ or $q \prec p$; it is an *antichain* if $p \parallel q$ for any pair $\{p,q\} \subseteq P$.

A point p is called *maximal* in P if $p \in P$ and no $q \in P$ exists such that $p \prec q$. We denote the set of maxima by $\mathcal{M}(P)$.

The *maxima problem* consists of retrieving the set of maxima of P, and has been studied in several frameworks. In the most general one points can be both inserted in and deleted from P, and the corresponding data structure has to be accordingly re-arranged in order to efficiently retrieve the maxima. Here we consider the boundary update framework that is a restriction of fully dynamic one: a point p can be inserted in P if and only if $x(p) < x(\text{left}(P))$ or $x(p) > x(\text{right}(P))$, and only points left$(P)$ and right(P) can be deleted.

3 The data structure

Given a set of points P and two points $p, q \in P$ we say that q *leftmost covers* p ($p \prec:_{\mathrm{L}} q$) if and only if $q = \text{left}\big(\{q' \in P \mid p \prec: q'\}\big)$. Note that the set of points $\text{cov}_{\mathrm{L}}(q) = \{p \in P \mid p \prec:_{\mathrm{L}} q\}$ is an antichain; in the sequel this set is also informally referred to as a *covered antichain.* If $p \in \text{cov}_{\mathrm{L}}(q)$ we use next(p) (resp. prev(p)) for denoting the possible point in $\text{cov}_{\mathrm{L}}(q)$ immediately to the right (left) of p; we use the same notation also for $\mathcal{M}(P)$.

We define the partial order relation $\prec_{\mathrm{L}}$ as the transitive closure of $\prec:_{\mathrm{L}}$, i.e. q *leftmost dominates* p ($p \prec_{\mathrm{L}} q$) if there is a sequence of points $p = p_1, p_2, \ldots, p_k = q$ such that $p_i \prec:_{\mathrm{L}} p_{i+1}$, for $i = 1, 2, \ldots, k-1$. The $\prec_{\mathrm{L}}$ relation plays a central role in our data structures.

Relation $\prec:_{\mathrm{L}}$ defines a rooted forest, in fact each point except maximal ones has exactly one parent, and there are no cycles since a parent is always to the right of its children. We refer to the forest defined by relation $\prec_{\mathrm{L}}$ as the *cover forest.*

The queries described in the next section can be answered by navigating in the cover forest. However, we do not explicitly represent the cover forest. In contrast, we maintain the following sets, which constitute the *cover structure* of P:

- For each point $p \in P$, its covered antichain $\text{cov}_{\mathrm{L}}(p)$ (the children of p in the forest), sorted by abscissae.
- For each point $p \in P$ such that either next(p) exists or p is maximal, the associated *maximal right chain* rc(p) in the cover forest, sorted by abscissae, where the maximal right chain is determined by the longest sequence of vertices $p = p_1, p_2, \ldots, p_k$ such that $p_{i+1} = \text{right}(\text{cov}_{\mathrm{L}}(p_i))$, for $i = 1, 2, \ldots, k-1$. For each non maximal point q for which next(q) does not exist, $\text{rc}(q) = \emptyset$.
- For each point $p \in P$ such that either prev(p) exists or p is maximal, the associated *maximal left chain* lc(p) in the cover forest, sorted by abscissae, where the maximal left chain is determined by the longest sequence of vertices $p = p_1, p_2, \ldots, p_k$ such that $p_{i+1} = \text{left}(\text{cov}_{\mathrm{L}}(p_i))$, for $i = 1, 2, \ldots, k-1$. For each non maximal point q for which prev(q) does not exist, $\text{lc}(q) = \emptyset$.
- $\mathcal{M}(P)$, sorted by abscissae.

Fig. 1 depicts the $\prec:_{\mathrm{L}}$ relation (a) and the corresponding cover structure (b).

Assuming that all the sets in the cover structure are represented by data structures whose space requirements are linear, the following holds:

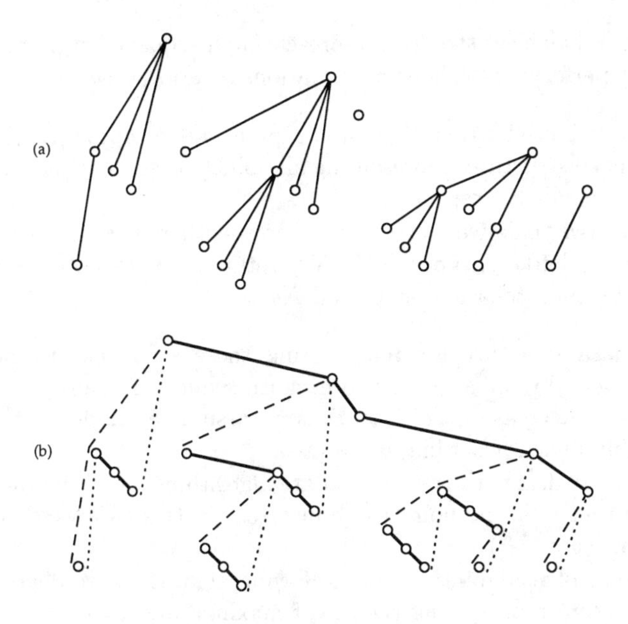

Fig. 1. The cover structure.

Lemma 1. *Given a set of points P, its cover structure can be represented by covered antichains and maximal left and right chains in $O(|P|)$ space.*

Proof. It directly derives from the observation that each point appears in exactly one maximal left chain and one maximal right chain. Moreover each point either appears in $\mathcal{M}(P)$ or in one covered antichain.

We assume in the sequel that covered antichains, maximal right and left chains, and maxima are represented by balanced search trees. In the next section we see how binary search trees allow to support boundary updates by means of split and merge operations.

Since the set of maxima points is explicitly stored in the cover structure, all the k maxima points can be reported in $O(k)$ worst case time. Moreover it is also possible to answer a search query on the maxima set in $O(\log k)$, like for example deciding whether a query point (x, y) is dominated or not by any point in the structure.

4 Updates

The cover structure of a set of points P can be maintained under under boundary updates in logarithmic worst case time, i.e. it is possible to insert a point such that either $x\,(p) < x\,(\text{left}(P))$ or $x\,(p) > x\,(\text{right}(P))$, and it is also possible to delete points left(P) and right(P).

This can be done if the cover structure is implemented by means of balanced structures supporting the split and merge operations in logarithmic time, i.e., starting from a balanced structure representing a sequence of points $K \subseteq P$, it is possible to perform the following operations in worst case time $O(\log |P|)$:

split given a balanced structure K and a point $q \in K$ `split`(K, q) splits K in two balanced structures representing all point preceding q and the remaining points (q included), respectively;

merge given two balanced structures representing sets K_1 and K_2, with $\text{right}(K_1) < \text{left}(K_2)$, `merge`$(K_1, K_2)$ returns a balanced structure representing the concatenation of K_1 and K_2.

A suitable data structure for representing these sequences is the red-black tree [GS78]. Actually, we also need to perform sequential scans on the sequence by increasing or decreasing abscissa, thus we assume all nodes in the red-black tree are doubly linked according to abscissa.

We separately deal with the four update algorithms. Let P be the current set of points, and let p be the point to be inserted (in this case we assume $p \notin P$) or deleted (hence $p \in P$).

For the sake of readability, in the following algorithms we denote by $\mathcal{M}(P)$ the data structure representing the set of maxima, and by $\text{cov}_\text{L}(\cdot)$ the covered antichains.

4.1 `InsertRight`

The algorithm for inserting a point p at the right of the current point set P is described in Fig. 2. The new inserted point p clearly is a maximal point, while the portion of $\mathcal{M}(P)$ possibly dominated by p is no longer in the set $\mathcal{M}(P \cup \{p\})$. Though the algorithm is straightforward, it is worth to point out how the maximal chains have to be updated. Concerning the assignments in the last two lines of Fig. 2, these are actually performed by detaching from a the balanced trees representing $\text{lc}(a)$ and $\text{rc}(a)$, and associating them with p after adding $\{p\}$. The behaviour of the algorithm is shown in Fig. 3.

4.2 `DeleteRight`

The algorithm for deleting point $\text{left}(P)$ is very simple, and is shown in Fig. 4. Note that also in this case the assignments between maximal chains are intended as simple assignments to pointers (see Fig. 5).

4.3 `InsertLeft`

The more involved case occurs when p is not a maximal point. In this case, as shown in Fig. 6, we have to split the maximal left chain associated with $\text{left}(\mathcal{M}(P))$ at $y(p)$, creating the two maximal left chains associated to $\text{left}(\mathcal{M}(P))$ and b, respectively (see Fig. 7).

procedure `InsertRight`(p)
input: a point p such that $x(p) > x(\text{right}(P))$
begin
  comment: p is always a maximal point
  if p does not dominate any point in $\mathcal{M}(P)$ **then**
    insert p into $\mathcal{M}(P)$
    $\text{cov}_{\text{L}}(p) := \emptyset$
    $\text{lc}(p) := (p)$
    $\text{rc}(p) := (p)$
  else
    $a := \text{left}\big(q \in \mathcal{M}(P) \;\big|\; q \prec_{\text{L}} p\big)$
    $b := \text{right}\big(q \in \mathcal{M}(P) \;\big|\; q \prec_{\text{L}} p\big)$
    `split`$(\mathcal{M}(P), p)$; let $\mathcal{M}(P)$ be the left part and $\text{cov}_{\text{L}}(p)$ the right part
    $\mathcal{M}(P) := \mathcal{M}(P) \cup \{p\}$
    $\text{lc}(p) := \text{lc}(a) \cup \{p\}$; $\text{lc}(a) := \emptyset$
    $\text{rc}(p) := \text{rc}(b) \cup \{p\}$; $\text{rc}(a) := \emptyset$
  end if
end

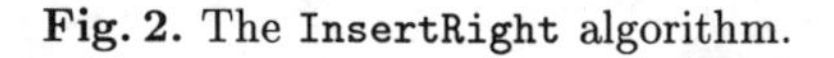

Fig. 2. The `InsertRight` algorithm.

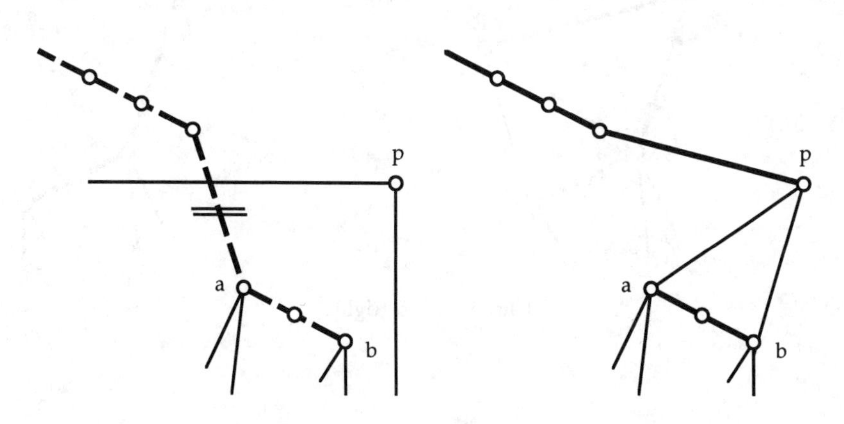

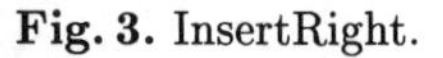

Fig. 3. InsertRight.

```
procedure DeleteRight
begin
    p := right(P)
    comment: the rightmost point is always a maximal point
    M(P) := M(P) − {p}
    lc(left(cov_L(p))) := lc(p) − {p}
    rc(right(cov_L(p))) := rc(p) − {p}
    M(P) := merge (M(P), cov_L(p))
end
```

Fig. 4. The `DeleteRight` algorithm.

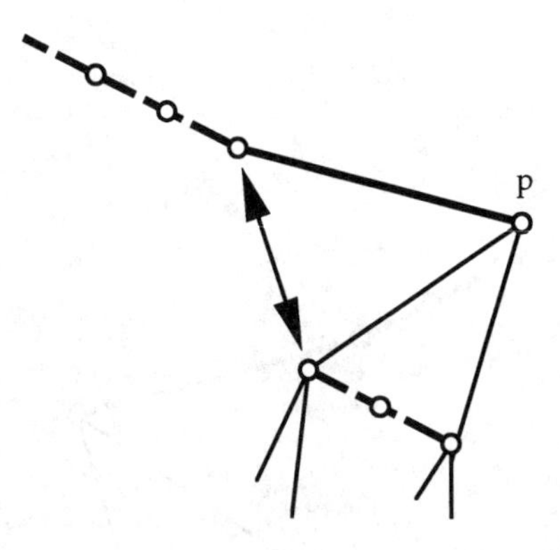

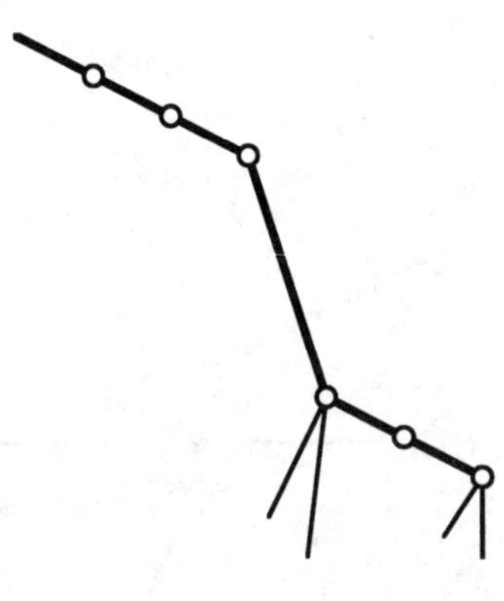

Fig. 5. DeleteRight.

procedure `InsertLeft`(p)
input: a point p such that $x(p) < x(\mathrm{left}(P))$
begin
  $\mathrm{cov_L}(p) := \emptyset$
  $q := \mathrm{left}(\mathcal{M}(P))$
  if $q \parallel p$ **then**
    $\mathcal{M}(P) := \mathcal{M}(P) \cup \{p\}$
    $\mathrm{lc}(p) := (p)$
    $\mathrm{rc}(p) := (p)$
  else
    $a :=$ the point in $\mathrm{lc}(q)$ such that $p \prec\!:_{\mathrm{L}} a$
    if $\mathrm{cov_L}(a) \neq \emptyset$ **then**
      $b := \mathrm{left}(\mathrm{cov_L}(a))$
      `split`$(\mathrm{lc}(q), a)$; let$\mathrm{lc}(b)$ be the left part and $\mathrm{lc}(q)$ the right part
      $\mathrm{rc}(p) := (p)$
    else
      $\mathrm{rc}(a) := \mathrm{rc}(a) \cup \{p\}$
    end if
    $\mathrm{cov_L}(a) := \mathrm{cov_L}(a) \cup \{p\}$
    $\mathrm{lc}(q) := \mathrm{lc}(q) \cup \{p\}$
  end if
end

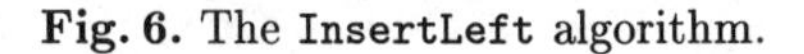

Fig. 6. The `InsertLeft` algorithm.

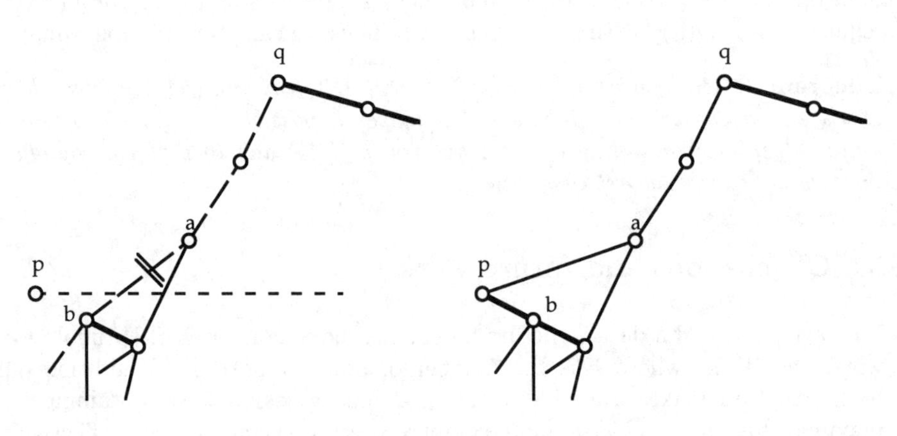

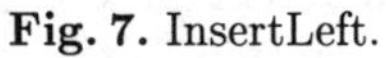

Fig. 7. InsertLeft.

4.4 DeleteLeft

Also in this update, the more involved case occurs when left(P) is not a maximal point. If this happens, after the point has been deleted we have to merge the maximal left chains associated with q and b (see Fig. 8) in a single maximal chain (see Fig. 9).

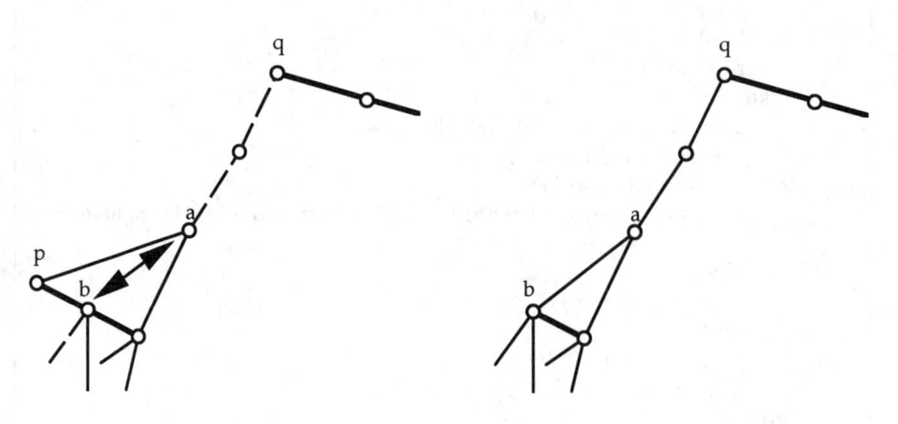

Fig. 8. DeleteLeft.

4.5 Complexity

The update algorithms only perform a constant number of split and merge operations on the lc(·), rc(·) and $\text{cov}_\text{L}(\cdot)$ representing the cover forest. Its overall space requirement is $O(n)$, as shown in Lemma 1. Thus we can state the following:

Theorem 2. *Given a point set $P \subset \mathbb{R}^2$, with $|P| = n$ and $|\mathcal{M}(P)| = k$, there exists a $O(n)$ space data structure that makes it possible to report a balanced search structure containing maxima queries in $O(1)$, and to perform boundary updates in $O(\log n)$ worst case time.*

5 Conclusions and future work

We have presented a data structure for solving maxima queries in optimal $O(k)$ worst case time, where k is the number of points reported. In addition our data structure makes available a balanced binary search tree containing the maxima: this allows to solve more complex queries on maxima very effectively. Also, we have shown how to maintain the data structure under boundary update operations in $O(\log n)$ time per update, where boundary updates are insertions/deletions occurring at the extremities of the point set. This is equivalent to have a moving and resizable vertical window on the point set.

procedure `DeleteLeft`
begin
    $p := \text{left}(P)$
    if $p \in \mathcal{M}(P)$ **then**
        $\mathcal{M}(P) := \mathcal{M}(P) - \{p\}$
    else
        $q := \text{left}(\mathcal{M}(P))$
        comment: p is the lowest element in $\text{lc}(q)$
        $\text{lc}(q) := \text{lc}(q) - \{p\}$
        $a := \text{left}(\text{lc}(q))$
        $\text{cov}_\text{L}(a) := \text{cov}_\text{L}(a) - \{p\}$
        if $\text{cov}_\text{L}(a) \neq \emptyset$ **then**
            $b := \text{left}(\text{cov}_\text{L}(a))$
            $\text{lc}(q) :=$ `merge` $(\text{lc}(b), \text{lc}(q))$
        else
            $\text{rc}(q) := \text{rc}(q) - \{p\}$
        end if
    end if
end

Fig. 9. The `DeleteLeft` algorithm.

Future research on the topics presents several challenging aspects. First of all how to solve maxima queries in optimal time allowing both insertion and deletions in optimal time. It would also be interesting to extend our technique to $\mathbb{R}^3$, and to apply similar techniques to other fundamental problems, such as dominance and convex hulls.

References

[dG96] F. d'Amore and R. Giaccio. Simplified hive-graphs with boundary updates. Technical report, Univ. "La Sapienza," Roma, 1996. Submitted.

[FR90] G. Frederickson and S. Rodger. A new approach to the dynamic maintenance of maximal points in a plane. *Discrete Comput. Geom.*, 5:365–374, 1990.

[GNO89] R. H. Güting, O. Nurmi, and T. Ottmann. Fast algorithms for direct enclosures and direct dominances. *J. Algorithms*, 10:170–186, 1989.

[GS78] L. J. Guibas and R. Sedgewick. A dichromatic framework for balanced trees. In *Proc. 19th Annu. IEEE Sympos. Found. Comput. Sci.*, Lecture Notes in Computer Science, pages 8–21, 1978.

[Jan91] R. Janardan. On the dynamic maintenance of maximal points in the plane. *Inform. Process. Lett.*, 40:59–64, 1991.

[Kap94] S. Kapoor. Dynamic maintenance of maximas of 2-d point sets. In *Proc. 10th Annu. ACM Sympos. Comput. Geom.*, pages 140–149, 1994.

[OvL81] M. H. Overmars and J. van Leeuwen. Maintenance of configurations in the plane. *J. Comput. Syst. Sci.*, 23:166–204, 1981.

[OW88] M. H. Overmars and D. Wood. On rectangular visibility. *J. Algorithms*, 9:372–390, 1988.

An Optimal Algorithm for One-Separation of a Set of Isothetic Polygons

(Extended Abstract)

Amitava Datta* Kamala Krithivasan† Thomas Ottmann‡

Abstract

We consider the problem of separating a collection of isothetic polygons in the plane by translating one polygon at a time to infinity. The directions of translation are the four isothetic (parallel to the axes) directions, but a particular polygon can be translated only in one of these four directions. Our algorithm detects whether a set is separable in this sense and computes a translational ordering of the polygons. The time and space complexities of our algorithm is $\Theta(n \log n)$ and $\Theta(n)$ respectively, where n is the total number of edges of the polygons in the set. The best previous algorithm in the plane for this problem had complexities of $O(n \log^2 n)$ time and $O(n \log n)$ space.

1 Introduction

The problem of collision-free motion of objects in the presence of obstacles has many applications in *computer graphics*, *CAD-CAM systems* and *robotics*. There has been considerable work on moving line segments, polygons and polyhedra in two and three dimensional spaces in the presence of obstacles. These problems are collectively called *motion planning* problems. In this paper, we are interested in a different kind of motion planning problem called *separability* problem. This class of problems is interesting from the point of view of two potential application areas. In the first case, we have a collection of objects and we are interested in finding out whether the objects can be separated one by one from the collection by collision-free motion. In the second case, the problem arises in machine assembly. A robot has to assemble or disassemble a composite machine part by moving elementary machine parts through collision-free motion. Reif [10] has proved that this problem is PSPACE-hard in its most general form. However, when only translational motion is allowed, this problem is tractable

*Department of Mathematics, Statistics and Computing Science, University of New England, Armidale, NSW 2351, Australia. e-mail : datta@neumann.une.edu.au

†Department of Computer Science and Engineering, Indian Institute of Technology, Madras 600 036, India. e-mail: kamala@iitm.ernet.in

‡Institut für Informatik,Universität Freiburg, am Flughafen 17, 79110 Freiburg, Germany. e-mail : ottmann@informatik.uni-freiburg.de

and has been studied extensively in recent years. Usually, two related problems are considered in this context. The first is the *detection* problem, where it is to be decided whether a composite part can be disassembled or not. In the *determination* problem, a disassembly sequence is computed such that the composite part can be disassembled following this sequence of translations.

Guibas and Yao [6] have shown that a disassembly sequence for a set of M rectangles (resp. convex n-gons) can be computed in $O(M \log M)$ (resp. $O(Mn + M \log M)$) time. Ottmann and Widmayer [8] gave an algorithm for translating a set of line segments. They also gave an alternate and simple solution for the problem studied in [6]. For convex composite parts, a disassembly sequence exists for any fixed direction. However, in case of non-convex parts, there may not exist any direction of disassembly. This happens if a subset of the parts are interlocked. Nussbaum and Sack [7] have presented matching lower and upper bounds for detecting and determining a disassembly sequence in a single direction. Their algorithm runs in $\Theta(Mn + M \log M)$ time. Recently, Agarwal *et al* [1] have presented an algorithm for separation of n polyhedra in 3-space when k directions of separation are allowed and each polyhedron can be translated in one of these k directions. The algorithm in [1] runs in $O(km^{4/3+\epsilon})$ time where, m is the total number of vertices of the polyhedra. See also [2] for a similar problem.

In the isothetic domain,i.e., when the polygons have sides parallel to the axes, the first work on separation is due to Guibas and Yao [6]. Chazelle *et al* [3] first considered two problems related to disassembling a set of isothetic polygons. They considered separating the polygons in directions parallel to the axes (i.e., isothetic directions). The first problem is called the *iso-separability* problem. In this case, the direction of translation is unique. Chazelle *et al* [3] presented an optimal $O(n \log n)$ time and $O(n)$ space algorithm for the iso-separability problem. Here, n is the total number of vertices of the polygons. In the more general *one-separability* problem, all the four isothetic directions of translation are allowed. However, a particular polygon can be translated only once in one of these four directions. Chazelle *et al* presented an $O(n \log^2 n)$ time and $O(n \log n)$ space algorithm for deciding whether a collection of isothetic polygons is one-separable. They could also compute a separating sequence within the same complexity. Recently, Devine and Wood [5] have investigated this one-separability problem in 3 and higher dimensions. In dimensions $d \geq 3$, their algorithm runs in $O(dn^{d/2} \log^3 n)$ time and $O(dn^{d/2} \log n)$ space.

In this paper, we present the first optimal algorithm for deciding one-separability and computing a separation order of the polygons in two dimensions. Our algorithm runs in $\Theta(n \log n)$ time and $\Theta(n)$ space. Though the basic strategy of our algorithm is similar to that in Chazelle *et al* [3], our data structure is completely different and we expect that this data structure will be useful in other applications. If the input scene is not one-separable, our algorithm finds a maximum subset of the polygons which is one-separable.

The rest of this abstract is organized as follows. In Section 2, we discuss some preliminaries. We describe our data structure in Section 3. The algorithm for one-separability is presented in Section 4.

2 Preliminaries

We consider m isothetic polygons $P_1, P_2, \ldots, P_m$. There are totally n edges $e_1, e_2, \ldots, e_n$. If e_i is a vertical (resp. horizontal) edge, its x (resp. y) coordinate is represented by x_i (resp. y_i). The left and right end points of a horizontal edge e_i is represented by l_i and r_i respectively. Similarly, the top and bottom end points of a vertical edge e_j are represented by t_j and b_j respectively. We sometime represent a horizontal (resp. vertical) edge e_i by the interval $[l_i, r_i]$ (resp. $[t_i, b_i]$).

We set up four data structures called *separation trees* in the four isothetic directions. These four trees are called the *eastern (E)*, *western (W)*, *northern(N)* and *southern (S)* trees. The E (resp. W) tree keeps the freedom information of vertical polygon edges in the $+x$ (resp. $-x$) direction and the N (resp. S) tree keeps freedom information of horizontal polygon edges in the $+y$ (resp. $-y$) direction. In other words, if an edge is free in the $+x$ direction, it will be free in the E tree. For the N (resp. S) tree, a horizontal edge e_i *blocks* another horizontal edge e_j in the $+y$ (resp. $-y$) direction, if e_i and e_j overlap and $y_i > y_j$ (resp. $y_i < y_j$). Similarly, for the E (resp. W) tree, a vertical edge e_i *blocks* another vertical edge e_j in the $+x$ (resp. $-x$) direction if e_i and e_j overlap and $x_i > x_j$ (resp. $x_i < x_j$). If an edge e_i is not blocked by any other edge (in a fixed direction), we say that e_i is *unblocked* in that direction.

For every polygon P_i, we keep four counters called eastern, western, northern and southern counters and denoted by EC_i, WC_i, NC_i and WC_i. EC_i keeps the count of the number of free edges in the eastern ($+x$) direction. The meanings of the other counters are similar. The total number of horizontal (resp. vertical) edges of a polygon P_i are represented by $hcount(P_i)$ (resp. $vcount(P_i)$).

We discuss our algorithm briefly and informally in this section. The details are given in Section 4. We first set up the four data structures. The E (resp. W) tree is set up by sweeping the vertical polygon edges in the decreasing (resp. increasing) x direction. Similarly, the N (resp. S) trees are constructed by sweeping the horizontal polygon edges in the decreasing (resp. increasing) y direction. After this, we try to peel the trees and as a result, try to determine the freedom of the polygons. Suppose, we start with the E tree. We check whether any edge is free in the eastern ($+x$) direction and if we find an edge $e_i \in P_j$, we increase EC_j by 1. This way, if EC_j becomes equal to the number of edges of the polygon P_j, P_j is free in the eastern ($+x$) direction. We delete all the edges of P_j from all the four trees and continue our peeling process with the E tree. If we are stuck with the E tree, i.e., there is no free polygon in the $+x$ direction, we go to one of the other trees and repeat the same process. If the scene is one-separable, all the trees will become empty at the end of this process. Otherwise, we will not be able to free any polygon from any of the trees at some stage and we can infer that the scene is not one-separable.

3 Separation tree

In this section, we describe our data structure called *separation tree* which we will use to solve the one-separation problem. We discuss the tree N (as mentioned in Section 2) in this section and the details of the other three trees are similar. N is a balanced binary tree which stores horizontal edges of the isothetic polygons. Once we construct N and insert all the edges, we only delete the edges during the execution of our algorithm. As a result, we ensure that the data structure is balanced when we insert all the edges and there is no need of rebalancing afterwards.

We store the end points of the edges at the leaves of N. Let n_i be an internal node of N and $N(n_i)$ be the subtree of N rooted at n_i. The two subtrees rooted at the two children of n_i are denoted as $left(n_i)$ and $right(n_i)$. We denote the parent of n_i by $parent(n_i)$. The additional fields at n_i are the following.
$min(n_i)$: Stores the coordinate of the least x coordinate leaf in $N(n_i)$.
$max(n_i)$: Stores the coordinate of the highest x coordinate leaf in $N(n_i)$.
$L\_high(n_i)$: Stores an edge which is completely contained and unblocked in $left(n_i)$.
$R\_high(n_i)$: Stores an edge which is completely contained and unblocked in $right(n_i)$.
$LL\_cross(n_i)$: Stores an edge e_j which is unblocked in $left(n_i)$ and l_j is in $left(n_i)$, but r_j is not in $left(n_i)$, i.e., $min(left(n_i)) \leq l_j \leq max(left(n_i)) < r_j$.
$LR\_cross(n_i)$: Stores an edge e_j which is unblocked in $left(n_i)$ and r_j is in $left(n_i)$, but l_j is not in $left(n_i)$, i.e., $l_j < min(left(n_i)) \leq r_j \leq max(left(n_i))$.
$RL\_cross(n_i)$: Stores an edge e_j which is unblocked in $right(n_i)$ and l_j is in $right(n_i)$, but r_j is not in $right(n_i)$, i.e., $min(right(n_i)) \leq l_j \leq max(right(n_i)) < r_j$.
$RR\_cross(n_i)$: Stores an edge e_j which is unblocked in $right(n_i)$ and r_j is in $right(n_i)$, but l_j is not in $right(n_i)$, i.e., $l_j < min(right(n_i)) \leq r_j \leq max(right(n_i))$.

It is easy to see that in all the above cases, only one edge can be the candidate for a particular field except the fields R_high and L_high. If there are multiple candidate edges for these two fields, we choose the highest y coordinate edge. Also, if no such edge exists, the field is assigned a value $null$.

Lemma 3.1 *The separation tree as described above can be constructed in $O(n \log n)$ time and $O(n)$ space for a set of n edges.*

A simple scene of isothetic polygons and the initial tree N is shown in Figure 1.

We now discuss how to update the fields in N when an edge is deleted. To delete an edge e_j, we delete the two leaves containing l_j and r_j. If these two leaves are occupied by other end points from some other edges, we keep the two leaves but delete the information regarding e_j. We first delete l_j and then r_j. We assume that the fields upto the node n_i have already been updated. We discuss how to update the fields in $parent(n_i)$. Let $High(e_m, \ldots, e_n)$ be a function which returns the edge with highest y coordinate among $e_m, \ldots, e_n$. We only consider the case when l_j is deleted. The other case is similar. Also, we only indicate how to update the fields when n_i is the left child of $parent(n_i)$. We can update the fields in the other case (when n_i is the right child of $parent(n_i)$) in a similar way. By $r(.)$ (resp. $l(.)$) we mean the right (resp. left) endpoint of the edge in the argument.

$L\_high(parent(n_i))$:
If ($LL\_cross(n_i) = RR\_cross(n_i)$) then
$L\_high(parent(n_i)) := High(L\_high(n_i), LL\_cross(n_i), R\_high(n_i))$

else $L\_high(parent(n_i) := High(L\_high(n_i), R\_high(n_i))$;
$LL\_cross(parent(n_i))$:
If $r(LL\_cross(n_i)) > max(n_i)$ then
$LL\_cross(parent(n_i)) := High(LL\_cross(n_i), RL\_cross(n_i))$
else $LL\_cross(parent(n_i)) := RL\_cross(n_i)$
$LR\_cross(parent(n_i))$:
If $l(RR\_cross(n_i)) < min(n_i)$ then
$LR\_cross(parent(n_i)) := High(LR\_cross(n_i), RR\_cross(n_i))$
else $LR\_cross(parent(n_i)) := LR\_cross(n_i)$
$min(parent(n_i)) := min(n_i)$.

$R\_high(parent(n_i))$, $RL\_cross(parent(n_i))$, $RR\_cross(parent(n_i))$ and $max(parent(n_i))$ remain unchanged. The fields in the leaf nodes can be updated in a similar way and the details are omitted from this version. It is easy to see that all the nodes along the two deletion paths can be updated in $O(\log n)$ time since we spend $O(1)$ time for each node. It is easy to prove that this updation takes $O(\log n)$ time since we spend $O(1)$ time for each node.

Lemma 3.2 *If two edges e_i and e_j ($r_i < l_j$) are unblocked in the $+y$ direction, in $O(\log n)$ time we can find a free edge e_k (if it exists) such that $r_i < l_k < r_k < l_j$.*

Proof: We first consider the path from the leaf containing r_i to the root. We call this path as $\mathcal{P}$. We store the possible candidate for the edge e_k in a variable $free\_edge$. While traversing $\mathcal{P}$, at every internal node $n_i \in \mathcal{P}$ (including $root(N)$), we do the following computation. There are two different cases, depending on whether the path $\mathcal{P}$ has come upto n_i from the left or the right subtree of n_i. We consider the case when $\mathcal{P}$ has come to n_i from its left subtree. The other case is similar and the details are omitted.

We first check whether the edge currently present in *free_edge* is blocked in the $+y$ direction by the edges present in the fields at n_i. In case $free\_edge$ is blocked, we assign $null$ to $free\_edge$. Suppose, n_i has an edge e_m which has the highest y coordinate among the following fields, **(i)** $L\_high(n_i)$, **(ii)** $LL\_cross(n_i)$ and $RR\_cross(n_i)$ (i.e., both these fields contain the same edge) and **(iii)** $R\_high(n_i)$, and such that e_m is unblocked and completely contained in the subtree rooted at n_i. Similarly, assume that $free\_edge$ currently contains an edge e_n which is unblocked and completely contained in the subtree rooted at $left(n_i)$ as well as in the subtree rooted at n_i. In this case, if e_m is different from e_j and $y_m > y_n$, we delete the edge e_n from $free\_edge$ and include e_m instead. The reason is that if e_m is blocked by some other larger edge e_l, the end points of e_l are outside the subtree rooted at n_i. In that case, e_l will also block the edge e_n in the $+y$ direction. If e_m is same as e_j, we keep $free\_edge$ unchanged. Following this scheme, when we reach $root(N)$ along $\mathcal{P}$, $free\_edge$ will contain an edge like e_k if it exists. The complete computation takes $O(\log n)$ time, since the length of $\mathcal{P}$ is $O(\log n)$ and we do a constant number of comparisons at every node along $\mathcal{P}$. □

Corollary 3.3 *We can find all edges which are free in the $+y$ direction in $O(k \log n)$ time by repeatedly applying the method in Lemma 3.2, where k is the number of horizontal edges currently free in the $+y$ direction.*

We now discuss how to find new edges which are freed due to the deletion of an edge e_i. We refer to an edge which has got freedom due to the deletion of e_i as a *freed edge*. The freed edges fall into several categories. **i.** A freed edge e_j has a partial overlap with the deleted edge e_i. **ii.** A freed edge e_j completely covers the deleted edge e_i. **iii.** A freed edge e_j is completely covered by the deleted edge e_i.

Lemma 3.4 *If an edge e_i is deleted from the eastern tree, in $O(\log n)$ time we can find at least one edge e_j (if it exists) which is freed due to the deletion of e_i.*

Proof: In all the three cases we first delete the edge e_i (i.e., the leaves containing the end points) and update the fields in all the internal nodes along the deletion paths. After this, we again traverse the deletion paths from the leaf to the root to find an unblocked edge.

First we consider the case when the deleted edge e_i has partial overlap with the freed edge e_j. We assume w.l.o.g. that $l_i < l_j < r_i < r_j$ i.e., the left end point of e_j falls inside e_i. In the following, by $lca(n_i, n_j)$ we mean the least common ancestor of two nodes n_i and n_j. Suppose, the leaves containing r_i and r_j are n_i and n_j respectively. Suppose n_k is the node $lca(n_i, n_j)$. n_i and n_j are respectively in the left and right subtree of n_k. If e_j is free in the right subtree of n_k, it will be present in the field $RR\_cross(n_k)$, since the fields in the internal nodes have been already updated after the deletion of e_i. Similarly, if e_j is free in the left subtree of n_k, it will be present in the field $LL\_cross(n_k)$. While traversing the path from n_i to root, we check at every internal node n_l along this path whether $LL\_cross(n_l)$ and $RR\_cross(n_l)$ stores the same edge or not. If we find such an edge, we store this as a candidate for e_j. Note that, e_j cannot be blocked by some edge e_m either completely contained in e_j or having partial overlap with e_j. Since in both of these cases, at least one end point of e_m will be present in the subtree rooted at $lca(n_i, n_j)$ and hence e_j cannot be present in the LL_cross and RR_cross fields of $lca(n_i, n_j)$. It may happen that such an edge e_j is blocked by some larger edge e_n whose end points are outside the subtree rooted at $lca(n_i, n_j)$. In this case, e_n will be present in the fields of some node along the path from $lca(n_i, n_j)$ to $root(N)$. We can easily check this while traversing the path from n_i to $root(N)$.

The second case is similar to the first case. If we traverse one of the deletion paths, we will encounter such an edge e_j in an internal node n_k in both the fields $LL\_cross(n_k)$ and $RR\_cross(n_k)$.

In the third case, the deleted edge e_i completely covers an edge e_j which has been freed due to the deletion of e_i. Note that, many small edges may be freed due to the deletion of e_i. If we can find at least one such edge, we can find the rest of the freed edges by repeated application of the method discussed in Lemma 3.2. So, in the following, we discuss how to find an unblocked edge e_j with the highest y coordinate among all such edges.

We traverse a path $\mathcal{P}$ from the leaf containing l_i upto the root. Since e_j is completely contained in e_i, there exists a node n_k on $\mathcal{P}$ with the following property. $\mathcal{P}$ reaches n_k from $left(n_k)$ and e_j is completely or partially contained in the subtree rooted at $right(n_k)$. If e_j is really free after the deletion of e_i, e_j must be present either in $R\_high(n_k)$ or in $RL\_cross(n_k)$. Hence at every internal node along $\mathcal{P}$, we check for such an edge e_j according to the above observation. There is a possibility that such an edge is blocked by some other edge e_l. Note that, such an edge e_l cannot have a partial overlap with e_j, since in that case e_j cannot be present in $R\_high(n_k)$ or $RL\_cross(n_k)$. This is because, one end point of e_l will be present in the subtree rooted at $right(n_k)$. Hence, if such an edge e_l exists, it will have both its end points outside the subtree rooted at $right(n_k)$. Hence, we will encounter e_l in the part of $\mathcal{P}$ from n_k upto $root(N)$. e_l will be present in one of the fields of a node in this part of $\mathcal{P}$. We can check this while we continue traversing $\mathcal{P}$. At the end of traversal of $\mathcal{P}$, if such an unblocked edge e_j exists, we can find it. □

4 Algorithm for one-separation

1. We sort the vertical (resp. horizontal) edges of the polygons in ascending and descending order according to x (resp. y) coordinates and construct the four separation trees E, N, W and S.
2. We try to detect the freedom of the polygons by peeling the four trees. We arbitrarily start with the tree E and check whether any polygon is free in E. Whenever we find a free polygon, we delete all its edges from all the four trees. If we cannot find any free polygon in E (i.e., in the $+x$ direction), we visit the other three trees in a round robin fashion and repeat the same process. At the end of this process, either all the trees will be empty or no polygon is free in any of the trees. In the first case, the set of polygons is one-separable and a separation order can be determined from the sequence of freedom of the polygons. In the second case, the set is not one-separable, but we can find a maximum subset of polygons which is one-separable.

Algorithm 1: The main steps in the algorithm for one-separation.

We have already discussed how to implement Step 1 in Section 3. In this section, we explain Step 2. We will explain this step through some examples. We first introduce some more definitions. In E, if the $-x$ side of an edge $e_i \in P_j$ is inside (resp. outside) the polygon P_j, we call e_i as an *inside* (resp. *outside*) edge. Similarly, inside and outside edges are defined for the other three trees. We keep four free sets EFS, NFS, WFS and SFS associated with the four trees E, N, W and S. EFS is used to store the edges which are free in the eastern ($+x$) direction. The other three free sets are used similarly. Notice that, the vertical edges in EFS cannot have any overlap since all of them are free in the $+x$ direction. Suppose, $e_i, e_{i+1}, e_{i+2}, \ldots, e_j$ is a set of contiguous edges of the polygon P_k in EFS such that $b_i < t_i = b_{i+1} < t_{i+1} = b_{i+2} < t_{i+2} = \ldots < t_j$. We call a set of such edges as a *group*. The vertical interval $[b_i, t_j]$ is called the *span* of the set of edges $e_i, e_{i+1}, e_{i+2}, \ldots, e_j$.

There are two important issues in Step 2. **(a)** While deleting the free edges from one of the trees, we should not wrongly give freedom to some polygon which is not free in that direction. **(b)** We should spend only $O(\log n)$ time per deleted edge to achieve the overall time complexity of $O(n \log n)$.

We first explain how we maintain the correct freedom information in each tree. Assume that we start the peeling process with E. Initially, at most three edges may be free at $root(E)$. These are the edges present in the fields L_high, LL_cross and RR_cross (if these two fields contain the same edge) and R_high. These are the starting members of EFS. After this, we can find all the other free edges (we assume this number is k) initially present in E in $O(k \log n)$ time by the method in Lemma 3.2 and Corollary 3.3. We include all these edges in EFS. All contiguous free edges in EFS are grouped together. The jth group for polygon P_i is denoted by P_{ij}.

Example : In Figure 2, inside edges e_1, e_3, e_4 and e_5 of polygon P_2 are free in the $+x$ direction, but e_2 is not free. There are two groups of contiguous edges namely $P_{21} = \{e_1\}$ and $P_{22} = \{e_3, e_4, e_5\}$.

While adding the free edges in EFS, we ensure that they are in a bottom to top order in terms of the y coordinates of their bottom end points. When we find a new free edge by the method in Lemma 3.2, we insert that edge in EFS by doing a binary search in additional $O(\log n)$ time. We delete the edges in EFS in two stages. First, we mark the edges as *inside* or *outside* and also mark an edge as deleted if it is free in the $+x$ direction. When we delete the edges in a particular group, say, P_{ij}, more outside and inside edges of P_i may get freedom and we can find all such edges by Lemma 3.4 and Lemma 3.2. All the edges which get freedom due to the deletion of edges in a group P_{ij} are called $free(P_{ij})$. Notice that, if $EC_i \neq vcount(P_i)$ after all these deletions, P_i is not free in the $+x$ direction. Hence, the deletion of all these edges may wrongly give freedom to some other polygon which is blocked by the edges of P_i. To prevent this, in the second stage we delete the individual free edges but replace them by the span of each group in both E and EFS.

Example : In Figure 2, $free(P_{21}) = \{e_1\}$ and $free(P_{22}) = \{e_3, e_4, e_5, e_6, e_7\}$, but deletion of all these edges does not give freedom to P_2 in $+x$ direction. Hence, we replace all these edges by $span(P_{21}) = \{e_1\}$ and $span(P_{22}) = [t_3, b_5]$ in both E and EFS.

The following lemma ensures the correctness of this process.

Lemma 4.1 *If some polygon P_k was blocked in the $+x$ direction by the edges in $free(P_{ij})$, P_k remains blocked in the $+x$ direction after insertion of $span(P_{ij})$. Moreover, no new polygon is blocked in the $+x$ direction due to the insertion of $span(P_{ij})$.*

Now, we explain how to spend only $O(\log n)$ time per deleted edge. We consider the case when we are visiting the tree E either first time or some other time during the execution of our algorithm. The members of EFS are either inside edges of some polygons or span edges, since we do not keep any outside edge in EFS. In our algorithm, we want to check whether the deletion of some edge $e_k \in P_i$ gives freedom to some other edge e_m of a polygon P_j such that e_m is only blocked by a span edge of P_j after the deletion of e_k. Notice that, we can only spend $O(\log n)$ time for each edge in EFS to achieve our claimed time complexity of $O(n \log n)$. Also, whenever we find an inside edge in EFS we delete it from E and from EFS and possibly replace it by its span (or the span of the group to which it belongs). The main difficulty is that when we visit EFS several time during the execution of our algorithm, the same span edge may be present in EFS. We should delete such a span edge only if its deletion gives freedom to other edges of the same polygon. We cannot spend $O(\log n)$ time for such a span edge every time and we want to spend only $O(\log n)$ time overall for such a span edge. We need some extra properties of our data structure to ensure this.

Suppose, we have found a free polygon P_k in the $-y$ direction. We delete all the edges of P_k from all the four trees. In the following lemma, we consider the deletion of vertical edges of P_k from E and show through an example how we spend $O(\log n)$ time for each edge in EFS.

Lemma 4.2 *If we delete the edges of P_k one by one from higher x to lower x direction, each vertical edge of P_k can unblock a set of edges of atmost one polygon P_m such that, this set of edges is blocked in the $+x$ direction by a span edge of the same polygon P_m after the deletion of the edges of P_k.*

Example : See Figure 3. Suppose, we started the peeling process with the tree E and after removal of P_{11}, i.e., e_1 and the edges in $free(P_{11})$(i.e., the edges e_{11}, e_9 and e_3), we inserted $span(P_{11})$ i.e., e_1 in the eastern tree. Since, no polygon is free in the $+x$ direction, we go to the tree S and find that P_2 is free in the $-y$ direction. Now, we have to delete the vertical edges of P_2 from E. Notice that, the vertical edge $e_5 \in P_1$ is blocked by the edges e_{13} and e_{19} of P_2. We delete the vertical edges of P_2 from E in a right to left order. After deletion of e_{13}, e_{19} blocks the edge $e_5 \in P_1$ such that, after deletion of e_{19}, e_5 will be blocked in the eastern direction by $span(P_{11})$, i.e., e_1. The claim of Lemma 4.2 is that there cannot be another edge (like e_5) from another polygon which is freed due to the deletion of e_{19}. Suppose, there is another edge $e_m \in P_j$ like e_5, such that after the deletion of e_{19}, e_m will be blocked in the $+x$ direction by some span edge of P_j. Then the polygon P_j to which e_m belongs should have some part of it behind e_{19} (i.e., towards the less x coordinate side of e_{19}). If P_j has some part behind e_{19}, there are two possibilities. In the first case, e_5 is blocked by some span edge of P_j even after the deletion of e_{19}. In the second case, $span(P_{11})$ blocks the part of P_j which is behind e_{19} even after the deletion of e_{19}. In either case, a set of edges of only one polygon (P_1 or P_j) remains blocked only by some span edge of the same polygon.

We ensure that whenever we delete a vertical edge from the eastern tree, if we find some hidden edge e_m of a polygon P_i such that e_m is blocked by some span edge of the same polygon P_i, we can indicate this in EFS by marking the corresponding span edge. Notice that, Lemma 4.2 implies that there can be at most one such span edge. This is because, if there are more than one such span edges, they are part of the same polygon and as a result they have been already merged to form a single span edge. This follows from the definition of a span edge. When we visit EFS next time, we only delete the span edge in EFS if it is marked, otherwise we ignore the span edge in that particular visit to E. This way, we invest $O(\log n)$ time for an edge in EFS if the deletion of that edge really gives freedom to some other edge of that same polygon. This is done in the following way.

When we insert a span edge $e_k \in P_i$ in E, suppose the two end points of this span edge are at the leaves n_i and n_j. We mark the node $lca(n_i, n_j)$ in E with the identity of the span edge e_k. Suppose, we are deleting a vertical edge e_l from E which has an overlap with e_k (e.g., like an edge e_{19} in Figure 3, which has an overlap with $span(P_{11})$). Notice that one end point of e_l is in the subtree rooted at $lca(n_i, n_j)$. Suppose this end point of e_l is at the leaf n_p. Before deletion of e_l, we traverse the path from n_p to root and discover the node $lca(n_i, n_j)$. If a marked node like $lca(n_i, n_j)$ exists, while deleting the end point of e_l from the leaf n_p we search for an edge of P_i (to which e_k belongs) to update the internal fields of E along the deletion path. If such an edge of P_i exists, it will ultimately be brought to one of the internal fields of either the left or the right child of $lca(n_i, n_j)$. In that case, we mark the span-edge e_k in EFS as a candidate for future deletion from EFS. We omit the details of this process from this version, but this can be done in the same way as in Lemma 3.2. This whole process takes $O(\log n)$ time for each edge we actually delete and hence we can charge this time to each deleted edge.

Finally, we illustrate the algorithm in the following example.

Example : We illustrate the algorithm in Figure 3. We start the peeling with E and initially, $EFS = \{e_1\}$. Since e_1 is an inside edge, we can delete e_1 from E. After the deletion of e_1, e_{11} becomes free and when e_{11} is deleted, e_9 gets freedom. After the deletion of e_9, e_3 gets freedom. So, $free(e_1) = \{e_1, e_{11}, e_9, e_3\}$. Since the polygon P_1 does not get freedom (EC_1 is less than the number of vertical edges) in the $+x$ direction, we delete all the edges in $free(e_1)$ and insert $span(e_1)$, i.e., e_1 in E and EFS. At this point, no polygon is free in the $+x$ direction. So, we go to tree N. Initially, $NFS = \{e_{12}, e_{10}, e_8, e_{16}\}$. Since all these edges are inside edges with respect to the $+y$ direction, we delete all of them. As a result, e_2, e_4 and e_6 get freedom in the $+y$ direction and the polygon P_1 is free in the $+y$ direction. We delete all the horizontal edges of P_1 from the N and S trees and all its vertical edges from the E and W trees. Now e_{20} and e_{18} also get freedom and $NFS = \{e_{18}, e_{20}\}$. Since both these are inside edges, we delete them and as a result, e_{14} gets freedom and the polygon P_2 gets freedom in the $+y$ direction. We delete all the edges of P_2 from the other three trees and all the four trees become empty.

Theorem 4.3 *The two-dimensional one-separability of a set of isothetic polygons can be decided in $\Theta(n \log n)$ time and $\Theta(n)$ space, where n is the total number of edges of the polygons in the set. A translational ordering of the polygons can also be derived within the same time and space bounds.*

Proof: The correctness of the algorithm follows from the description above. If the scene is separable, we are never stuck with the peeling process and the trees become empty at the end. Sorting of the edges and the construction of the four trees take $O(n \log n)$ time. With every edge, we associate the identity of the polygon to which the edge belongs. We assume that the polygons are given as a clockwise list of edges. In $O(n)$ time, we can mark each edge as inside or outside. When we visit a particular tree and the corresponding free set, we delete a span-edge only if it is already marked. For any other edge e_i in the free set, we either delete it completely (if the corresponding polygon is free in that direction) or replace it by its span. Hence, we spend only $O(\log n)$ time for each edge in the scene. Hence, the total execution time is $O(n \log n)$. The space requirement is $O(n)$ since each internal node in a tree stores only a constant number of fields. Also the four freedom counters EC_i, WC_i, NC_i and SC_i for each polygon P_i takes $O(1)$ space.

The lower bound on space is obvious. The lower bound on the time complexity can be proved from the fact that the *iso-separation* problem (i.e., separating in a single axis-parallel direction) is a special case of the one-separation problem. The iso-separation problem is equivalent to detection of intersection of rectangles. This problem is in turn equivalent to the *element uniqueness* problem which has a lower bound of $\Omega(n \log n)$ [9]. □

Acknowledgements: The authors would like to thank one of the referees for extensive comments which improved the presentation of the paper considerably.

References

[1] P. K. Agarwal, M. de Berg, D. Halperin and M. Sharir. "Efficient generation of k-directional assembly sequences", *Proc. 7th ACM-SIAM Symp. Discrete Algorithms*, (1996), pp. 122-131.

[2] M. de Berg, H. Everett and H. Wagener. "Translation queries for sets of polygons", *Internat. J. Comput. Geom. Appl.*, **Vol. 5**, (1995), pp. 221-242.

[3] B. Chazelle, Th. Ottmann, E. Soisalon-Soininen and D. Wood. "The complexity and decidability of $SEPARATION^{TM}$", *Proc. International Colloquium on Automata, Languages and Programming*, (1984), *LNCS* **172**, pp. 119-127.

[4] F. Dehne and J. -R. Sack. "Translation separability of sets of polygons", *The Visual Computer*, **3**, (1987), pp. 227-235.

[5] M. Devine and D. Wood. "$SEPARATION^{TM}$ in d dimensions or strip mining in asteroid fields", *Comput. and Graphics*, Vol. **13**, No. **3**, (1989), pp. 329-336.

[6] L. J. Guibas and F. F. Yao. "On translating a set of rectangles", in *Advances in Computing Research, Volume I: Computational Geometry*, Ed. F. P. Preparata, JAI Press Inc., (1983), pp. 61-77.

[7] D. Nussbaum and J. -R. Sack. "Disassembling two-dimensional composite parts via translation", *International Journal of Computational Geometry & Applications*, Vol. **3**, No. **1**, (1993), pp. 71-84.

[8] Th. Ottmann and P. Widmayer, "On translating a set of line segments", *Computer Vision, Graphics and Image Processing*, Vol. **24**, (1983), pp. 382-389.

[9] F. P. Preparata and M. I. Shamos. *Computational Geometry : an Introduction*, Springer-Verlag, 1985.

[10] J. Reif. "Complexity of the mover's problem and generalizations", *Proc. 20th IEEE Symposium on Foundations of Computer Science*, (1979), pp. 560-570.

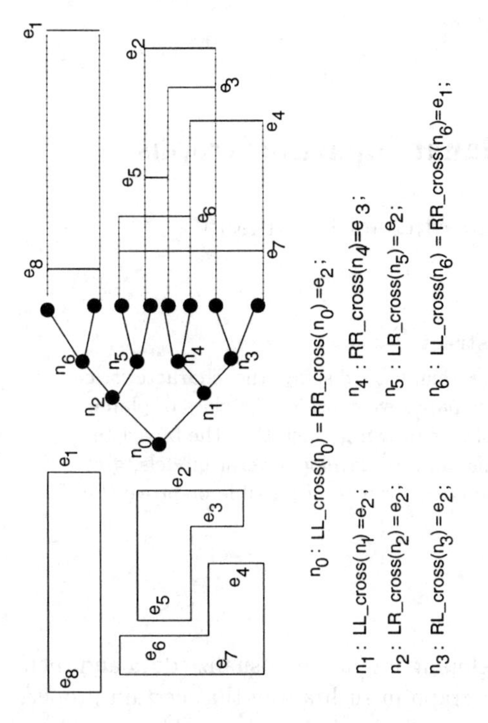

Figure 1: A simple scene of isothetic polygons and the initial N_tree. The non-empty fields at the internal nodes are shown. max and min fields are not shown.

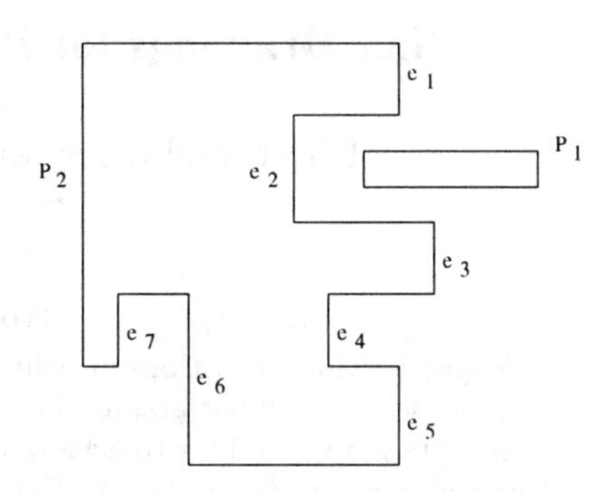

Figure 2: Some inside edges of P_2 are free in the eastern direction and some are not free.

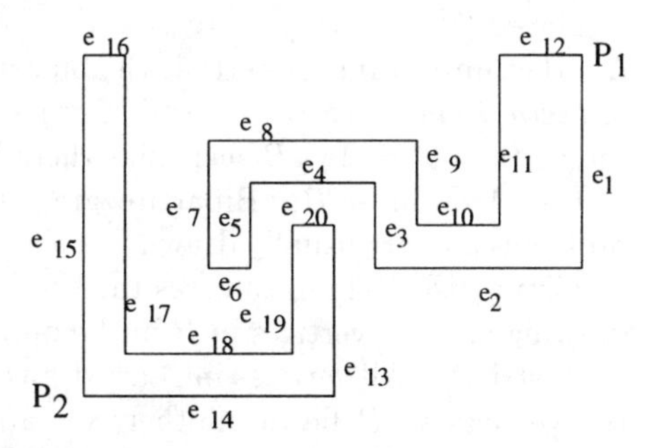

Figure 3: Illustration of the peeling process (Step 4 in the informal description). The polygon edges are numbered in the clockwise order from e_1 to e_{12} for P_1 and from e_{13} to e_{20} for P_2.

Nice Drawings for Planar Bipartite Graphs

Ulrich Fößmeier* and Michael Kaufmann*

Abstract

Graph drawing algorithms usually attempt to display the characteristic properties of the input graphs. In this paper we consider the class of planar bipartite graphs and try to achieve planar drawings such that the bipartiteness property is cleary shown. To this aim, we develop several models, give efficient algorithms to find a corresponding drawing if possible or prove the hardness of the problem.

1 Introduction

Graph drawing is a more and more developing method to visualize data and their relations. The main goal is to draw the graph in such a way that certain properties are clearly displayed: Planar graphs should be drawn planar [6], symmetries should be displayed [3, 11], if the graph is directed and acyclic then it should be drawn 'upward' [12], cliques should be easily recognized. There are many more properties developed in graph theory and graph algorithms that are worth to be displayed [1].

Another important property is the bipartiteness. A graph $G = (V, E)$ with a set V of *vertices* and a set $E \subseteq V \times V$ of *edges* is called *bipartite* if there is a partition of the vertices $V = A \dot{\cup} B$ such that there is no edge $e = (u, v)$ with both endpoints $u, v \in A$ or in $\in B$. Bipartite graphs often arise in assignment or matching problems and are usually drawn in such a way that the two sets A and B can easily be distinguished; in most cases the vertices in A are drawn in the lower part of the drawing and the vertices in B in the upper part. Often this 'layout rule' is defined more strictly: All vertices in A are drawn in a row (having the same y-coordinate), the vertices in B lie in another row above them and every edge is drawn as a straight line. We will refer to this model as the *full-constrained model.* Related work is on the crossing minimization problem for arbitrary bipartite graphs in this model. Garey and Johnson [7] show the NP-completeness of this problem and in [5] and many other papers heuristical approaches are given. Recently, Jünger and Mutzel [9] presented sophisticated combinatorial optimization techniques to solve such problems. Many more references can be found in the survey [1].

In this paper we incorporate the additional planarity requirement. Hence the goal is to produce crossing-free drawings that also reflect in some way the bipartiteness

*Wilhelm-Schickard-Institut für Informatik, Universität Tübingen, Sand 13, 72076 Tübingen, Germany, `foessmei / mk @informatik.uni-tuebingen.de`

property. A characterization of planar bipartite graphs has been given in [2], where it has been shown that all such graphs have an upward drawing. But the resulting drawings do not directly reflect the bipartiteness property in the desired sense. For the full-constrained model, necessary and sufficient conditions for the drawability of the graphs are known and can be found in [13, 4]. We will review this model, analyse various other models and give algorithms to decide whether a graph can be drawn without an edge crossing in the specified model or not.

Models. Since the class of graphs that can be drawn without edge crossings in the full-constrained model is very restricted (see Section 2), we define some other models being more powerful: In the *two-sides-constrained model* we allow the vertices in B to have two different y-coordinates such that these vertices define two rows having the row of the A-vertices between them. In the *one-side-free model* the vertices in A lie in a row and the vertices in B may have any position at one side of this row, i.e. each y-coordinate of a vertex in B must be larger than the y-coordinate of the vertices in A. Finally in the *two-sides-free model* the vertices in A must lie in a row, whereas there are no restrictions to the positions of the vertices in B.

We analyse the classes of graphs which can be drawn without edge crossings in these four models and call them B_{fc} for the full-constrained model, B_{2c} for the two-sides-constrained model, B_{1f} for the one-side-free model and B_{2f} for the two-sides-free model. It is clear that B_{2f} contains all the other classes and that B_{fc} is contained in all the other classes. A simple example shows that the classes B_{1f} and B_{2c} are not comparable: The graph $K_{3,2}$ belongs to B_{2c}, but not to B_{1f}, and the graph $K_{2,3}$ belongs to B_{1f} but not to B_{2c}. Note that we implicitely use an ordering of sets A and B here to keep the example simple. In the rest of the paper we will show that checking whether a given graph belongs to some class is easy for B_{fc}, B_{2c} and B_{1f} and difficult for B_{2f}.

Throughout this paper, we assume that the graphs are connected, otherwise we can handle the connected components separately.

2 The Full-Constrained Model

Definition 2.1 *A vertex will be called* small *if it has degree one, and* large *otherwise.*

Definition 2.2 *A connected graph G is called a* caterpillar *if G is a tree and each vertex in G has at most two large neighbours.*

Theorem 2.1 *[13, 4] A connected graph G is in B_{fc} if and only if G is a caterpillar. A corresponding drawing can be found in linear time.*

The proof is simple. It consists of a straightforward algorithm and an easy contradiction argument. Details are omitted [13, 4].

The full-constrained model allows a planar drawing only for a very restricted class of graphs, so we turn to more powerful models to represent the bipartiteness property.

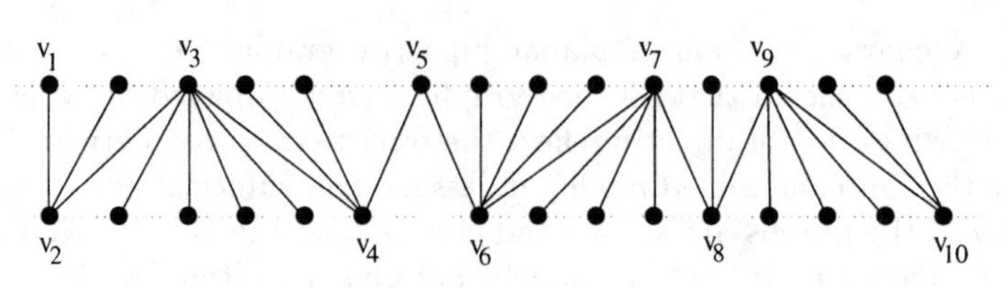

Figure 1: A caterpillar drawn in the B_{fc} model

3 The Two-Sides-Constrained Model

In the two-sides-constrained model we again force the vertices in set A to have a unique y-coordinate, but we allow the vertices in B to have one of two distinct y-coordinates, such that they define two rows, one above and one below the vertices in A.

Our test whether a graph G admits such a drawing is a recursively defined procedure with two parameters G and *borders*; G is the actual graph which should be tested by the procedure and *borders* may contain up to two vertices: A vertex p in *borders* must lie at the border of the drawing of G, it will be either the leftmost or the rightmost vertex of G in the drawing. If *borders* contains two vertices, they must lie at opposite sides of the drawing.

Definition 3.1 *A vertex v is called a* separator vertex, *if removing v and all incident edges from G leads to a non-connected graph.*
An edge $e = (u, v)$ is called a separator edge, *if removing u, v and all incident edges from G leads to a non-connected graph.*

The procedure tries to split the graph at separator vertices or at separator edges into components that will be handled separately. We have to guarantee that the components can be put together at the split vertices; consequently these vertices cannot be placed in the interior of one of the components. The split vertices will denote the parameter *borders*. Note that we can freely choose on which side we place the common vertex since we always can reflect the drawing of the actual graph if the border vertex is at the left side but it should be at the right side or vice versa. Obviously, the number of such split vertices must be restricted to two for any component (one at the left side and one at the right side of the drawing); otherwise the graph does not admit a drawing obeying our rules. The procedure *test(G)* returnes true if the graph G can be drawn without edge crossings in the two-sides-constrained model and false otherwise. We show only the interesting cases; the cases are checked in the given order.

The procedure *test* is called with the parameters G and *borders* initialized as an empty set, *test* mainly consists of a case analysis on the vertices in A:

(a) (* Simple checks *)

(1) **if** $|A| \leq 1$ **then** return true;

(2) **if** there is a small vertex $v \in A$ **then**
if $v \in borders$ **then** insert v's neighbour into $borders$ **fi**;
call test$(G \backslash \{v\}, borders \backslash \{v\})$;

(3) **if** there is a vertex $v \in A$ with > 4 large neighbours, **then** return false;

(* the non-trivial cases *)

(b) there are vertices in A with four large neighbours; let v be such a vertex; **if** v is a separator vertex **and** all (at most four) connected components of $G \backslash \{v\}$ are drawable in the two-sides-constrained model (where v gets a new border) **and** at most two of the components are no-caterpillar components **then** return true **else** return false;

(c) there are some vertices in A with three large neighbours; let v be such a vertex;

(1) **if** v is a separator vertex **then** we make the same test as in case (b)

(2) **else** we need a more special case analysis involving separator edges which is omitted;

(d) (* all vertices in A have at most 2 large neighbours *)

(1) **if** there is a separator vertex $v \in A$
then call *test* for the two connected components of $G \backslash \{v\}$ after inserting v into $borders$ for these procedure calls;

(2) **if** there is a separator vertex $w \in B$ **then**
if there are at most two no-caterpillar components among the connected components of $G \backslash \{w\}$ **then** call *test* for the no-caterpillar components after inserting w into $borders$ for these calls **else** return false;

(3) (* no separator vertices at all, vertices in A have degree 2 *)
let L be the graph obtained by replacing all $v \in A$ by edges between its neighbours. **if** L is a ladder graph (an outerplanar graph with completely nested shortcut edges) **then** return true **else** return false;

Theorem 3.1 *The procedure* test *exactly identifies bipartite graphs that can be drawn in the two-sides-constrained model without edge crossings.*

Proof: We shortly discuss the most important cases: the first interesting case is (a3): If v has five large neighbours, at least three of them lie at the same side; the rightmost and the leftmost of them are connected to v by straight edges e_1 and e_2; so any straight edge between the middle neighbour at this side and any point on the middle row must intersect e_1 or e_2.

Case (b): Let $v \in A$ be a vertex with four large neighbours.
Obviously, the four large neighbours of v have to be placed two on each side of v. Let v_{ru} be the rightmost large neighbour of v in the upper row and v_{lu} denote the other large neighbour in the upper row; the neighbours in the lower row are

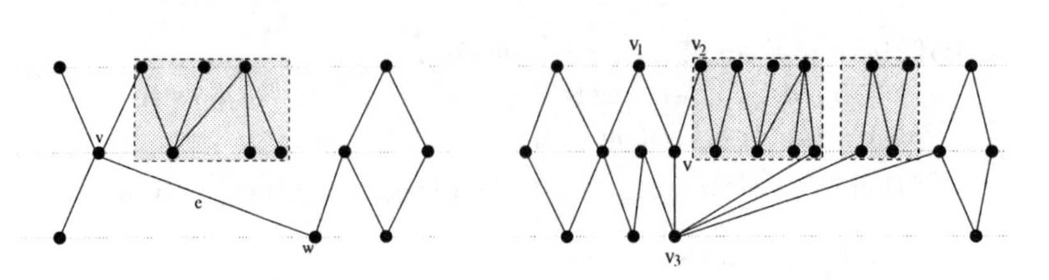

Figure 2: Typical configurations for Cases (b) and (c2) (caterpillars in shaded areas)

called v_{rl} and v_{ll}. If v is a separator, there may be two, three or four connected components of $G \backslash \{v\}$. If three of them are no-caterpillar components, two of these no-caterpillar components contain w.l.o.g. v_{ru} and v_{rl} (the other case, that two of them contain v_{lu} and v_{ll} is symmetric). Since a no-caterpillar component uses all three rows, there is a path from v to a vertex at the lower row within the component that contains v_{ru}; moreover (v, v_{ll}) is an edge of the graph; the component that contains v_{rl} must cross this path (because it is a no-caterpillar component and must use a vertex at the upper row); since the two components are disjoint we conclude that an edge crossing is necessary.
On the other hand, if at most two of the components are no-caterpillar components, a drawing can be produced in the following way: Draw the component recursively with v at the left border, then shift the drawing to the right by stretching the edge being incident to v; in the free area above of this edge there is enough space for a caterpillar component (see Fig. 2 for illustration); a second caterpillar component can be drawn at the left side of v.
We omit the case that v is not a separator vertex. It is relatively easy to see, that a planar drawing is impossible in this case.

Case (c): The case of v being a separator vertex is very similar to case (b), the other case is not very complicated, and is omitted here; Fig. 2 shows an example.

Case (d): If there is a separator vertex $v \in A$ we apply the same recursion as in the cases (b) and (c) with no restriction to the number of no-caterpillar components because there are only two connected components and we place one of them to the left of v and the other one to the right. If we divide the problem at a separator vertex $w \in B$ we may draw two no-caterpillar components (at the extreme left and at the extreme right of the current graph) and have place for caterpillar components in the middle as shown in Fig. 3.
Case (d3) is most interesting: Since previous cases do not apply, G is biconnected and $deg(v) = 2$ for all $v \in A$. By replacing the vertices $v \in A$ by an edge each, we construct the induced graph $L(G)$. If $L(G)$ is a ladder graph then from the natural drawing of this ladder a drawing for G can be obtained as indicated by Fig. 4. On the other side, each drawing of a biconnected graph $G \in B_{2c}$ with $deg(v) = 2$ for all $v \in A$ induces a ladder graph $L(G)$. This completes the proof of Theorem 3.1. □

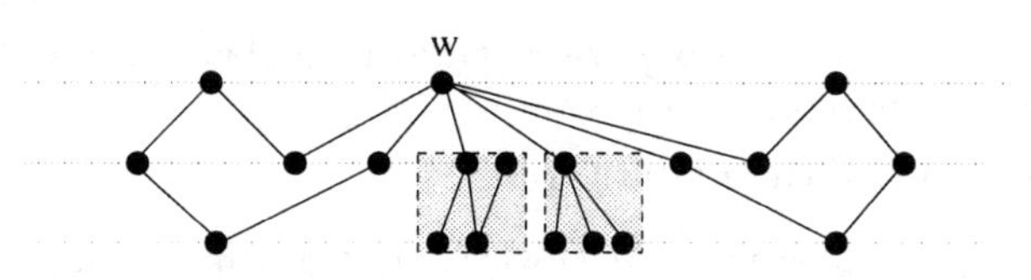

Figure 3: A separator vertex in B

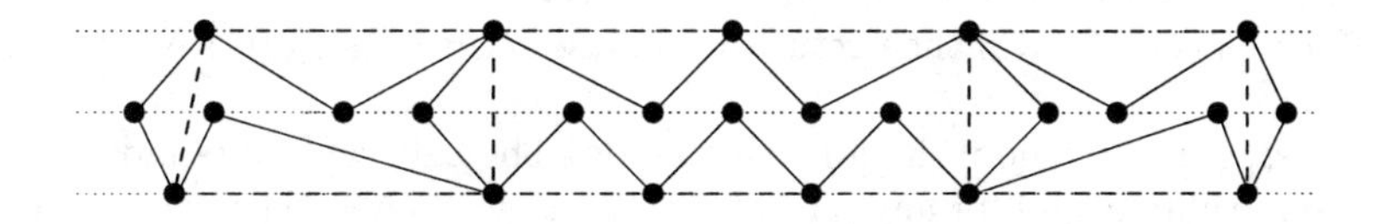

Figure 4: The ladder graph and the resulting drawing

Corollary 3.2 *Implementing the details described in the proof of Theorem 3.1 leads to a linear time algorithm that computes a drawing in the two-sides-constrained model, if there is a solution.*

Proof: The correctness of the algorithm follows directly from the correctness of the proof of Theorem 3.1. Analysing the running time we observe that all tests in the algorithm concerning the separators can be performed by a single computation of biconnected components. In case (d3) we apply a modified version of the well-known linear-time algorithm for recognition of outerplanar graphs [10]. Summing up the running times for the single components gives the linear time bound. □

4 The One-Side-Free Model

In the one-side-free model the vertices in A again must have a unique y-coordinate, and the vertices in B may have arbitrary positions *above* this row, i.e. the y-coordinate of any vertex in B must be larger than the y-coordinate of the vertices in A.
Next we give a procedure that returns true if and only if the input graph G can be drawn without edge crossings in the one-side-free model.

(a) (1) **if** there is a vertex $v \in B$ with degree one, **then** G is drawable if and only if $G\backslash\{v\}$ is drawable. **if** two vertices in B have three common neighbours, **then** $G \notin B_{1f}$.

(2) Partition the vertices in B into groups, such that all vertices belonging to the same group have exactly the same (two) neighbours. $G \in B_{1f}$

if and only if $G^* \in B_{1f}$ where G^* is the subgraph of G that contains exactly one representative out of each group.

(3) **If** G is a tree **then** return true.

(b) Let $C = v_1, \ldots, v_k$ be a cycle in G where (v_i, v_j) is an edge if $|i-j| \bmod k = 1$ and $v_i \in B \Leftrightarrow i$ is even. C defines two sets of vertices IN_C and $OUT_C \subseteq G \backslash C$: $v \in IN_C$ if there are edges (v, v_i) and (v, v_j) where v_i and $v_j \in C$ such that i and j are odd and $|i-j| \bmod k \geq 4$; $v \in OUT_C$ if there is an edge (v, w) with $w \notin C$. Let C be any cycle.
If $IN_C \cap OUT_C \neq \emptyset$ then return false; else return $test_{IN}(C) \wedge test_{OUT}(C)$ where
- $test_{IN}(C)$ = true if the following holds: The subgraph of G induced by C is outerplanar and for any pair of vertices $u, w \in IN_C$: The neighbours of u have a common adjacent face of this subgraph, the neighbours of w also have such a common face, and there is a pair v_{i_1}, v_{i_2} of consecutive neighbours of u (i.e. the interval $[i_1, i_2] \bmod k$ of the cycle C does not contain any other neighbours of u), such that all neighbours of w are in the interval $[i_1, i_2]$.
- Let $G_C^i = OUT_C \cup \{v_i\}$ and compute $R_1^i, \ldots R_r^i$, the connected components of G_C^i containing v_i; $test_{OUT}(C) = true$ if the following conditions hold:

- $R_j^i \in B_{1f}$ for all i and j;
- there is no triple $R_x^{i_1}$, $R_y^{i_2}$, $R_z^{i_3}$ which pairwise have common vertices;
- if $R_x^{i_1} = R_y^{i_2}$ then $|i_1 - i_2| \bmod k = 2$ and the subgraph $R_x^{i_1} \cup \{v_{i_1}, v_{i_2}, v^*\}$ can be drawn where v^* is the common neighbour of v_{i_1} and v_{i_2} in C.

Lemma 4.1 *The procedure given above decides whether a graph can be drawn without edge crossings in the one-side-free model or not.*

Proof: The simple cases of part (a) are omitted. When arriving at case (b), the graph contains a cycle C and does not contain any two vertices with the same adjacencies.

We draw the cycle C as shown in Fig. 5; this is obviously the only way to draw a cycle, only the choice which vertex in B stands at the top of the large triangle is open. This drawing of C together with the horizontal line through the vertices in A divide the plane into several regions, one region inside of C and the other regions outside of C each of them having exactly two vertices in $A \cap C$ at its border. Choosing another vertex to be the top of the large triangle does not change anything topologically because the resulting regions are the same.
The characterization of the regions given above directly implies that every vertex in B not being part of C and having two non consecutive neighbours in C must be placed in the region inside of C. On the other hand, for any edge $e = (v_i, v_j)$ where v_i and v_j are not being part of C both vertices have to be placed in one of the regions outside of the cycle (since inside there is no place for any other vertex in A). Thus if $IN_C \cap OUT_C$ is not empty then $G \notin B_{1f}$.

Otherwise we have to check if the subgraph inside of C and the subgraph outside of C can be drawn without edge crossings. At the inside we start with drawing

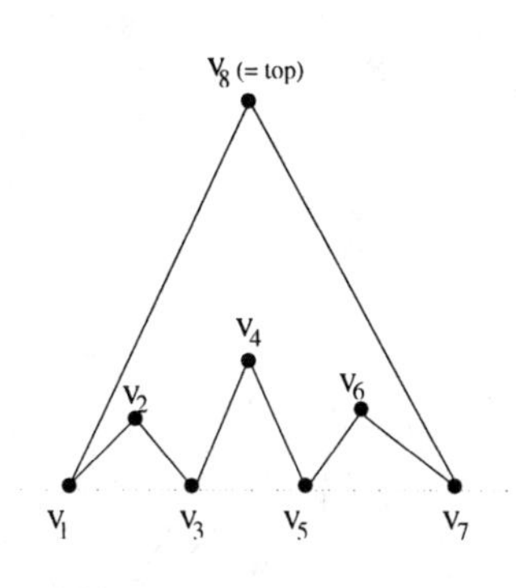

Figure 5: A cycle in the one-side-free model

edges $e = (v_i, v_j)$ where $v_i \in A$ and $v_j \in B$ are both part of the cycle. This is easy if v_j is the top. Otherwise we place v_j 'far away' from the row of vertices in A such that e does not cross other edges of C; eventually also the top must be placed higher in order to avoid edge crossings. This can always be done if the subgraph induced by C is outerplanar. If there is a vertex $u \in IN_C$, u together with its incident edges and a part of C defines a region of the plane; with the same argument as before we state that no vertex $w \in IN_C$ may have neighbours at the inside and at the outside of this region. The characterization given in the test procedure exactly describes this fact.
For the outside of C similar arguments work, they are omitted here. Parts of the graph which are not connected to some vertex of $A \cap C$ but to some vertex $v \in B \cap C$ can be handled very easily: Let w_1 and w_2 be the two neighbours of v in C; if $R_j^{w_1} = R_k^{w_2}$ for some j, k (thus there exists a path between w_1 and w_2 outside of C) then such a part cannot be drawn because an edge connecting v with a vertex in A between w_1 and w_2 would intersect the path. Otherwise this part is a connected component of $G \backslash \{v\}$ and is drawn recursively.
Fig. 6 shows an example for the different cases arising in the algorithm: The cycle C is indicated by the thick points, e is an edge between two vertices of C, u is a vertex inside of C and the regions between v_1 and v_3 respectively between v_7 and v_9 show examples for components which are drawn recursively. □

Lemma 4.2 *The tests described in this section lead to a drawing, if one exists and they can be implemented in quadratic time.*

Proof: We sketch the running time proof only. Most of the operations applied in the procedure are elementary computations or computations of (bi-)connected components which can be done in linear time. The recursion of depth $\Theta(n)$ leads to a quadratic running time. Only the operations for drawing the inside of a cycle in part (b) may require more than linear time because we have to sort the vertices in the inside of a cycle in the following way: If the interval defined by

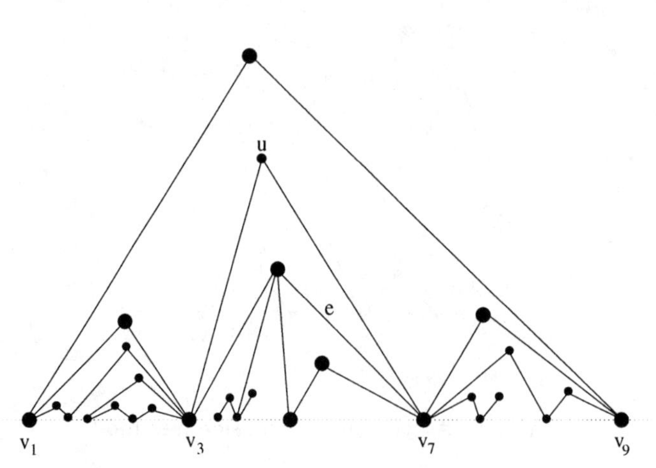

Figure 6: An example for a drawing in the one-side-free model

the neighbours of a vertex v contains the interval defined by the neighbours of a vertex w then we have to draw w below v. If there is a series of such vertices each of them containing the interval of the next one, they must be sorted. But this does not concern the total running time because there are no recursive calls in this part of the procedure. □

Note that the graphs in B_{1f} remind of outerplanar graphs when the vertices in B are neglected. Their adjacencies can be interpreted as hyperedges on the A-vertices. This viewpoint, to see the problem as a recognition problem of outerplanar graphs with hyperedges might lead to a future linear time algorithm.

5 The Two-Sides-Free Model

In this section, we consider the most general natural model that we have in mind: Let $G = (A \cup B, E)$ be the planar bipartite graph, such that there are only edges between vertices of A and B. The model requires that the set A of vertices should be arranged on a (horizontal) line while the vertices of B can be placed freely such that there are no edge crossings when we represent the edges by straight lines.
In the next subsection, we show that the problem to decide if a given planar bipartite graph can be represented within the two-sides-free model is hard. After that we give an approach for such drawings that works quite well in practice.

5.1 Drawing in the two-sides-free model is hard.

Even this model is not powerful enough to enclose all planar bipartite graphs: Let u, v and w be three vertices in B having pairwise three common neighbours, then no two of the vertices can be placed at the same side (see Fig. 7).
We show the following theorem on the membership problem in B_{2f}.

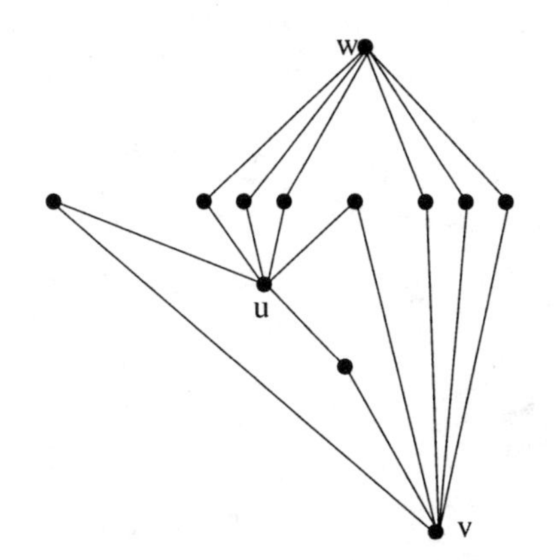

Figure 7: A planar bipartite graph $\notin B_{2f}$.

Theorem 5.1 *Given a planar bipartite graph $G = (A \cup B, E)$, deciding whether G can be represented within the two-sides-free model is NP-complete.*

Proof: The problem is in NP: Compute an arbitrary embedding for G such that the vertices in A lie on a horizontal line and the vertices of B are placed freely. Check the edges for crossings. Both can be done in polynomial time.

Let G be an arbitrary planar bipartite graph which is w.l.o.g connected. Consider any planar drawing Γ of G. From this drawing we can obtain a drawing in the two-sides-free model in the following way: We define a new embedded graph $G' = (A, F)$ on the set A of the vertices, and $F = \{\{a, b\}$, such that a and b lie on the same face in $\Gamma\}$. Let C be a hamiltonian cycle in G' with respect to Γ. Assume w.l.o.g that C connects two vertices on the outer face of Γ, otherwise choose another outer face for Γ. If C exists and it is not selfintersecting, it induces an order of the vertices in A. Together with the drawing Γ this order gives the drawing in the two-sides-free model. Cycle C broken on the outer face corresponds exactly to the horizontal line where the vertices in A are placed.

On the other side, in any drawing in this model the order of the A-vertices on the horizontal line induces a non-selfintersecting hamiltonian path in the corresponding graph G'. This relation between drawings of G in the two-sides-free model and a certain hamiltonian path in some corresponding graph G' will be used in the following construction:

Lemma 5.2 *Given a planar cubic tri-connected graph H, deciding whether H has a hamiltonian cycle is NP-complete [8].*

Let H be an arbitrary planar cubic tri-connected graph $H = (A, F)$. Since H is tri-connected, the embedding is unique and the faces are well-defined. From H we construct a planar bipartite graph $G = (A \cup B, E)$ in the following way: In each face f of H, insert one vertex v_f of B. Let E be the set $\{(v_f, w)$, for any face f such that w is a vertex in A lying on face $f\}$.

Obviously G is bipartite and the edges do not cross. Note that all faces of G are quadrangles by construction. Hence the corresponding graph G', as defined

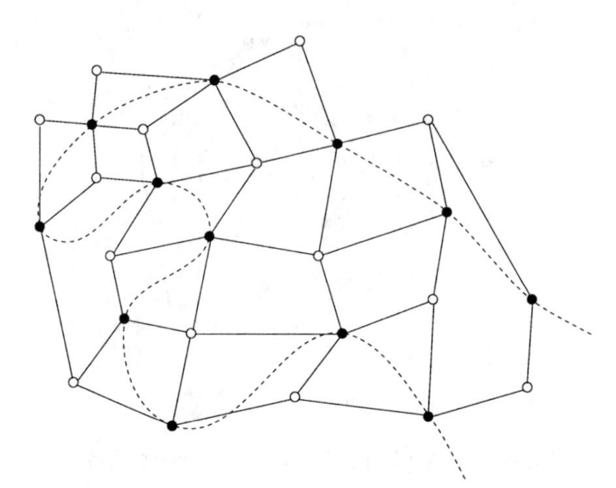

Figure 8: The planar hamiltonian cycle in G' induces a drawing of G.

above, is planar. As already observed, a drawing of G in the two-sides-free model induces a hamiltonian path in the corresponding graph G'. The observation that G' and H are identical finishes the reduction. □

5.2 A closer look to the two-sides-free model

Theorem 5.3 *Let $G = (A \cup B, E)$ be a planar bipartite graph. If the partition of B in B_u and B_l is given, such that the vertices in B_u must be placed on the upper side of the horizontal line induced by the A-vertices and B_l must be place below, a drawing of G in the two-sides-free model can be found efficiently, if it exists.*

Proof: Analysing the algorithm described in Section 4 (one-side-free model) we see that the main part of the algorithm is deterministic; we are only free to choose the order of some connected components when G is a tree or when OUT_C contains several separated parts. Thus we start drawing the vertices in B_u using the algorithm of Section 4 until we have such a choice; then we adjourn drawing B_u and continue with drawing the other part (B_l) where an order for some vertices in A is already given; if there are no more deterministic steps for the subproblem B_l we resume drawing B_u and so on. If there is no deterministic step at all (at both sides) we choose a new vertex arbitrarily and continue. Since we only make free choices if there are no restrictions for any vertices there is no need to make any backtracking; therefore a drawing is impossible if once a contradiction to the part of the graph drawn so far arises. □

The theorem implies that the hard part is the decision on which side of the horizontal line each vertex in B should go. To this end, we propose a back-track approach which is inspired by the proof of the last theorem: For a large number of pairs of vertices $u, v \in B$ it is easy to find out that they cannot be drawn at the same side (e.g. if u and v have more than two common neighbours). Thus if we place a vertex v at one fixed side (e.g. in B_u) then there may be some other vertices $w_1, \ldots, w_k$ that have to be put into B_l; these vertices force some other vertices to be in B_u and so on. We use some more involved conditions for vertices that have to be put on different sides and observe that in practice there remain only few vertices which yet have a choice. We assign them arbitrarily and use backtracking, if we find out that a drawing is not possible with the actual choice.

Our algorithm can even draw graphs not being in B_{2f}: If it does not find a legal drawing after a suitable amount of time the algorithm deletes some vertices and draws the rest. The deleted vertices can be reinserted by hand at positions where they do not disarrange the aesthetic too much (e.g. such that the drawing remains planar). We found that this approach works quite well in practice. As an example we add a drawing of a quite large graph that arised in an application of a telephone company. Our algorithm deleted four vertices from the graph, then it found a drawing. The four vertices are also drawn in the figure.

References

[1] Di Battista G. , P. Eades, R. Tamassia and I.G. Tollis, 'Algorithms for Drawing Graphs: an Annotated Bibliography', *Computational Geometry: Theory & Applications*, (1994), pp. 235–282.

[2] Di Battista G. , W.-P. Liu and I. Rival, 'Bipartite graphs, upward drawings, and planarity', *Information Processing Letters*, 36 (1990), pp. 317-322.

[3] Eades P., 'A heuristic for graph drawing', *Congr. Numerant.*, (1984), pp. 149–160.

[4] Eades P., B.D. McKay and N. Wormald, 'On an edge crossing problem', *Proc. Australian Computer Science Conf.*, Austr. Nat. Univ. (1986), pp. 327–334.

[5] Eades P. and N. Wormald, 'The Median Heuristic for Drawing 2-layered Networks', Technical Report 69, Dept. of Comp. Science, Univ. of Queensland, 1986.

[6] Fraysseix, H. de, J. Pach and R. Pollack, 'How to draw a planar graph on a grid', *Combinatorica* 10 (1990), pp. 41–51.

[7] Garey, M.R. and D.S. Johnson, 'Crossing Number is NP-complete', *SIAM J. Algebraic and Discrete Methods*, 4(3) (1983), pp. 312–316.

[8] Garey, M.R., D.S. Johnson and R.E. Tarjan, 'The Planar Hamiltonian Circuit Problem is NP-complete', *SIAM J. Comput.* 5 (1976), pp. 704–714.

[9] Jünger M. and P. Mutzel, 'Exact and Heuristic Algorithms for 2-Layer Straightline Crossing Minimization', *Proc. Graph Drawing*, (1995), LNCS 1027, pp. 337–349.

[10] Mitchell S.L. 'Linear algorithms to recognize outerplanar and maximal outplanar graphs', *Information Processing Letters* 9 (1979), pp. 229–232.

[11] Kamada T. and S. Kawai, 'An algorithm for drawing general undirected graphs', *Information Processing Letters*, 31 (1989), pp. 70–15.

[12] Sugiyama K. , S. Tagawa and M. Toda, 'Methods for visual understanding of hierarchical Systems', *IEEE Trans. on Systems, Man and Cybern.*, 11 (1981), pp.109-125.

[13] Tomii, N., Y. Kambayashi, and Y. Shuzo, 'On Planarization Algorithms of 2-Level Graphs', *Papers of tech. group on elect. comp.*, IECEJ, EC77-38 (1977), pp. 1–12.

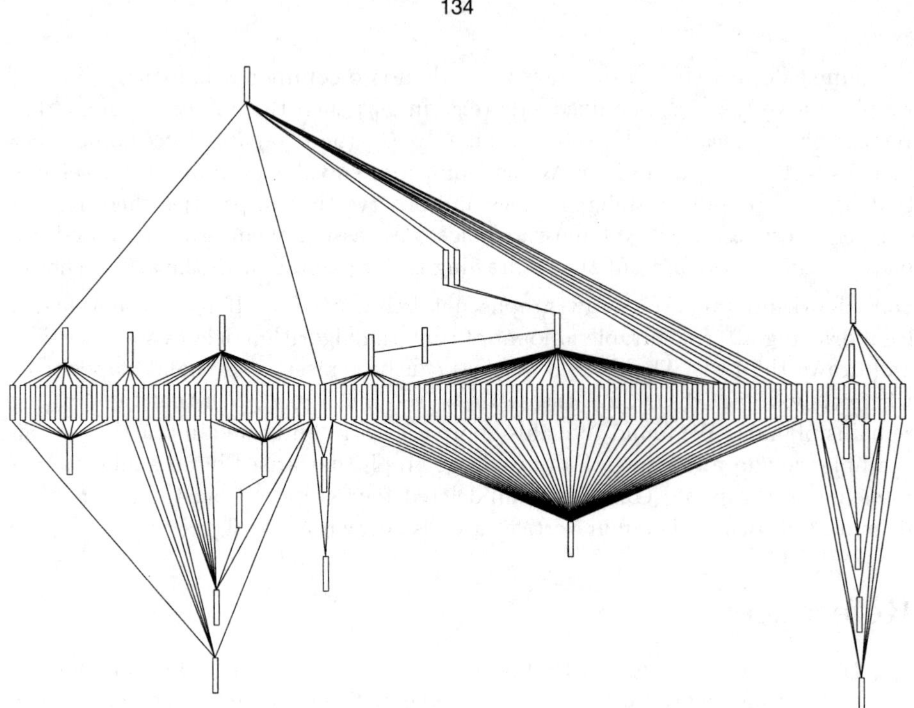
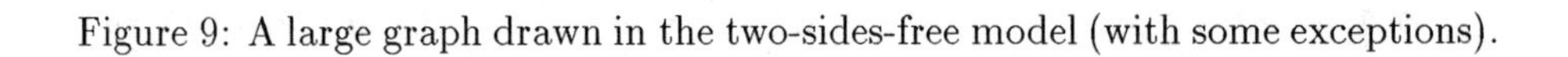

Figure 9: A large graph drawn in the two-sides-free model (with some exceptions).

Area Requirement of Gabriel Drawings * (extended abstract)

G. Liotta[1], R. Tamassia[2], I. G. Tollis[3], P. Vocca[4]

[1] Dipartimento di Informatica e Sistemistica, Universitá di Roma "La Sapienza", via Salaria 113, 00198 Rome, Italy
liotta@dis.uniroma1.it

[2] Center for Geometric Computing, Department of Computer Science, Brown University, Providence, RI 02912–1910, USA
rt@cs.brown.edu

[3] Department of Computer Science, The University of Texas at Dallas, Richardson, TX 75083–0688, USA
tollis@utdallas.edu

[4] Dipartimento di Matematica, University of Rome "Tor Vergata", via della Ricerca Scientifica, 00133, Rome, Italy
vocca@utovrm.it

Abstract. In this paper we investigate the area requirement of proximity drawings and we prove an exponential lower bound. Our main contribution is to show the existence of a class of Gabriel-drawable graphs that require exponential area for any Gabriel drawing and any resolution rule. The result is further extended to an infinite class of proximity drawings.

1 Introduction.

Proximity drawings of graphs have received increasing attention recently in the computational geometry and graph drawing communities due to the large number of applications where they play a crucial role. Such applications include pattern recognition and classification, geographic variation analysis, geographic information systems, computational geometry, computational morphology, and computer vision (see, e.g. [23, 21, 26]).

A *proximity drawing* is a straight–line drawing where two vertices are adjacent if and only if they are *neighbors* according to some definition of *neighborhood.*

* Work supported in part by the US National Science Foundation under grant CCR–9423847, by the US Army Research Office under grant DAAH04–96–1–0013, by the NATO Scientific Affairs Division under collaborative research grant 911016, by EC ESPRIT Long Term Research Project ALCOM-IT under contract no. 20244, by Progetto Finalizzato Trasporti 2 (PFT 2) of the Italian National Research Council (CNR), and by the NATO–CNR Advanced Fellowships Programme. This research was done when the first author was with the Center for Geometric Computing, Department of Computer Science, Brown University,Providence, RI 02912–1910, USA. This research started while the fourth author was visiting the Computer Science Department of Brown University.

One way of defining a neighborhood constraint between a pair of vertices is to use a *proximity region*, that is a suitable region of the plane having the two points on the boundary. Two vertices are adjacent if and only if the corresponding proximity region is *empty*, i.e., it does not contain any other vertex of the drawing (however, an edge of the drawing may cross the proximity region). For example, two vertices u and v are considered to be neighbors if and only if the closed disk having u and v as antipodal points, is empty. Proximity drawings that obey this neighborhood constraint are known in the literature as *Gabriel drawings* ([13, 21]) and the closed disk is called *Gabriel disk*. A different notion of proximity region is based on lunes instead of disks. In a *relative neighborhood drawing* ([25, 27]) two vertices u and v are adjacent if and only if the intersection of two open disks, one having center at u and the other at v, and with radius the distance between u and v, is empty. Gabriel drawings and relative neighborhood drawings are just two examples of an infinite family of proximity drawings called β-drawings that have been first introduced by Kirkpatrick and Radke [16, 23] in the computational morphology context.

A *β-drawing* is a straight–line drawing such that there is an edge between a pair of vertices u and v if and only if the corresponding *β-lune* is empty. The β-lune is defined as the intersection of two disks whose radius depends on the value of the parameter β. For $\beta \geq 1$, the β-lune is the intersection of the two disks of radius $\beta d(u,v)/2$, where $d(u,v)$ is the distance between u and v, and centered at the points $(1-\beta/2)u+(\beta/2)v$ and $(\beta/2)u+(1-\beta/2)v$. In particular, for $\beta = 1$, the β-lune coincides with the Gabriel disk. Figure 1 depicts a set of β-lunes.

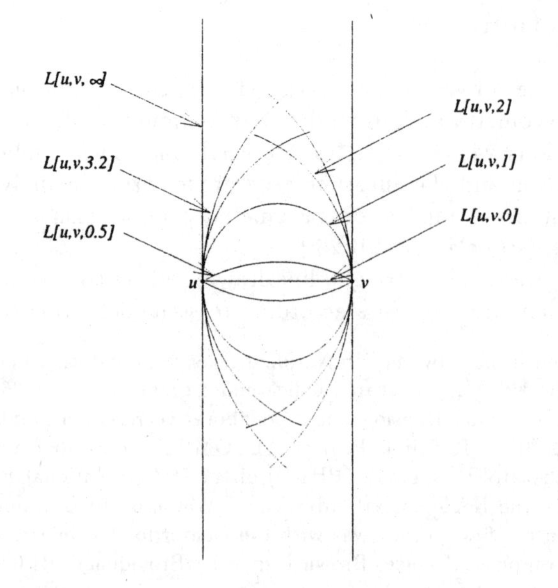

Fig. 1. A set of β-lunes between vertices u and v.

A different definition of proximity drawing is given in [8]. A *weak proximity drawing* is a straight–line drawing such that if there is an edge between a pair of vertices u, v then the proximity region of u and v is empty. This definition relaxes the requirement of classical β-drawings, that for each pair of non-adjacent vertices the β-lune is not empty. In other words, if (u, v) is *not* an edge, then no requirement is placed on the proximity region of u and v. Several papers have been recently published that characterize proximity drawings and show algorithms to construct proximity drawings of different classes of graphs and different definitions of proximity [2, 20, 12, 21, 10].

In [1, 2], the problem of characterizing β–drawable trees has been addressed and an algorithm to compute Gabriel drawings and relative neighborhood drawings of trees in the plane is given. The 3–dimensional version of the same problem has been studied in [19]. Lubiw and Sleumer [20] proved that maximal outerplanar graphs are both relative neighborhood and Gabriel drawings. This result has been extended in [18] to all biconnected outerplanar graphs. Also, in [8] several algorithms to construct weak proximity drawings of different families of graphs are given. For a survey on proximity drawings see [7].

In this paper, we investigate the area requirement of proximity drawings. The finite resolution of display and printing devices requires that some constraints be placed on the drawing so that its dimensions cannot be arbitrarily scaled down. Any constraint which imposes bounds on the minimum distance between vertices and (non–incident) edges in the drawing is called a *resolution rule*. Typical resolution rules are [4, 9]: The *vertex resolution rule* which requires that any two vertices have distance at least (a constant) δ, (typically, $\delta = 1$); the *edge resolution rule* which requires that the minimum distance between any vertex and a non–incident edge is at least δ; the *angular resolution rule* which states that the vertex resolution rule is verified, and that the minimum angle between two edges incident at the same vertex is at least $\alpha(d)$, where $\alpha(d)$ is a predefined function of the maximum degree of the graph. Once the resolution rule is given, the problem of evaluating the area of a drawing naturally arises [3, 9, 15, 14, 5, 17, 24]. In fact, any resolution rule implies a finite minimum area for a drawing of a graph.

All known algorithms that compute (weak) proximity drawings produce representations whose area increases exponentially with the number of vertices. As a consequence, the problem of constructing proximity drawings of graphs that have small area is considered a very challenging one by several authors [2, 10, 22]. Additionally, the importance of this question arises in practice, by observing that several heuristics for drawing graphs often produce proximity drawings [11].

We present the first results on area requirements of proximity drawings. Namely, we exhibit a class of graphs whose proximity drawings require exponential area under several different definitions of proximity. The main contributions of the paper are listed below.

- We describe a class of planar triangular graphs, such that any Gabriel drawing of a graph in the class has area that is exponential in the number of vertices of the graph. Planar triangular graphs are a classical field of study in graph drawing, because of the many applications where such graphs arise (see, e.g. [6]).

- We extend the above result to weak Gabriel drawings.
- We show an exponential lower bound on the area of an infinite class of β–drawings, for $1 \leq \beta < \frac{1}{1-\cos 2\pi/5}$.

Some of the proofs are omitted in this extended abstract, because of page limits.

2 Preliminaries.

A *graph* $G = (V, E)$ consists of a finite non empty set $V(G)$ of *vertices*, and a set $E(G)$ of unordered pairs of vertices known as *edges*. Given an edge $e = (u, v)$, u and v are the *endpoints* of e and are said to be *adjacent* vertices. A *simple path* of *length* k in a graph is a finite sequence $P = v_1 v_2 \ldots v_k$, where $v_i \neq v_j$, for $1 \leq i < j \leq k$, and $(v_i, v_{i+1}) \in E(G)$, for $i \in \{1, \ldots, k-1\}$. The vertices v_1 and v_k are the *endpoints* of the path. A k*–cycle* $C_k = v_1 v_2 \ldots v_k$ is a sequence of vertices such that $P = v_1 v_2 \ldots v_k$ is a simple path and $(v_k, v_1) \in E(G)$. A *drawing* Γ of a graph $G = (V, E)$ is a function which maps each vertex of G to a distinct point of the plane and each edge $e = (u, v)$ in G to a simple Jordan curve with endpoints the points of the plane corresponding to u and v. Γ is a *straight–line* drawing if each edge is a straight–line segment; Γ is *planar* if no two edges intersect, except possibly at their endpoints. In this paper, when it does not give rise to ambiguities, we refer to a drawing of a graph as the graph itself. A *planar triangular graph* is an embedded planar graph so that every internal face is a 3-cycle, a *triangle*. The *area* of a drawing Γ can be defined in several ways depending on whether we evaluate lower or upper bounds. In this paper, we define the area of Γ as the area of the smallest polygon covering Γ [9].

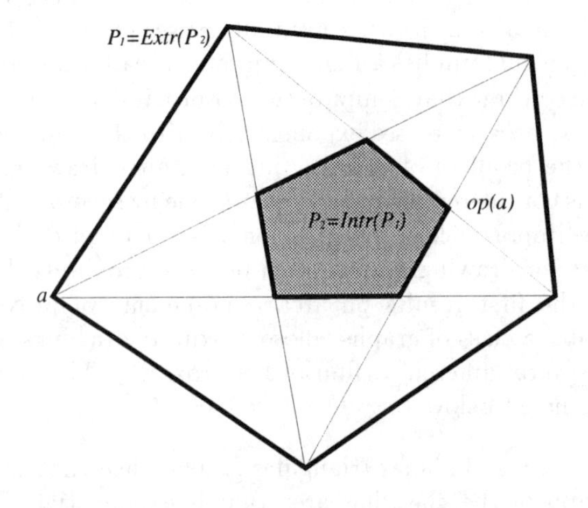

Fig. 2. Extruded and Intruded Pentagons.

A *Gabriel drawing* is a planar straight–line drawing such that there is an edge between two vertices u and v if and only if the closed disk having u and v as antipodal points is empty. The closed disk is denoted as $D[u,v]$. A given graph G is *Gabriel drawable* if it admits a Gabriel drawing. A *maximal Gabriel graph* is a Gabriel drawable graph with the maximum number of edges. A *weak Gabriel drawing* is a planar straight–line drawing such that if there is an edge between two vertices u and v then the closed disk having u and v as antipodal points is empty. A weak Gabriel drawing is contained in or coincides with a Gabriel drawing.

Property 1. *A Gabriel drawing of a planar triangular graph is such that all internal faces have acute angles.*

Property 2. [21] *In a Gabriel drawing every 3–cycle and every chordless 4–cycle is an internal face.*

Property 3. *A weak Gabriel drawing of a maximal Gabriel graph is also a Gabriel drawing.*

In our proofs, we will use several geometric objects. Let $\mathbb{R}^2$ denote the euclidean plane. Given any three distinct points $a, b, c \in \mathbb{R}^2$, $\angle abc$ denotes the counterclockwise angle between line segments $\overline{ab}$ and $\overline{bc}$; $\Delta(abc)$ denotes the triangle whose vertices are a, b, and c. Let P_1 be a convex pentagon. The intersections of the five diagonals of P_1 define a pentagon P_2 inside P_1 (see Figure 2). We call P_1 the *extruded pentagon* of P_2 and denote it with $Extr(P_2)$. Conversely, P_2 is the *intruded pentagon* of P_1, denoted with $Intr(P_1)$. Let a be a vertex of P_1. Its *opposite vertex*, $op(a)$, is the vertex of $Intr(P_1)$ in the region of the plane delimited by the two diagonals outgoing from a.

3 Description of the Class of Graphs

Our class of graphs is recursively defined as follows. Graph G_1 (Figure 3 (a)) is a wheel graph with six vertices (the center of the wheel is v^*; the other vertices are $v_0^1, v_1^1, v_2^1, v_3^1$, and v_4^1; the edges are (v^*, v_i^1) and $(v_i^1, v_{(i+1)\mathrm{mod}5}^1)$, for $i \in \{0,\ldots,4\}$). For $n \geq 2$, G_n is constructed from G_{n-1} by adding five vertices (denoted as $v_0^n, v_1^n, v_2^n, v_3^n, v_4^n$), and the edges $(v_i^n, v_{(i+1)\mathrm{mod}5}^n)$, (v_i^n, v_i^{n-1}), and $(v_i^n, v_{(i+1)\mathrm{mod}5}^{n-1})$, as shown in in Figure 3(b). F_5^n denotes the 5-cycle composed by $v_0^n v_1^n v_2^n v_3^n v_4^n$.

Property 4. *G_n is a planar triangular graph, with $5n+1$ vertices, and $15n-5$ edges. Also, G_n is triconnected.*

Lemma 5. [21] *G_n is Gabriel drawable.*

Figure 4 shows a Gabriel drawing of G_2.

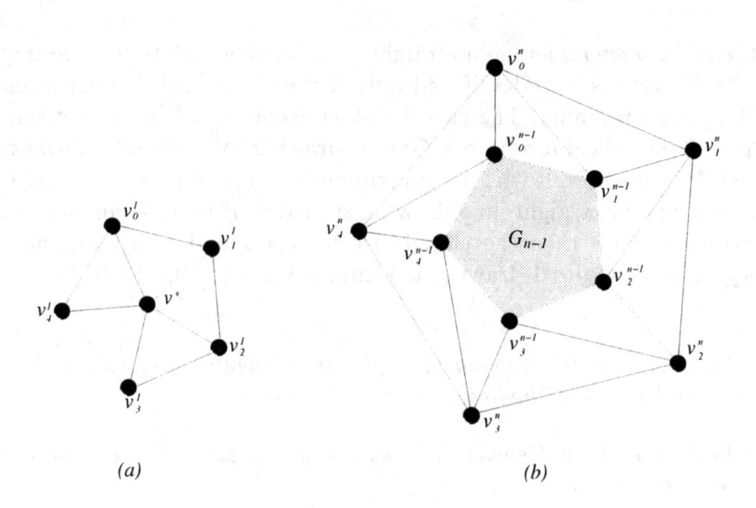

Fig. 3. A class of graphs that require exponential area.

Lemma 6. [21] *G_n is a maximal Gabriel graph.*

Lemma 7. *In a Gabriel drawing of G_n the external face is a 5-cycle.*

Sketch of Proof. As G_n is triconnected, its planar embedding is fixed for any given choice of the external face. From Property 2, the external face of a Gabriel drawing of G_n can only be the 5–cycle $F_5^n = v_0^n v_1^n v_2^n v_3^n v_4^n$.

Lemma 8. *In a Gabriel drawing of G_n ($n \geq 2$), for all $1 \leq i \leq n-1$, F_5^i is a strictly convex pentagon.*

Sketch of Proof. We prove the lemma by contradiction. Suppose F_5^{n-1} is drawn as a concave pentagon (see Figure 5). Let $\angle v_0^{n-1} v_1^{n-1} v_2^{n-1} \geq \pi$. Since the sum of the external angles of a pentagon is equal to 7π, then there exists at least one external angle, say $\angle v_1^{n-1} v_0^{n-1} v_4^{n-1}$, such that $\angle v_1^{n-1} v_0^{n-1} v_4^{n-1} \geq \frac{3}{2}\pi$. Hence, at least one of the angles $\angle v_1^{n-1} v_0^{n-1} v_0^n$, $\angle v_0^n v_0^{n-1} v_4^n$, and $\angle v_4^n v_0^{n-1} v_4^{n-1}$ is greater than or equal to $\frac{\pi}{2}$. Thus, at least one of the triangles $\Delta(v_1^{n-1} v_0^{n-1} v_0^n)$, $\Delta(v_0^n v_0^{n-1} v_4^n)$, $\Delta(v_4^n v_0^{n-1} v_4^{n-1})$ has a non acute angle. This contradicts Property 1.

4 Area Requirement

In this section, we prove that any Gabriel drawing of G_n requires exponential area. Because of Property 3, the result is immediately extended to weak Gabriel drawings.

Before showing the main result, we need a preliminary technical lemma. Let two straight lines in general position (not parallel) l_1 and l_2 be given; let a and

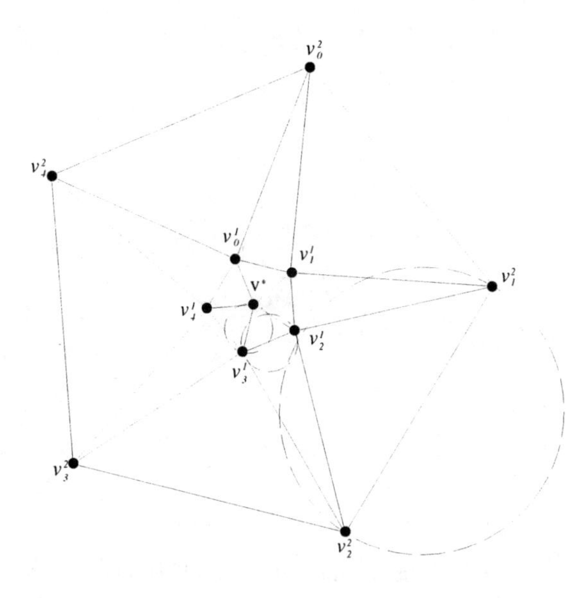

Fig. 4. A Gabriel drawing of G_2.

b two points on l_1, and c and d two points on l_2, such that a is to the left of b and c is to the left of d (see Figure 6). Suppose that the crossing point of l_1 and l_2 lays to the left of a and c. Let o be the crossing point of the line through a and d and of the line through c and b. Then we have:

Lemma 9. *The area of the triangle $\Delta(aoc)$ is smaller than the area of triangle $\Delta(bod)$.*

Theorem 10. *A Gabriel drawing and a weak Gabriel drawing of graph G_n with $5n+1$ vertices require area $\Omega(3^n)$, under any resolution rule assumption.*

Sketch of Proof. Let A_n be the area of a Gabriel drawing Γ_n of G_n and let A_{n-1} be the area of be a Gabriel drawing $\Gamma_{n-1} \subset \Gamma_n$ of G_{n-1}. We prove that $A_n \geq 3A_{n-1}$. Since $A_1 \geq c$, for some constant c depending on the resolution rule, the theorem follows by induction. For the sake of simplicity, we assume in the following that the index $i \in \{0, \ldots, 4\}$ and all the operations on the indexes are modulo 5.

We start by studying how to construct Γ_n from Γ_{n-1}. We adopt the notation of Figure 7: B_i denotes the region of the plane delimited by the two lines p_i and p_{i+1} through v_i^{n-1} and v_{i+1}^{n-1}, perpendicular to the edge $(v_i^{n-1}, v_{i+1}^{n-1})$; $HP(v_i^{n-1})$ is the the half-plane delimited by the line through edge $(v_{i-1}^{n-1}, v_i^{n-1})$ and not containing Γ_{n-1}; $C_i = B_i \cap HP(v_i^{n-1}) \cap HP(v_{i+2}^{n-1})$.

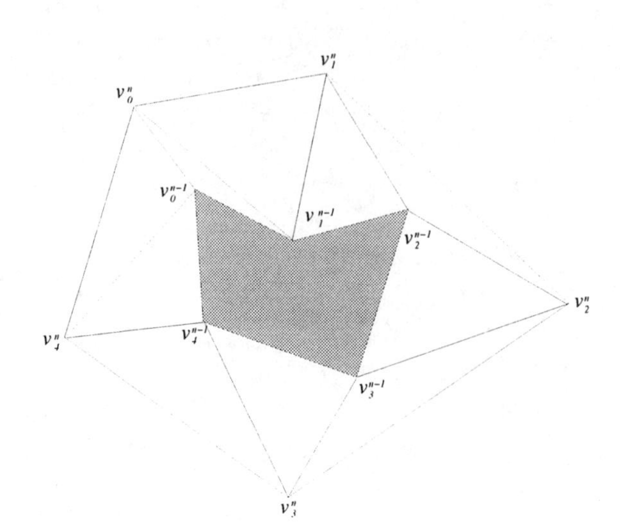

Fig. 5. Illustration for Theorem 8.

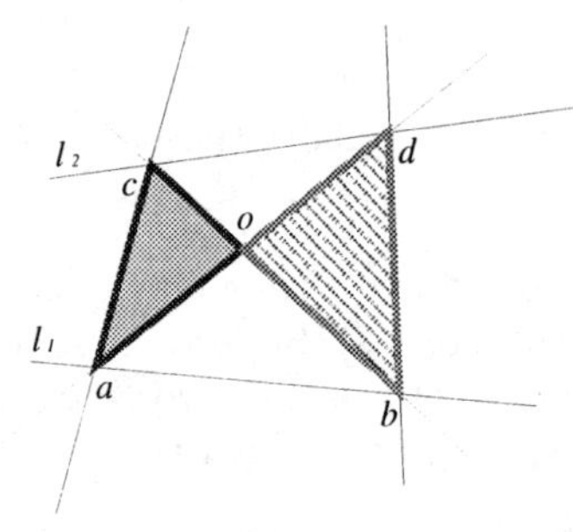

Fig. 6. Illustration for Lemma 9.

Claim 1: In order to construct Γ_n from Γ_{n-1}, v_i^n must be placed inside C_i.

Proof of Claim 1: Suppose, for a contradiction, that $v_i^n \notin C_i$. Three cases are possible. (i) $v_i^n \notin B_i$. Let v_i^n be placed to the right of p_i. Then $\angle v_{i+1}^{n-1} v_i^{n-1} v_i^n > \frac{\pi}{2}$. Hence, from Property 1, edge (v_{i+1}^{n-1}, v_i^n) cannot be drawn. Similarly, v_i^n cannot be placed to the left of p_{i+1}. (ii) $v_i^n \notin HP(v_i^{n-1})$. According to the above condition, each vertex v_i^n is placed in the corresponding B_i region. Thus, in this case, $v_i^{n-1} \in D[v_i^n, v_{i-1}^n]$. Hence, edge (v_i^n, v_{i-1}^n) cannot be drawn. (iii) $v_i^n \notin HP(v_{i+2}^{n-1})$. Similar to the previous case, $v_{i+1}^{n-1} \in D[v_i^n, v_{i+1}^n]$. Hence, edge (v_i^n, v_{i+1}^n) cannot be drawn.

Claim 1 implies that a Gabriel drawing Γ_n of G_n must strictly contain $Extr(F_5^{n-1})$ (see Figure 8), where $Extr(F_5^{n-1})$ is the extruded pentagon of the external face F_5^{n-1} of G_{n-1}.

We are now ready to show that $Area(Extr(F_5^{n-1}))$ is at least three times

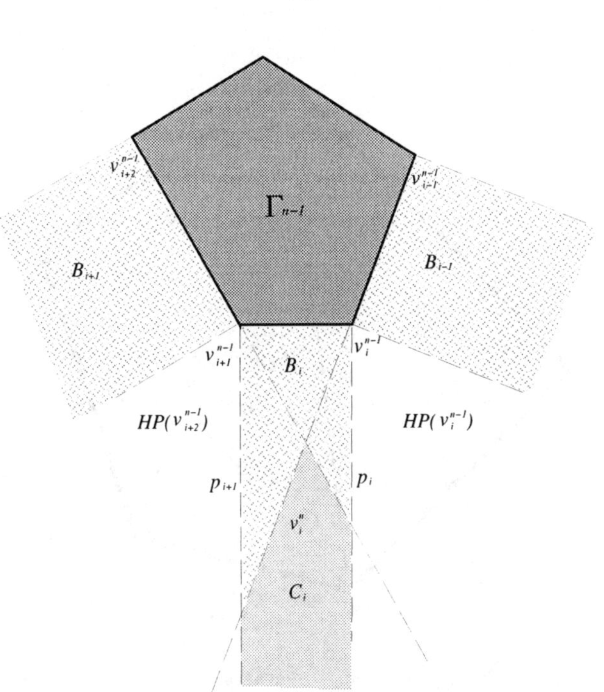

Fig. 7. Construction of Γ_n starting from Γ_{n-1}.

$Area(\Gamma_{n-1})$, which implies that $Area(\Gamma_n) > 3Area(\Gamma_{n-1})$. The proof proceeds in two steps. In the first step we prove that $Area(Extr(F_5^{n-1})) \geq 2Area(\Gamma_{n-1})$; in the second step we prove that $Area(Extr(F_5^{n-1})) \geq 3Area(\Gamma_{n-1})$. We refer to Figure 8: $w_0^n, w_1^n, w_2^n, w_3^n, w_4^n$ denote the vertices of $Extr(F_5^{n-1})$, such that $v^{n-1}_{(i+3)\mathrm{mod}5}$ is the opposite vertex of w_i^n; P_i^n is the triangle $\Delta(v_i^{n-1}, v_{i+1}^{n-1}, w_i^n)$; W_i^n is the triangle $\Delta(v_i^{n-1}, w_{i-1}^n, w_i^n)$.

The first step of the proof is based on the following claim.

Claim 2: $Area(\bigcup_{i=0}^4 P_i^n) > Area(\Gamma_{n-1})$.

Proof of Claim 2: Let us consider the line segments having as endpoints vertices of $Extr(F_5^{n-1})$ and the corresponding opposite vertices of F_5^{n-1}. These segments subdivide each P_i^n in two triangles. With reference to Figure 9(a), let us consider the two triangles $\Delta(COD)$ and $\Delta(AOB)$. Due to Lemma 8, line segments ED and EB and points C, D, A, and B satisfy the conditions of Lemma 9, hence, $Area(\Delta(COD))$ is greater than $Area(\Delta(AOB))$. A similar reasoning holds for all P_i^n. It is well known that in any arrangement of lines, every internal region is a simply connected region. Thus, $Area(\bigcup_{i=0}^4 P_i^n) > Area(\Gamma_{n-1})$.

The second step of the proof is based on the following claim.

Claim 3: $Area(\bigcup_{i=0}^4 W_i^n) > Area(\Gamma_{n-1})$.

Proof of Claim 3: Consider the pentagon $Intr(F_5^{n-1})$. We extend the notation to the intruded pentagon, so defining the P_i^{n-1}'s and W_i^{n-1}'s regions. From Lemmas 8 and 9, we derive that $Area(W_i^n) > Area(W_i^{n-1}) + Area(P_i^{n-1}) +$

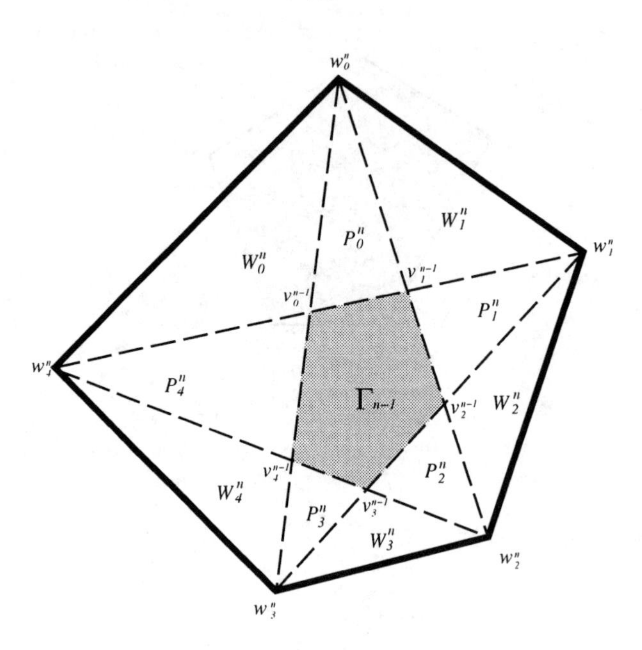

Fig. 8. Extruded pentagon of F_5^{n-1}.

$Area(W_{i+1}^{n-1})$. Thus, $Area(\bigcup_{i=0}^{4} W_i^n) > Area(\bigcup_{i=0}^{4} P_i^{n-1}) + 2Area(\bigcup_{i=0}^{4} W_i^{n-1})$. As $\bigcup_{i=0}^{4} W_i^{n-1}$ is covered twice, we can recursively apply the above strategy, to $Intr(Intr(F_5^{n-1}))$. Because of Claim 2, $Area(Extr(F_5^{n-1}))$ is at least twice $Area(\Gamma_{n-1})$. Thus, the recursion ends after a finite number of steps, proving that $Area(\bigcup_{i=0}^{4} W_i^n) > Area(\Gamma_{n-1})$.

5 Extensions and Open Problems

The result of Theorem 10 can be extended to an infinite family of (β)–drawings.

Theorem 11. *A β-drawing and a weak β-drawing of graph G_n with $5n+1$ vertices require area $\Omega(3^n)$, under any resolution rule assumption, for $1 \leq \beta < \frac{1}{1-\cos 2\pi/5}$.*

Several problems remain open in this area: (1) Study the area of proximity drawings using different definitions of proximity. For example, we find interesting to investigate the area of relative neighborhood drawings, and minimum spanning trees. For what concerns minimum spanning trees, note that Monma and Suri [22] conjectured an exponential lower bound on the area requirement. (2) Motivated by our exponential lower bounds, it is interesting to investigate classes of graphs that admit a Gabriel drawing (or a weak Gabriel drawing) with polynomial area.

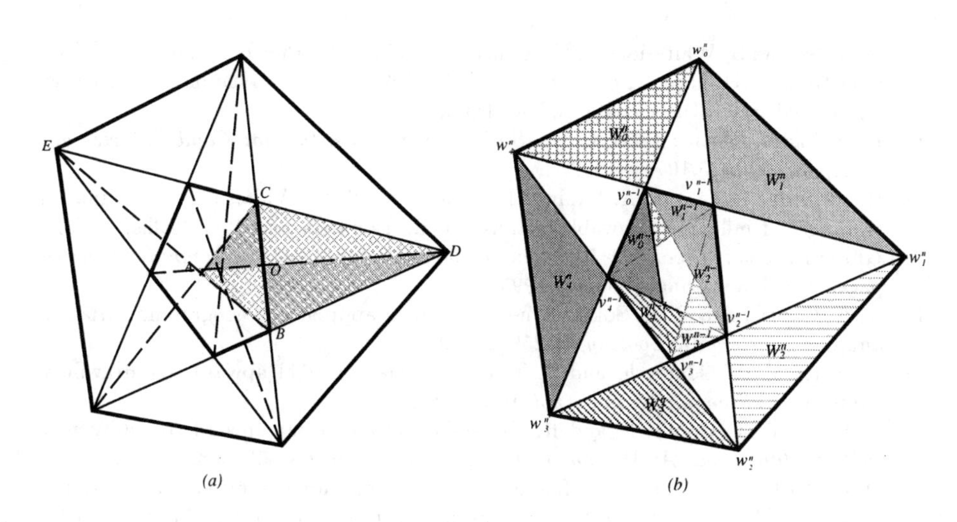

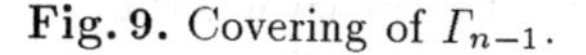
Fig. 9. Covering of Γ_{n-1}.

References

1. P. Bose, G. Di Battista, W. Lenhart, and G. Liotta. Proximity constraints and representable trees. In R. Tamassia and I. G. Tollis, editors, *Graph Drawing (Proc. GD '94)*, volume 894 of *Lecture Notes in Computer Science*, pages 340–351. Springer-Verlag, 1995.
2. P. Bose, W. Lenhart, and G. Liotta. Characterizing proximity trees. *Algorithmica*, 16:83–110, 1996. (special issue on Graph Drawing, edited by G. Di Battista and R. Tamassia, to appear).
3. R. P. Brent and H. T. Kung. On the area of binary tree layouts. *Inform. Process. Lett.*, 11:521–534, 1980.
4. Marek Chrobak, Michael T. Goodrich, and Roberto Tamassia. Convex drawings of graphs in two and three dimensions. In *Proc. 12th Annu. ACM Sympos. Comput. Geom.*, pages 319–328, 1996.
5. P. Crescenzi, G. Di Battista, and A. Piperno. A note on optimal area algorithms for upward drawings of binary trees. *Comput. Geom. Theory Appl.*, 2:187–200, 1992.
6. G. Di Battista, P. Eades, R. Tamassia, and I. G. Tollis. Algorithms for drawing graphs: an annotated bibliography. *Comput. Geom. Theory Appl.*, 4:235–282, 1994.
7. G. Di Battista, W. Lenhart, and G. Liotta. Proximity drawability: a survey. In R. Tamassia and I. G. Tollis, editors, *Graph Drawing (Proc. GD '94)*, volume 894 of *Lecture Notes in Computer Science*, pages 328–339. Springer-Verlag, 1995.
8. G. Di Battista, G. Liotta, and S. H. Whitesides. The strength of weak proximity. In F. J. Brandenburg, editor, *Graph Drawing (Proc. GD '95)*, volume 1027 of *Lecture Notes in Computer Science*, pages 178–189. Springer-Verlag, 1996.
9. G. Di Battista, R. Tamassia, and I. G. Tollis. Area requirement and symmetry display of planar upward drawings. *Discrete Comput. Geom.*, 7:381–401, 1992.

10. P. Eades and S. Whitesides. The realization problem for Euclidean minimum spanning trees is NP-hard. *Algorithmica*, vol. 16, 1996. (special issue on Graph Drawing, edited by G. Di Battista and R. Tamassia).
11. P. D. Eades. Drawing free trees. *Bulletin of the Institute for Combinatorics and its Applications*, 5:10–36, 1992.
12. H. ElGindy, G. Liotta, A. Lubiw, H. Meijer, and S. H. Whitesides. Recognizing rectangle of influence drawable graphs. In R. Tamassia and I. G. Tollis, editors, *Graph Drawing (Proc. GD '94)*, volume 894 of *Lecture Notes in Computer Science*, pages 352–363. Springer-Verlag, 1995.
13. K. R. Gabriel and R. R. Sokal. A new statistical approach to geographic variation analysis. *Systematic Zoology*, 18:259–278, 1969.
14. A. Garg, M. T. Goodrich, and R. Tamassia. Area-optimal upward tree drawings. *Internat. J. Comput. Geom. Appl.* to appear.
15. A. Garg, M. T. Goodrich, and R. Tamassia. Area-efficient upward tree drawings. In *Proc. 9th Annu. ACM Sympos. Comput. Geom.*, pages 359–368, 1993.
16. D. G. Kirkpatrick and J. D. Radke. A framework for computational morphology. In G. T. Toussaint, editor, *Computational Geometry*, pages 217–248. North-Holland, Amsterdam, Netherlands, 1985.
17. C. E. Leiserson. Area-efficient graph layouts (for VLSI). In *Proc. 21st Annu. IEEE Sympos. Found. Comput. Sci.*, pages 270–281, 1980.
18. W. Lenhart and G. Liotta. Proximity drawings of outerplanar graphs. In *Graph Drawing (Proc. GD '96)*, Lecture Notes in Computer Science. Springer-Verlag, 1996.
19. Giuseppe Liotta and Giuseppe Di Battista. Computing proximity drawings of trees in the 3-dimemsional space. In *Proc. 4th Workshop Algorithms Data Struct.*, volume 955 of *Lecture Notes in Computer Science*, pages 239–250. Springer-Verlag, 1995.
20. A. Lubiw and N. Sleumer. Maximal outerplanar graphs are relative neighborhood graphs. In *Proc. 5th Canad. Conf. Comput. Geom.*, pages 198–203, Waterloo, Canada, 1993.
21. D. W. Matula and R. R. Sokal. Properties of Gabriel graphs relevant to geographic variation research and clustering of points in the plane. *Geogr. Anal.*, 12:205–222, 1980.
22. C. Monma and S. Suri. Transitions in geometric minimum spanning trees. *Discrete Comput. Geom.*, 8:265–293, 1992.
23. J. D. Radke. On the shape of a set of points. In G. T. Toussaint, editor, *Computational Morphology*, pages 105–136. North-Holland, Amsterdam, Netherlands, 1988.
24. C. D. Thompson. Area-time complexity for VLSI. In *Proc. 11th Annu. ACM Sympos. Theory Comput.*, pages 81–88, 1979.
25. G. T. Toussaint. The relative neighbourhood graph of a finite planar set. *Pattern Recogn.*, 12:261–268, 1980.
26. G. T. Toussaint. A graph-theoretical primal sketch. In G. T. Toussaint, editor, *Computational Morphology*, pages 229–260. North-Holland, Amsterdam, Netherlands, 1988.
27. R. B. Urquhart. Graph theoretical clustering based on limited neighbourhood sets. *Pattern Recogn.*, 15:173–187, 1982.

Design of Reliable Combinatorial Algorithms Using Certificates*

Fabrizio Luccio[1] and Alberto Pedrotti[2]

[1] Dipartimento di Informatica, Università di Pisa, Italy
[2] Scuola Normale Superiore, Pisa, Italy

Abstract. Assuming random computational errors in the execution of a RAM program, we discuss how to obtain a correct result with a given confidence bound. An algorithm ALG (requiring A operations) is used in connection with a certificate CERT (requiring C operations, $C \leq A$). CERT assesses the result of ALG, however, it may fail as well. Each operation of ALG (resp. CERT) is repeated k (resp. h) times, and the result of the operation is established by majority voting. Letting π be the success probability of an operation, and $1 - \delta$ be the desired confidence on the overall result, k and h are evaluated as functions of A, C, π and δ. The resulting running time is reduced by a factor of up to A/C over the time required without using CERT, for the same value of δ.
Two case studies with $C < A$ are examined, their expected behavior is analyzed, and computational results on the running times are presented.

1 Formulation of the problem

The problem of computing in the presence of errors has been carefully considered in different areas. In this work we take an algorithmic view of the problem on an arbitrary abstract machine, where some instructions may randomly fail, giving wrong results, while a program is running. In fact we refer to a RAM machine, but our results can be easily transferred to other computational models. We pose a confidence bound for the correctness of the results, and study how to design the algorithm to obtain such a bound. In this sense we may speak of a fault-tolerant algorithm, where the amount of faults is probabilistically predicted, yet not limited a priori, and the final result is correct with arbitrarily large probability, given as a design parameter.

Several theoretical lines of research may fall into algorithmic fault-tolerance. The studies opened by Blum show how to overcome the effect of computational errors by randomization techniques [2, 3, 4]. In this method, the computation is organized in two interactive phases: a computational phase that may provide a wrong result, and a randomized checking phase that works on a fully reliable medium. The results are carried out within a given confidence bound.

---

* Supported in part by MURST of Italy. E-mail: `luccio@di.unipi.it`, `pedrotti@cibs.sns.it`

Another approach considers a bounded number of errors of specific type. Then, for given problems, it is shown how to design an algorithm yielding a result that is certainly correct. Information-theoretic problems dealing with "computation against a liar" typically fall in this line [5, 8].

In hardware design, errors have physical origin; circuit redundancy is commonly used to ensure reliable computation. Among the huge number of contributions in this area, the mathematical analysis of logical circuits done by Pippenger shares some ideas with our approach [10, 11, 12]. The model considers transient random failures, under which a computational step may fail with fixed, or at least bounded, probability. No subsidiary reliable hardware is available, and thus the results are certified to be correct only with given confidence. This is achieved by appropriate circuit redundancy. More recently, a similar approach has been adopted for decision trees, to compute Boolean functions, and for comparison trees, to deal with searching and sorting problems [6]. In all these studies, a family of "algorithms" (logic circuits or trees) is considered for a given problem and varying input size.

In this work, we consider a new approach for designing combinatorial algorithms in the presence of random computational errors. As already pointed out, we code our algorithms as RAM programs, as formally defined, for example, in [1]. Then, compared to the studies on circuits and decision trees, we have a single algorithm for any given problem and all possible input sizes. We introduce redundancy by repeating k times each machine instruction of the algorithm ALG. Moreover, we use an additional algorithm CERT to certify the correctness of the result devised by ALG. Since CERT is in turn a RAM algorithm, its operations are repeated h times. We have to determine the values of k and h which minimize the expected processing time, maintaining the probability to obtain an incorrect final result less than a given δ. Due to the novelty of this approach, we cannot directly compare our results with previous ones.

The paper is organized as follows. In this section we define the computational strategy. Then we explain how the overall error probability is related to the error probabilities of ALG and CERT. Next, we pose a minimization problem on the overall computational time; this problem is solved in the Section 2, where we find the optimal values of k and h. For simplicity, the study is carried out under the assumption that the most common outcome among the k (or h) outcomes of a RAM operation is selected by an infallible mechanism, requiring zero time. In Section 3 we remove this assumption, showing that this does not affect the previous results substantially. Finally (Section 4), we apply our results to two classical problems, namely, sorting with Quicksort, and searching with Binary search.

Let A and C denote the number of RAM operations (without repetitions) performed by ALG and CERT, respectively, and let $1 \leq C \leq A$. Every RAM operation produces a correct result with probability $\pi > \frac{1}{2}$; let us call *step* the repetition of a single operation plus the choice of the most common outcome as the result. We make two hypotheses, the second of which will be later removed.

1. A RAM operation may produce an incorrect result, but the destination of such a result is always correct, in order to protect previous results. We store the k (resp. h) results in dedicated registers, to perform the majority voting on the content of these registers, and to transfer the most common result into the (certainly correct) assigned destination. An extension where also the destinations may be incorrect is presently under study.
2. The implementation of the majority voting is infallible, and has zero cost. This hypothesis is posed to focus on the analysis of our computational strategy, and will be removed in Section 3, where we show that the analysis remains valid even if the voting is fallible.

If many RAM operations within a single step fail, the step itself may yield an incorrect result. We then say that the step *fails.* If none of the steps of ALG fail, we say that ALG *succeeds* (namely, its result is guaranteed to be correct). Otherwise, we say that ALG *fails.* Similarly, if some step of CERT fails, we say that CERT itself fails; namely, it may erroneously accept an incorrect result, or reject the right one. On the other hand, if none of the steps of CERT fail, we say that CERT succeeds. Let us define:

Strategy S_1 :
repeat
  execute ALG, to get a result R;
  execute CERT on R;
until CERT says "yes";
claim that R is the right result.

Note that we do not compare the successive outcomes of ALG among themselves; we simply rely on the approval of CERT. Let q_a and q_c be the failure probabilities of ALG and CERT, respectively, and let $p_a = 1 - q_a$, $p_c = 1 - q_c$. Let us call *attempt* a single repetition of ALG + CERT. Within an attempt, we have four cases:

(i) ALG and CERT succeed (probability: $p_a p_c$). The right result is produced by ALG and approved by CERT.
(ii) ALG and CERT fail (probability: $q_a q_c$). ALG may produce a wrong result, and CERT may incorrectly accept it.
(iii) ALG fails and CERT succeeds (probability: $q_a p_c$). A wrong result may be produced by ALG, but is then refused by CERT.
(iv) ALG succeeds and CERT fails (probability: $p_a q_c$). The result returned by ALG, although correct, may be refused.

Denote with Γ the probability that strategy S_1 comes up with the right result; we want to give a lower bound to Γ. Let γ denote the probability that the *first* attempt claims the right result; since this happens at least in case (i), we have $\gamma \geq p_a p_c$ (Note, however, that in *all* four cases the right result may be generated and accepted, due to compensating or uninfluencing errors). Note also that, in cases (iii) and (iv), we either claim the right result, or enter another attempt. Thus, the probability that the first attempt is also the last does not exceed $\gamma + q_a q_c$. By these observations, we get $\Gamma \geq \gamma + (1 - \gamma - q_a q_c)\Gamma$, that is, $\Gamma \geq \gamma/(\gamma + q_a q_c)$. Using $\gamma \geq p_a p_c$, we get the following *sufficient* condition for the constraint $\Gamma \geq 1 - \delta$ to be satisfied:

$$1 - q_a - q_c \geq \left(\frac{1}{\delta} - 2\right) q_a q_c. \tag{1}$$

Let us turn to the execution time of strategy S_1; it is a random variable that we shall denote with $t(S_1)$. Let N_1 be the random variable counting the number of attempts executed; since in case (i) a result is always claimed, the probability that a given attempt is not the last cannot exceed $1 - p_a p_c$; hence

$$\mathbf{E}[N_1] \leq 1 + (1 - p_a p_c) + (1 - p_a p_c)^2 + \ldots = \frac{1}{p_a p_c} < \frac{1}{1 - q_a - q_c}. \tag{2}$$

Note that the same number of attempts is expected for an execution ending up with a wrong result as is for a successful one. Let us find out the relationship between k_1 (resp., h_1) and q_a (resp., q_c). Taking the majority among k outcomes, the probability of getting a wrong result does not exceed the probability that $k/2$ or more outcomes are wrong. Then, by Chernov bounds, the probability that a single step of ALG (resp., CERT) fails does not exceed $(4\pi - 4\pi^2)^{k_1/2}$, (resp., $(4\pi - 4\pi^2)^{h_1/2}$). Applying the union bound on all the steps performed by ALG and CERT, we obtain that $q_a \leq A(4\pi - 4\pi^2)^{k_1/2}$ and $q_c \leq C(4\pi - 4\pi^2)^{h_1/2}$. We then choose:

$$k_1 = \frac{2\ln(A/q_a)}{-\ln(4\pi - 4\pi^2)}, \quad h_1 = \frac{2\ln(C/q_c)}{-\ln(4\pi - 4\pi^2)}. \tag{3}$$

Since we have $t(S_1) = N_1(k_1 A + h_1 C)$, with the aid of equations (2) and (3) we obtain an upper bound $T_1(q_a, q_c)$ to $\mathbf{E}[t(S_1)]$, namely

$$T_1(q_a, q_c) = \frac{2}{-\ln(4\pi - 4\pi^2)} \frac{1}{1 - q_a - q_c} \left(A \ln \frac{A}{q_a} + C \ln \frac{C}{q_c}\right) \tag{4}$$

We then consider:

Problem 1. *Given A, C and δ, $1 \leq C \leq A$, $\delta < 1/2$, minimize $T_1(q_a, q_c)$, subject to the condition (1).*

Let us introduce two more strategies:

- **Strategy** S_2 : we do not compute certificates, that is, we execute ALG once, relying on its result. The number k_2 of operations within each step is chosen such that $A\,(4\pi - 4\pi^2)^{k_2/2} = \delta$. The execution time of S_2 is $T_2 = k_2 A$, that is:

$$T_2 = \frac{2A(\ln A + \ln(\frac{1}{\delta}))}{-\ln(4\pi - 4\pi^2)}. \tag{5}$$

  The performance of S_2 will be used as a benchmark for evaluating the convenience of using certificates.

- **Strategy** S_3 is nothing but S_1, considered in the abstract setting where CERT is an infallible oracular computation, working at zero cost. Note that, since the final result cannot be wrong, no confidence bound is needed in input. As we did for S_1, we can find an upper bound T_3 to the expected performance of S_3. It can be shown that T_3 is minimized by setting $q_a = \tilde{q}_a$, where

$$\tilde{q}_a = \frac{1}{\ln A + \eta \ln \ln A}, \qquad \eta \in (1, 2). \tag{6}$$

The value of $\tilde{q}_a$ will turn out to be useful for the analysis of S_1. Let us record the competing performance:

$$T_3(\tilde{q}_a) = \frac{2A(\ln A + \eta \ln \ln A)}{-\ln(4\pi - 4\pi^2)}. \tag{7}$$

Incidentally, we note that, if a coarse confidence parameter is given as an input to S_2, strategy S_3 is worse than S_2 (see (5)), although it uses the oracular power of CERT. This happens because S_3, unlike S_2, may execute ALG more than once. Clearly, the counterpart to this additional effort is that the result returned by S_3 cannot be wrong.

2 Solving Problem 1

Let us now return to the analysis of S_1. The following result may be proved by standard calculus.

Theorem 1. *The function $T_1(q_a, q_c)$ admits a unique minimum $(\overline{q_a}, \overline{q_c})$, for $0 < q_a + q_c < 1$. Moreover, there exists a $\theta \in (1, 2)$ such that*

$$\overline{q_a} = \frac{A}{A + C} \frac{1}{\ln(A + C) + \theta \ln \ln(A + C)}, \qquad \overline{q_c} = \frac{C}{A} \overline{q_a},$$

and

$$T_1(\overline{q_a}, \overline{q_c}) = \frac{2(A + C)}{-\ln(4\pi - 4\pi^2)} [\ln(A + C) + \theta \ln \ln(A + C)].$$

Using the relations (3) and $\overline{q_c} = C\overline{q_a}/A$, we see that $k_1 = h_1$; moreover, from the values of $\overline{q_a}$ and $\overline{q_c}$ we get that the expected number of attempts is nearly one (see (2)). Thus, the common value of k_1 and h_1 is approximately equal to $T_1(\overline{q_a}, \overline{q_c})/(A + C)$.

Let $\overline{\delta}$ be the confidence achieved by setting $q_a = \overline{q_a}$, $q_c = \overline{q_c}$. By (1), we get $1 - \overline{q_a} - \overline{q_c} = (1/\overline{\delta} - 2)\overline{q_a}\,\overline{q_c}$; hence, $\overline{\delta}$ depends only on A and C. We have to distinguish two cases, according to whether $\delta \geq \overline{\delta}$ or $\delta < \overline{\delta}$.

2.1 Loose confidence constraints

If $\delta \geq \overline{\delta}$, the choice $q_a = \overline{q_a}$, $q_c = \overline{q_c}$ is admissible; hence, Theorem 1 completely solves Problem 1. To estimate T_2/T_1 under this hypothesis, we point out two facts:

- for fixed A, it is easily checked that $\overline{\delta}$ decreases together with C ($1/\overline{\delta}$ ranges from roughly $4\ln^2 A$, for $C = A$, to roughly $A\ln^2 A$, for $C = 1$);
- if we let δ decrease from $1/2$ to $\overline{\delta}$, the performance of S_1 remains constant (and S_1 achieves a confidence maybe greater than required), while that of S_2 deteriorates.

By these observations, we realize that the ratio $T_1(\overline{q_a}, \overline{q_c})/T_2$ attains its maximum when $\delta = \overline{\delta}$ and $C = 1$; in this case we obtain

$$T_1 \approx \frac{2A\ln A}{-\ln(4\pi - 4\pi^2)}, \quad T_2 \approx \frac{4A\ln A}{-\ln(4\pi - 4\pi^2)}. \tag{8}$$

On the other hand, if $C = A$ we have, for any $\delta \in [\overline{\delta}, 1/2)$:

$$T_1 \approx \frac{4A\ln A}{-\ln(4\pi - 4\pi^2)}, \quad T_2 \approx \frac{2A\ln A}{-\ln(4\pi - 4\pi^2)}. \tag{9}$$

2.2 Tight confidence constraints

If $\delta < \overline{\delta}$, the choice $q_a = \overline{q_a}$, $q_c = \overline{q_c}$ is no longer valid. We have to seek for minimum points on the boundary of the admissible region of Problem 1; such region is shown in Figure 1. The equation $q_c = f(q_a)$ describing its upper boundary is easily obtained from (1). Problem 1 is then restated in a one-variable framework; let us summarize the main points of the subsequent analysis, whose proof we omit:

P1 T_1 admits a unique minimum $(q_a^*, f(q_a^*))$ on the boundary;
P2 in the special case $C = A$, the minimum can be found explicitly, and we get

$$T_1(q_a^*, f(q_a^*)) > 2A\frac{2\ln A + \ln\left(\frac{1}{\delta}\right)}{-\ln(4\pi - 4\pi^2)}; \tag{10}$$

P3 T_1 admits the following lower bound:

$$T_1(q_a^*, f(q_a^*)) > \frac{2}{-\ln(4\pi - 4\pi^2)}\left[A\ln A + C\ln\left(\frac{C}{\delta}\right)\right]; \tag{11}$$

P4 q_a^* can be bounded as follows: $\sqrt{\delta - \delta^2} - \delta \leq q_a^* < \overline{q_a}$;
P5 $\overline{q_a}$ is less than $\tilde{q}_a$ (see also Theorem 1 and (7), respectively);

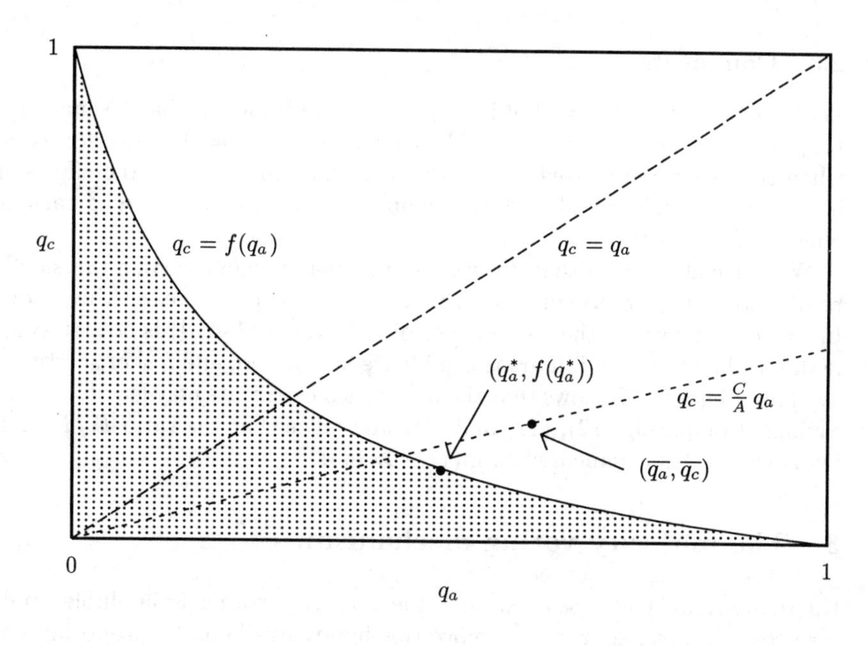

Fig. 1. Admissible region for Problem 1, in the case $\delta < \overline{\delta}$.

P6 note that the bound reported in **P4** is by no means tight; however, it will provide suggestions for the choice of a suitable approximation to q_a^*. Namely, let us think, for fixed A and δ, to let C decrease: we then expect S_1 to behave somewhat like S_3. Consequently, we expect q_a^* to approach $\tilde{q}_a$. Now, recall that, by **P4** and **P5**, we have $q_a^* < \overline{q_a} < \tilde{q}_a$; hence, we are led to try $\overline{q_a}$ as an approximation to q_a^*. We get:

$$T_1(\overline{q_a}, f(\overline{q_a})) < \frac{2}{-\ln(4\pi - 4\pi^2)}$$
$$* \left[A(\ln(A+C) + \theta \ln\ln(A+C)) + C\left(1 + \frac{1}{\ln(A+C)}\right)\ln\left(\frac{C}{\delta}\right)\right], \quad (12)$$

where $\theta \in (1,2)$ is the same as in Theorem 1. Note also that the competing k_1 is the same as in Section 2.1, while we now have

$$h_1 \approx \frac{2}{-\ln(4\pi - 4\pi^2)}(\ln C + \ln\left(\frac{1}{\delta}\right) - \ln\ln(A+C)).$$

Comparing the relations (12) and (11) with (5), we get:

Theorem 2. *For fixed A, C, and π, let $R(\delta)$ denote the ratio between the expected performance of strategy S_1, and the performance of S_2. Then, we have*

$$\frac{C}{A} \leq \lim_{\delta \to 0} R(\delta) \leq \frac{C}{A}\left(1 + \frac{1}{\ln(A+C)}\right).$$

2.3 Comments

Let us remark that, if $C = A$, it is better to work without certificates; namely, (9), (10) and (5) show that $T_1 > T_2$. The extremal case $C = A$ arises, for example, when the algorithm is used as its own certificate; in this case, the efforts spent in running it again could be better employed by repeating instead each RAM operation more times.

We have also shown that, for values of δ greater than a certain threshold, the result returned by S_1 is correct with a probability that is greater than required. In this case, however, the convenience of S_1 (even if $C \ll A$) is not striking (see (8)). On the other hand, if great confidence is required, and a short certificate is available, Theorem 2 shows that the use of such certificate may lead to significant savings. Comparing (12), (11) and (5), we realize that the ratio T_1/T_2 claimed by Theorem 2 is attained when $\ln(1/\delta) \gg \ln A$.

3 The majority voting mechanism.

Up to now, we have assumed that the majority voting is infallible, and has zero cost (Hypothesis 2). We remove this hypothesis here, by proposing a more feasible mechanism of majority voting in the RAM, and discussing its impact on the overall complexity of the computation.

Let us analyze a step of ALG. The same conclusions will hold for CERT, with k replaced by h. Let $k = 2r + 1$. In Section 1 (Hypothesis 1), we assumed that the k values, resulting from the k-fold repetition of the operation, were stored in k registers $R_1 \dots R_k$. As we shall see, it is not necessary to have k special registers at our disposal; we need only one dedicated register C. Let *op(arg)* be the operation realized in the given step, and X be the destination assigned to its result. The following program for a step embeds both the repeated operation and the voting mechanism:

```
repeat
    X := op(arg);
    t := 0;
    for j := 1 to k do
    begin
        C := op(arg);
        if C = X then t := t + 1
    end
until t ≥ r + 1;
```

Let us call *round* a single iteration of the **repeat** cycle. We suppose that the operations involving the counters i and t are free of error; namely, they can be performed with special pieces of hardware. Then, the following operations may fail within a round:

- the assignment $X := op(arg)$;

- each of the k assignments $C := op(arg)$;
- each of the k comparisons $[C : X]$.

Recall that the destination address of an operation is supposed to be always correct (Hypothesis 1). Thus, the assignment $X := op(arg)$, even if erroneous, cannot affect the value of registers different from X. Hence, the results of the previous steps cannot be corrupted. The same is true for the subsequent assignments $C := op(arg)$.

Let us give an upper bound to the probability that the step fails, i.e., that it ends up with X containing a wrong result.

Let us focus on the k iterations of the **for** cycle embedded in an arbitrary round. We say that the round is *clean* if the following conditions are simultaneously fulfilled in $r + 1$ or more iterations:

i) the assignment $C := op(arg)$ succeeds,
ii) the comparison $[C : X]$ succeeds.

Clearly, a clean round accepts the right value, and rejects a wrong one, while a round that is not clean may fail to do so. Note that the conditions i) and ii) are independent, and each of them is satisfied with probability π. Let us suppose that $\pi > \sqrt{2}/2$, such that the probability π^2 that i) and ii) are both satisfied exceeds 1/2. We denote with γ the probability that a round is not clean; using Chernov bounds, we obtain $\gamma \leq (4\pi^2 - 4\pi^4)^{k/2}$.

Let P_F (resp. P_S) be the probability that a given round claims a wrong result (resp., the right result). We want to give an upper bound to P_F. Note that a round may claim a wrong result if and only if the assignment $X := op(arg)$ fails, and the round itself is not clean. Since these events are independent, we have $P_F \leq \gamma(1 - \pi)$. Similarly, it is easy to see that $P_S \geq (1 - \gamma)\pi$.

The failure probability Q of a step is easily expressed in terms of P_F and P_S; namely, we have

$$Q = P_F + (1 - P_S - P_F)P_F + (1 - P_S - P_F)^2 P_F + \ldots = \frac{P_F}{P_S + P_F}$$
$$\leq \frac{(1 - \pi)\gamma}{\pi(1 - \gamma) + (1 - \pi)\gamma}.$$

From this, using the fact that $\pi > 1/2$, we get $Q < \gamma$; recalling the above bound on γ we get

$$Q \leq (4\pi^2 - 4\pi^4)^{k/2}. \tag{13}$$

Recall that, in the infallible case, the same result held with π^2 replaced by π.

To bound the expected number of operations that are executed in each step, we first consider the expected number α_1 of rounds. Since each round has probability $P_S + P_F$ of being the last one, we have

$$\alpha_1 = \frac{1}{P_S + P_F} \leq \frac{1}{P_S} = \frac{1}{\pi(1 - \gamma)}. \tag{14}$$

Note that γ depends on k; however, it can be shown that our choices of k_1, h_1 and k_2 are always such that γ is negligible; hence, we can safely ignore the dependence of α_1 on these parameters.

Next, let us note that we can find a constant α_2, independent of k, such that the number of operations performed in a round does not exceed $\alpha_2 k$. This constant may be determined by a straightforward analysis of the program describing a step.

As a conclusion, we may summarize the effects of having a fallible majority voting, embedded in the step, as follows:

- we have to take $\pi > \sqrt{2}/2$, rather than $\pi > 1/2$. Moreover, π has to be substituted with π^2 in all the formulas;
- the execution times $t(S_1), t(S_3), T_2$ introduced in Section 1 have to be further multiplied by the factor $\alpha_1 \alpha_2 = \alpha_2/\pi$.

Hence, we have:

Theorem 3. *Let T_1 and T_3 be the expected execution times, expressed in terms of RAM operations, of strategies S_1 and S_3, respectively, and let T_2 be the execution time of S_2. Let T_1, T_2 and T_3 be computed assuming the majority voting to be infallible and zero-cost. If $\pi^2 > 1/2$, an effective voting mechanism can be embedded in each computational step, such that T_1, T_2 and T_3 are evenly increased by a factor, that depends only on π.*

4 Two examples.

Our computing strategies will now be applied to two sample problems, solved with selected algorithms and certificates coded in the MIX language of Knuth.

1. Sorting a file of n records with Quicksort. For simplicity, we suppose that a new sorted file is created, while preserving the original one. Moreover, we suppose that each record $R(i)$ not only contains the information $K(i)$ to be sorted, but also keeps track of its original position; for example, $R(i) = (K(i), i)$. Quicksort requires an expected MIX time of $11.67\, n \ln n$ [7]. The certificate consists in a scan of the file, to check that the elements are in order, and that they have not been erroneously modified by the algorithm. It has a straightforward implementation requiring $7.5\, n$ MIX time.
2. Searching a sorted file for a record which is known to exist, using Binary search. The search returns the address a of the searched element e in the sorted file F; the MIX program requires $18 \log_2 n - 16$ time units [7]. The certificate amounts to comparing e with $F(a)$, and admits a MIX implementation requiring three time units.

In Figure 2, we plot the expected performance of the various strategies, in logarithmic scale, for increasing values of $1/\delta$. Recall that the ratio T_2/T_1 is expected to approach the value A/C (Theorem 2). For reference, we have reported also the time T_3 obtained with the aid of the oracle certificate.

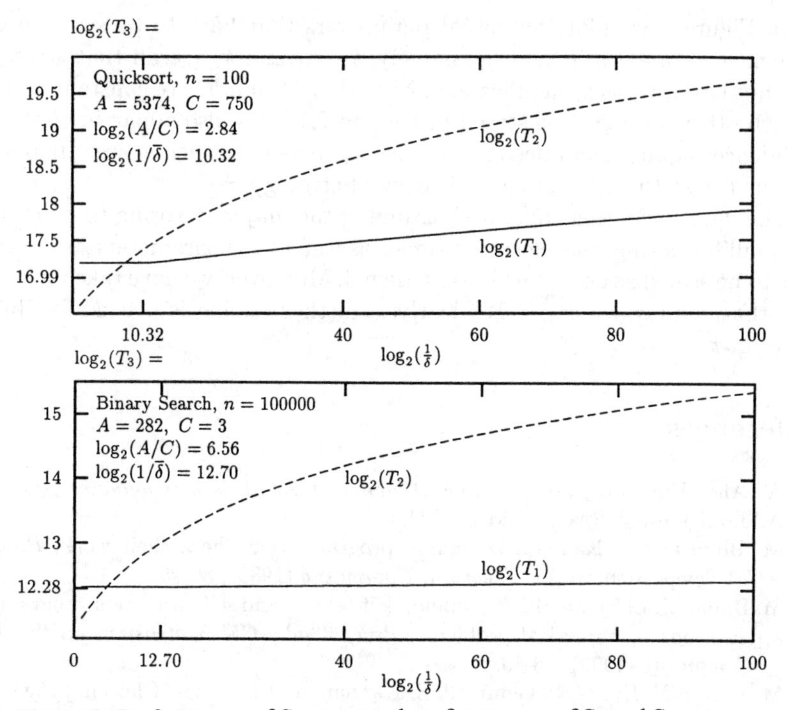

Figure 2: Performance of S_2; expected performances of S_1 and S_3

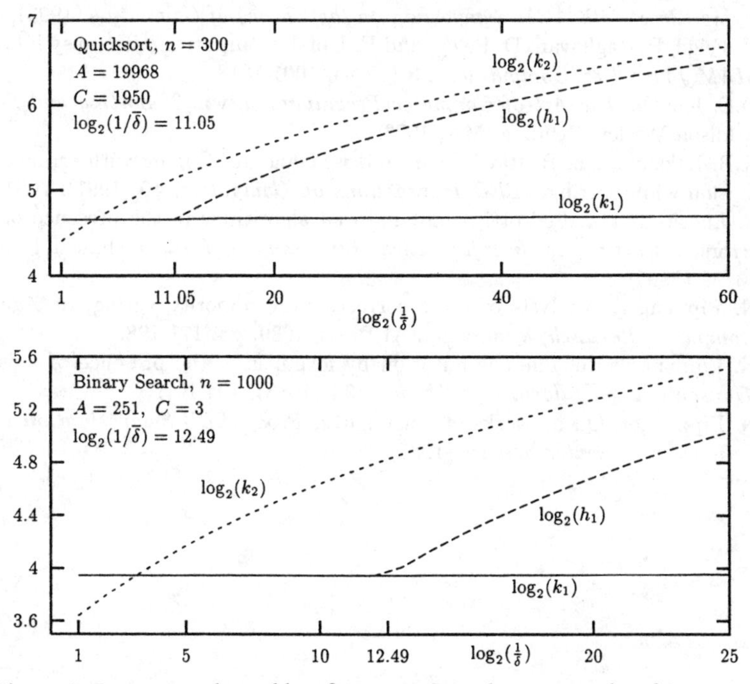

Figure 3: Parameters k_1 and h_1 of strategy S_1 and parameter k_2 of S_2

In Figure 3 we plot the actual parameters that have to be set in order to implement strategies S_1 and S_2, namely, k_1, h_1 and k_2 (recall that strategy S_3 does not have practical significance). Note that k_1 and h_1 are constant and equal to each other for $\delta \geq \overline{\delta}$, as stated in Section 2.1. It is also apparent that, as the confidence requirements become tighter, k_1 increases much slower than k_2 and h_1; this lies at the basis of the efficiency of strategy S_1.

The data have been computed assuming the majority voting to be infallible. With fallible voting, the execution times, as well as the parameters k_1, h_1 and k_2, have to be scaled as described by Theorem 3. Moreover, we have taken $\pi = 0.897$, such that $-\ln(4\pi - 4\pi^2) = 1$. Clearly, a further scaling is needed for different values of π.

References

1. A. Aho, Hopcroft, Ullman, *The Design and Analysis of Computer Algorithms,* Addison-Wesley, Reading, MA, 1974.
2. M. Blum and S. Kannan, Designing programs that check their work, *Proc. 21st ACM Symposium on the Theory of Computing* (1989), 86-95.
3. M. Blum, M. Luby, and R. Rubinfeld, Self-testing and self-correction programs with applications to numerical problems, *Proc. 22nd ACM Symposium on the Theory of Computing* (1990), 73-83.
4. M. Blum, W. Evans, P. Gemmell, S. Kannan, and M. Naor, Checking the correctness of memories, *Proc. IEEE Symposium on the Foundations of Computer Science* (1991), 90-99.
5. R.S. Borgstrom and S.R. Kosaraju, Comparison-based search in the presence of errors, *Proc. 25th ACM Symposium on the Theory of Computing* (1993), 130-136.
6. U. Feige, P. Raghavan, D. Peleg, and E. Upfal, Computing with noisy information, *SIAM Journal on Computing,* **23** (1994), 1001-1018.
7. D.E. Knuth, *The Art of Computer Programming,* vol. 3: *Sorting and Searching,* Addison-Wesley, Reading, MA, 1973.
8. K.B. Lakshmanan, B. Ravikumar, and K. Ganesan, Coping with erroneous information while sorting, *IEEE Transactions on Computers,* **40** (1991), 1081-1084.
9. F. Luccio and A. Pedrotti, Combinatorial algorithms in the presence of random errors, *Dipartimento di Informatica, University of Pisa,* Technical Report TR-13/95 (1995).
10. N. Pippenger, Analysis of error correction by majority voting, in *Advances in Computing Research,* Volume 5, JAI Press, 1989, pp. 171-198.
11. N. Pippenger, Reliable computation by formulas in the presence of noise. *IEEE Transactions on Information Theory,* **34** (1988), 194-197.
12. N. Pippenger, On networks of noisy gates, *Proc. IEEE Symposium on the Foundations of Computer Science* (1985), 30-38.

An Improved Deterministic Algorithm for Generalized Random Sampling

Amihood Amir* Georgia Tech and Bar-Ilan University

Emanuel Dar** Bar-Ilan University

Abstract. We consider the problem of *deterministically* selecting s uniformly random different m-element subsets of $\{1, ..., k\}$. The only known lower bound for the time to solve this problem is the trivial $\Omega(sm)$. The best two previously known solutions are of time $O(sm^3 \log m \log\log m)$ and $O(s(k+m))$, respectivly. In this paper we present an algorithm whose time comlexity is $O(s^2m^2 + sm^2 \log m \log\log m + sm \log sm)$. Thus, for $s < m \log m \log\log m$ this algorithm is the fastest known deterministic algorithm.
The main idea of the algorithm is using a uniform random number generator to efficiently construct biased random numbers.

1 Introduction

When writing heuristics or randomized algorithms one needs efficient souces of randomness. We distinguish between three possible levels:

1. Given an integer k, generate a random *integer* r, $1 \leq r \leq k$.
2. Given integers k, m, generate a *set* of m different integers chosen randomly between 1 and k.
3. Given integers k, s, m, generate a *set whose elements are s different m-element subsets of* $\{1, ..., k\}$.

Problem 1 is a clasic problem in computer science. The only use we make of it is the assumption that there exists a function `RandInt(x)` that returns a random integer between 1 and x. We will also assume that such a random integer is generated in unit time.

Problem 2 has proven very popular. Bentley devoted four *Programming Pearls* columns to it ([3, 6, 5, 4]), culminating with an algorithm by Floyd ([4]). Floyd's

* Department of Mathematics and Computer Science, Bar-Ilan University, 52900 Ramat Gan, ISRAEL; (972-3)531-8770; amir@cs.biu.ac.il; Partially support by NSF grant CCR-92-23699 and Israel Ministry of Science and Arts grant 6297.

** Department of Mathematics and Computer Science, Bar-Ilan University, 52900 Ramat Gan, ISRAEL; dar@macs.biu.ac.il;

algorithm generates a random subset of $\{1, ..., k\}$ using m calls to `RandInt`. Its deterministic time is $O(m \log m)$ but can be done in time $O(m)$ using hashing. Algorithms with the same time complexity but that produce a random sample (i.e. the elements in each set are given in random order) rather than a random set appeared independently in [1, 7]. A parallel algorithm appeared in [13].

We are interested in an efficient *deterministic* solution to problem 3. There are many applications that require an efficient generation of random sets. We were motivated by robust estimators for surface fitting (e.g. [11, 8, 14, 16]).

Our solution is based on a novel idea of using a uniform random number generator to produce biased number generation. We provide efficient algorithms for this task.

Problem Definition:

Let us formally define problem 3. Let $P_a(X)$ be the set of a-element subsets of X. and let $N_k = \{1, \ldots, k\}$. We are interested in generating a random, uniformly chosen element of $P_s(P_m(N_k))$. Our algorithm will, in fact, produce both the elements of $P_m(N_k)$ and the elements of $P_s(P_m(N_k))$ in a random order, i.e. it will produce a *random sample*.

The previously mentioned algorithms for problem 2 ([4, 1, 7, 13]) can also be used to solve a special case of the problem. By successive application of the algorithm, mutually disjoint random m-element sets can be chosen.

This paper presents a solution to the more generalized problem where the sets are different but not necessarily disjoint.

A straight-forward way of solving the problem is by repeatedly generating random elements of $P_m(N_k)$ and adding them to our sample as long as there is no conflict (*i.e.* the added sample did not appear previously). This algorithm is randomized and works well with high probabilty, especially where k is very large relative to s and m. Since sm numbers need to be generated, it is an interesting question whether such a sample can be generated *deterministically* in time $O(sm)$, assuming that `RandInt` operates in constant time. This problem is still open, but deterministic solutions exist.

All previous solutions (see [2]) involved combining efficient methods for generating random sets ([4, 1, 7]) with set *unranking* algorithms ([12, 9, 10]). A *ranking* is a bijection from the set of m-element subsets of $\{1, \ldots, k\}$ to the set of numbers $\{1, \ldots, \binom{k}{m}\}$. *Unranking* is the inverse function, i.e. recovering the set from the number.

Other than the fact that, for large k, all above algorithms are far from the trivial lower bound of $\Omega(sm)$, they use the powerful *numerical analysis* model where arithmetic operations $(+, -, \times, /, \exp, \log)$ are done in unit cost for numbers of all sizes[3].

Let us translate the complexity from the numerical analysis model to the standard *work model* where each of the four basic arithmetic operations takes unit time for words of $O(\log k)$ bits. Addition of $O(m \log k)$ bit numbers can be done in time $O(m)$ and multiplication of such numbers can be achieved in time $O(m \log m \log\log m)$ using Schönhage-Strassen techniques [15]. There are sm^2 multiplications in the algorithm of [2], thus the time that algorithm spends on multiplication alone is $O(sm^3 \log m \log\log m)$ operations. The log function is used sm times, but simulating it costs $O(m^2 \log m \log\log m)$ basic operations. Thus the total algorithm time of the algorithm in [2] is $O(sm^3 \log m \log\log m)$.

We present here a simpler algorithm. Our algorithm does not require powerful unit-cost operations such as *log* or *exp*. The *main contribution* of this paper is a *novel idea* for generating random sets that does not involve unranking. We show how to efficiently generate biased random numbers using an unbiased random number generator.

Our algorithm's running time is $O(sm \log sm + s^2 m)$ on the numerical analysis model. In the work model it is $O(sm^2 \log m \log\log m + s^2 m^2)$. For cases where $s < m \log m \log\log m$ this is the *fastest known* algorithm for the generalized random sampling problem.

For simplicity of the exposition and for ease of comparison with previous algorithms, we will compute the time assuming unit cost for large number basic arithmetic operations. In section 4 we compute the time complexity in the conventional work model.

2 The Algorithm's Idea

The previous algorithms generate a number that represents a *set* at every random call, and their main task is deciphering that set from the number. To avoid unranking, we generate the *numbers* of each set sequentially. In order to assure that the generated sets are indeed uniformly randomly distributed, we need to generate the numbers within a set with a probability conditional on the previously generated numbers.

The following lemma tells us what probability each particular number should have at a given stage in the generation, in order to assure a uniformly random set distribution.

Before the lemma, let us proceed with some definitions and notations that will serve us throughout the paper.

[3] It is true that if $x^{1/m}$ can be computed in time $O(m)$ the exp and log operations are not needed in [2], however, we are not aware of such a method.

Definitions and Notations: Denote the generated m-element sets by $S_1, ..., S_s$. We will represent the sets by tuples, ordered chronologically by the sequence of their numbers' generation. Denote the first i elements of the jth set by $S_j^i = S_j[1], ..., S_j[i]$.

Let $E^{i,j+1} = \{S|\ S = S_l,$ for some $l \in \{1, ..., j\}$ and $S_{j+1}^i \subset S\}$.

For each $c \in \{1, \ldots, k\}$, let $P_c^{i,j+1} = \{S|\ S \in E$ and $c \in S\}$, and let $p_c^{i,j+1} = |P_c^{i,j+1}|$.

In the future, we always deal with the case where i and j are known and fixed. To simplify notation we will henceforth drop there indices from our notation and just refer to E, P_c, or p_c.

Lemma: Suppose that $S_1, ..., S_j, S_{j+1}^i$ have been generated.
In order to guarantee a uniformly random set distribution we should assign the following probability to generating the number $c \in \{1, ..., k\}$ in $S_{j+1}[i+1]$:
If $c \in S_{j+1}^i$ then that probability is 0, else it is

$$\frac{1}{m-i} \frac{\binom{k-i-1}{m-i-1} - |P_c|}{\binom{k-i}{m-i} - |E|}$$

Proof: Since we have already generated i numbers in the current tuple, there remain $m-i$ numbers (out of a $k-i$ size universe) that still need to be generated. This means that there are $\binom{k-i}{m-i}$ such sets. Each of them should have an equal probability, provided they had not appeared as an earlier tuple. $|E|$ is the number of such sets that have already been generated in earlier tuples, so the remaining $\binom{k-i}{m-i} - |E|$ sets should have equal probability of being generated.
We now return to the probability we should assign to a specific number c. If $c = S_{j+1}^i[p]$ for some $p, 1 \leq p \leq i$, then obviously the probability of generating c should be 0. Consider, then, $c \neq S_{j+1}^i[p]$, $p = 1, \ldots, i$.
We have already generated i numbers in the current tuple. Thus the number of sets that c can be in is: $\binom{k-i-1}{m-i-1}$. Of these sets, $|P_c|$ have already been generated, so they should not be repeated. This leaves us with $\binom{k-i-1}{m-i-1} - |P_c|$ possible tuples that $\{S_{j+1}[1], ..., S_{j+1}[i], c\}$ is a subset of.
The probability of generating c should therefore be:

$$\frac{\binom{k-i-1}{m-i-1} - |P_c|}{\binom{k-i}{m-i} - |E|}$$

However, we are generating tuples, and not sets. So we should multiply this probability by the probability that c will be at the $(i+1)$st place (out of the $m-i$ remaining positions). This gives us the desired value:

$$\frac{1}{m-i}\frac{\binom{k-i-1}{m-i-1}-|P_c|}{\binom{k-i}{m-i}-|E|}. \qquad \square$$

Armed with the Lemma, we now produce the following simple algorithm.

Algorithm Outline

1. Create an m-element subset of $\{1,...,k\}$.
2. for $j=2$ to s do
   (a) for $i=1$ to m do
      i. Generate a number c by an appropriately weighted random function. Take $S_j[i] \leftarrow c$.
      ii. If $\{S_j[1],...,S_j[i]\} \not\subset S_l$, for $l=1,...,j-1$ then generate a uniformly random $(m-i)$ element sample of $\{1,...,k\}-\{S_j[1],...,S_j[i]\}$ and exit loop (a).

      end

   end

End Algorithm

Correctness: Obvious from the construction. The lemma proves that there exists a weighted random function that ensures uniformly distributed m-element tuples.

Time: Step 1 can be done in time of $O(m)$.

We will show that Step 2a can be done in time of $O(sm+m\log sm)$. This step is executed $s-1$ times.

Total Algorithm Time: $O(s^2m+sm\log sm)$.

3 Implementation Issues

3.1 The Weighted Random Number Generator

The idea of the algorithm is to generate each tuple one number at a time in time $O(sm)$. This can only be done by a weighted random number generator. We assume that we have a uniform random number generator. Our task now is to efficiently use the uniformly random number generator for generating weighted random numbers.

Assume that we have already generated j tuples. We proved that the probability of generating the number c in the $i+1$ position of the $j+1$ tuple is: $\frac{1}{m-i}\frac{\binom{k-i-1}{m-i-1}-|P_c|}{\binom{k-i}{m-i}-|E|}$. Our strategy is to generate a random number r, $1 \leq r \leq (m-i)\left(\binom{k-i}{m-i}-|E|\right)$ and *translate* this random number to the correct number $c \in \{1,...,k\}$. The translation assumes a logical bin of size $\binom{k-i-1}{m-i-1}-|P_c|$ for each number c and returns the appropriate number c in whose "bin" r happened to fall.

It would seem like we would have too many different bin values, and then would not be able to find the bin efficiently. Fortunately, this is not the case. The bin size is dependent on $|P_c|$. By definition the range of $|P_c|$ is $\{1,\ldots,s\}$.

We implement the translation by constructing $s+1$ sets of numbers $A_0, A_1, ..., A_s$. $c \in A_{p_c}$ if $c \notin S^i_{j+1}$ and c is in p_c of the previously generated j tuples that are in E. We represent sets A_p as arrays.

Implementation of the translation

We separate the translation into two cases. In the first case, the generated number did not appear in an earlier tuple in E, and in the second case the number appears in an earlier tuple. If c has been chosen before, then, $c \in A_{p_c}$ for some $p_c \in \{1,...,j\}$. Let $P = \sum_{p=1}^{j}\left(\binom{k-i-1}{m-i-1}-p\right)|A_p|$. P is the sum of the sizes of all logical bins, thus if $r \leq P$ (r is the generated random number) then the number has been chosen before, else, the number has not appeared yet.

Implementation of the translation in case $r \leq P$

If $r \leq P$ then we have to find the number in the right A_p set.

1. Compute $N \leftarrow \binom{k-i-1}{m-i-1}$.
2. For $l = 1$ to j do
   (a) If $r \leq (N-l)|A_l|$ then
      i. $c \leftarrow A_l[\lceil r/(N-l)\rceil]$.
      ii. Exit.
   (b) Else $r \leftarrow r-(N-l)|A_l|$.

End

Notes:

1. $\binom{k-i-1}{m-i-1} = \binom{k-i}{m-i}\frac{m-i}{k-i}$, so we pay only a constant amount of time for every generated number.
2. $|A_i|$ can be saved in the header of each set, so it does not need to be computed in every iteration.

Time: The time of generating the number for $r \leq P$ is $O(s)$.

Implementation of the translation in case $r > P$.

$r > P$ means that the generated number c has not appeared before. Therefore $c \in A_0$ and all the logical bins are of the same size.

If all the numbers previously chosen were $\left\{k - \left|\bigcup_{S\in E} S\right| + 1, k - \left|\bigcup_{S\in E} S\right| + 2 \ldots, k\right\}$ then we would be able to compute c easily. Simply take $c = \left\lceil \frac{(r-P)}{\binom{k-i-1}{m-i-1}} \right\rceil$. However, since the previously chosen numbers appeared at random in the range, we need to adjust the c computed above for the c-th *unchosen* number.

Let us assume that we maintain a bijection $Find : \{1, \ldots, k\} - |\bigcup_{S\in E} S| \rightarrow \{1, \ldots, k\}$. Then our translation will be done as follows:

1. $c' \leftarrow \left\lceil \frac{(r-P)}{\binom{k-i-1}{m-i-1}} \right\rceil$
2. $c \leftarrow Find(c')$

In the next subsection we will show how to maintain such a bijection in a manner similar to [1, 7]. The data structure to maintain it is constructed as long as $P \leq r$, and used once when $r > P$ is encountered (once that happens the sequence is easily completed by any algorithm for constructing a single random sample). The deterministic time for computing $Find(c')$ is $O(\log|\bigcup_{S\in E} S|) = O(\log sm)$.

We conclude:

Time: The time of generating the number for $r > P$ is $O(\log(sm))$.

Note: This type of translation can happen at most once per tuple because once we generate a number that has not appeared before we continue creating an $(m-i)$-element sample of $\{1, ..., k\} - \{S_j[1], ..., S_j[i]\}$ in a uniformly random fashion.

Maintaining a Dynamic Bijection Consider the following problem. Given a universe $U = \{1, \ldots, k\}$ we would like a data structure that supports the following operations:

1. *Delete* a number i from the universe.
2. *Reinsert* a previously deleted number back into the universe.
3. *Find*. Let $B = \{b_1, \ldots, b_p\}$ be the deleted numbers. We want a bijection $Find : \{1, \ldots, k-p\} \rightarrow U - B$ that allows us to efficiently compute $Find(i)$.

A data structure that supports the *Delete* and *Find* requirements was described in [1, 7]. We briefly describe that data structure and show how it can support the *Reinsert* operation as well.

The idea in [1, 7] is to pair each deleted number with one of $\{k-p+1, \ldots, k\}$. We then define

$$Find(i) = \begin{cases} i, & \text{if } i \notin B; \\ i\text{'s pair}, & \text{if } i \in B. \end{cases}$$

The space necessary for the pairs is $O(p)$. If the pairs are kept in a balanced tree the worst case deterministic time for $Find$ is $O(\log p)$. If we use hashing we can compute $Find$ in constant time with high probability.

We define $Delete(b)$, $b \in \{1, \ldots, k-p\}$ to mean that the number $Find(b)$ is deleted from U. We proceed to show the implementation for *Delete* as described in [1, 7].

Implementation of $Delete(b)$, $b \in \{1, \ldots, k-p\}$

1. If there exists a pair $(k-p, l)$ then $pair \leftarrow l$ else $pair \leftarrow k-p$
   (If $k-p$ has been previously deleted, then b's pair will be $k-p$'s former pair. Otherwise $k-p$ itself will be b's pair.)
2. If there exists a pair (b, l') then delete it and create new pair $(l', \texttt{nil})$.
   (b had been previously deleted and its pair was l'. Thus the current delete request actually requests to delete l'. We therefore create a new pair with l' as the left element. Since $l' > k-p$ it will never appear in a $Find$ or *Delete* future request, then its pair may be chosen as empty (`nil`).)
3. If $pair \neq b$ then create pair $(b, pair)$ else create pair $(b, \texttt{nil})$.
4. If $pair \neq b$ and $pair = k-p$ then set a pointer from $k-p$ to the pair $(b, k-p)$.

Important Observations:

1. For any pair (x, y), if $x \in \{k-p+1, \ldots, k\}$ then y is `nil`.
2. For every $x \in \{k-p+1, \ldots, k\}$, either there is a pair $(x, \texttt{nil})$ or a pair (b, x), $b \leq k-p$. In the later case there is a pointer from x to the pair (b, x).

Implementation of $Reinsert(b)$, $b \in B = \{b_1, \ldots, b_p\}$

1. If there exists a pair $(k-p+1, \texttt{nil})$ then destroy pairs (b,l) and $(k-p+1, \texttt{nil})$ and create pair $(k-p+1, l)$.
2. else destroy pairs (b,l) and $(b', k-p+1)$ and create pair (b', l).

Time: The time for *Delete* and *Reinsert* is proportional to searching for a pair, which is $O(\log p)$ in a dynamic balanced tree.

3.2 Getting Ready for the Next Element

Suppose we chose element c for position $i+1$ of tuple S_{j+1}. To get ready for generating the element in position $i+2$ we need to update the set E (discard all tuples where c does not appear) and the A_i lists (to reflect the new E). The implementation below shows how to do these updates in total time $O(sm + m \log sm)$ per tuple S_{j+1}.

Finding the right A_{p_c} set

Updating the A_i sets means moving some elements from one A set to another. We need a fast connection between an element c and its A_{p_c} set. To do this we build a balanced binary search tree where a leaf is: an element c_i, its p_{c_i} number, and a pointer to the appropriate $A_{p_{c_i}}$ set. Every element in the j m-element tuples points directly to its leaf. A number that appears in more than one tuple has each of its occurrences pointing to the same leaf.

The time for finding the right A_{p_c} set for $c \in S_l$, $l = 1, ..., j$, and moving c to another set, if necessary, is $O(1)$.

The time for adding a new element c to the tree is $O(\log(sm))$, as is the time for a tree search. If c is already in the tree then we need to increment by 1 the number of times it appears. This is done by moving the pointer from A_{p_c} to A_{p_c+1}. If c is not in the tree then add a new leaf, set the pointer to A_1, and add c to A_1 set.

We only add a new element to the tree when a new number is generated. Since m numbers are generated every sequence, then the time of adding the m element to the tree is $O(mlogsm)$. For the entire algorithm (s samples) it is $O(sm \log(sm))$.

We now describe how the A_i sets are updated. Note that any changes to the tree that are necessary when a tuple is completed involve changing only values in the current tuple. Thus it no more than doubles the total time.

Updating the A_i sets and E vector.

To update the A_i and E sets effectively, we use lists of indices B_c. Each $c \in A_{p_c}$ points to a list of indices B_c. For $l = 1, ..., j$, $l \in B_c$ iff $c \notin S_l$. We also define a vector $E[1], \ldots, E[s]$ where $E[l] = 1$ iff $S_l \in E$.

Updating the A_i sets.

Suppose that the last element generated was c, and $S^i_{j+1} \cup \{c\} \subseteq S_l$, for some $l \in \{1, ..., j\}$. (If not then we continue generating the tuple randomly). We can reach A_{p_c} in constant time and remove c from A_{p_c}. For each $l \in B_c$, if $E[l] = 1$ then for each $c_i \in S_l$ move c_i from $A_{p_{c_i}}$ to $A_{p_{c_i}-1}$.

Every $c_i \in S_l$, $i = 1, .., m$, $l = 1, ..., j$ can be moved from $A_{p_{c_i}}$ to $A_{p_{c_i}-1}$ at most once in a tuple generation, making the time of updating the A_i sets in a tuple generation $O(sm)$.

Updating the E vector.

For each $l \in B_c$ that has $E[l] = 1$, $l \in B_c$ means that $c \notin S_l$ therefore $S^i_{j+1} \cup \{c\} \not\subset S_l$ therefore it should be deleted from E. Set $E[l] \leftarrow 0$.

Building and updating the B_c lists of indices.

The B_c lists contains information about previous tuples. Consequently, during the first tuple generation the B_c's are empty. We update at the end of each tuple generation and not after each element generation. Consequently, chasing through every element of B_c and comparing it to E adds a total of $O(s^2m)$ to our entire algorithm running time, which is dominated by its overall time complexity.

Suppose that we finished generating the $(j+1)$-st tuple. The updating is done for each $c \notin S_{j+1}$. Simply let $B_c \leftarrow B_c \cup \{j+1\}$. This updating takes a constant amount of work for each element and is done at the end of each tuple generation, thus the total time per tuple is $O(sm)$. The following code describes the process.

1. For each $c \in S_{j+1}$ follow the pointer to its location in the A_{p_c} set and tag it.
2. For each $c \in S_l$, $l = 1, ..., j$ follow the pointer to its location in the A_{p_c}set. If the location is not tagged then update B_c. $B_c \leftarrow B_c \cup \{j+1\}$
3. For each $c \in S_{j+1}$ follow the pointer to its location in the A_{p_c} set and untag it.

We have a direct pointer from each element to its place in the A_i set, so each access is done in constant time. The time of the first step and the last step is $O(m)$. The time of step 2 is $O(jm) = O(sm)$. The total time of one updating of theB_c's is $O(sm)$. We update the B_c's after each tuple generation, so, we update the B_c's s times. Therefore the total time for updating the B_c's in the algorithm is $O(s^2m)$.

4 Complexity in the Conventional Work Model

The largest number we may handle is $O(k^m)$, requiring m computer words.

Generating a large random number is equivalent to generating m small random numbers. Thus generating $O(sm)$ large numbers will take time $O(sm^2)$.

The only large number operations we do are multiplication and addition (as opposed to previous algorithms that require large number *log* and *exp* operations). The number of multiplications is $O(sm)$, which again brings the actual cost up to $O(sm^2 \log m \log\log m)$ and $O(s^2m)$ additions which ends up as $O(s^2m^2)$ word operations.

Finally, If one is willing to sacrifice absolute randomness for faster results and if $k \geq s + m$ then our algorithm gives rise to the following very simple idea:

1. Create the first random sample.
2. Create $(s - 1)$ $(m - 1)$–element samples.
3. For each of the $s-1$ $(m-1)$-tuples generate the last number, giving all numbers that will complete a previously existing sample probability 0.

The above algorithm can be easily implemented in time $O(s^2m + sm\log(sm))$ without any large word operations. The only problem is that the probability of generating a sample is not equal for all samples, there are some very slight differences in the probabilities (depending on the size of k).

References

1. A. Amir. A pearl diver deals a poker hand. *Intl. J. of Computer Mathematics*, 48(1+2):11–14, 1993.
2. A. Amir and D. Mintz. An efficient algorithm for generalized random sampling. *Pattern Recognition Letters*, 12:683–686, 1991.
3. J. Bentley. Programming pearls - a little program, a lot of fun. *Comm. of ACM*, pages 1179–1182, 1984.
4. J. Bentley and R. Floyd. Programming pearls - a sample of brilliance. *Comm. of ACM*, pages 754–757, 1987.
5. J. Bentley and D. Gries. Programming pearls - abstract data types. *Comm. of ACM*, pages 284–290, 1987.
6. J. Bentley and D. Knuth. Programming pearls - literate programming. *Comm. of ACM*, pages 364–369, 1986.
7. M. Chrobak and R. Harter. A note on random sampling. *Inform. Process. Lett.*, 29:255–256, 1988.
8. M. A. Fischler and R. C. Bolles. Random sampling consensus: a paradigm for model fitting with applications to image analysis and automated cartography. *Comm. of ACM*, June 1981.
9. G. M. Knott. A numbering system for combinations. *Comm. of ACM*, pages 45–46, 1974.
10. D. H. Lehmer. *Applied Combinatorial Mathematics*, chapter The Machine Tool of Combinatorics. Wiley, New York, 1964.
11. P. Meer, D. Mintz, and A. Rozenfeld. Least median of squares based robust analysis of image structure. In *Proc. Image Understanding Workshop*, Pittsburgh, PA, Sept. 1990.
12. A. Nijenhuis and H. Wilf. *Combinatorial Algorithms*. Academic Press, New York, 2nd edition, 1978.

13. V. Rajan, R. K. Ghosh, and P. Gupta. An efficient parallel algorithm for random sampling. *Inform. Process. Lett.*, 30:265–268, 1989.
14. P. J. Rousseuw and A. M. Leroy. *Robust Regression and Outlier Detection.* Wiley, New York, 1987.
15. A. Schönhage and V. Strassen. Schnelle multiplikation grosser zahlen. *Computing (Arch. Elektron. Rechnen)*, 7:281–292, 1971. MR 45 # 1431.
16. A. Tirumalai and B. G. Schunck. Robust surface approximation using least median squares regression. Technical Report CSE-TR-13-89, University of Michigan, 1989.

Polynomial Time Algorithms for Some Self-Duality Problems

Carlos Domingo*

Department LSI,
Universitat Politecnica de Catalunya,
Pau Gargallo, 5
08028-Barcelona, Spain

Abstract. Consider the problem of deciding whether a Boolean formula f is self-dual, i.e. f is logically equivalent to its dual formula f^d, defined by $f^d(x) = \bar{f}(\bar{x})$. This problem is a well-studied problem in several areas like theory of coteries, database theory, hypergraph theory or computational learning theory. In this paper we exhibit polynomial time algorithms for testing self-duality for several natural classes of formulas where the problem was not known to be solvable. Some of the results are obtained by means of a new characterization of self-dual formulas in terms of its Fourier spectrum.

1 Preliminaries and Definitions

The self-duality problem is a well-studied problem in several areas. It is of particular interest the monotone case, since the problem of deciding whether a monotone DNF formula is self-dual or not is known to be equivalent to several other problems in different fields like theory of coteries (used in distributed systems) [6, 7], database theory [13], hypergraph theory [4] and computational learning theory [2].

Before discussing previous and our results, we provide a more detailed definition of the problem and related concepts.

We consider Boolean concepts $f : \{0,1\}^n \mapsto \{0,1\}$ depending on n variables $\{v_1, \ldots, v_n\}$. We denote by x_i the ith bit of vector $x \in \{0,1\}^n$ and by $\bar{x}$ and $\bar{f}$ the complement of x and f, respectively. A literal is either a variable or its negation. A Boolean concepts is said to be *monotone* if for every pair of assignments $a, b \in \{0,1\}^n$, $a \leq b$ implies $f(a) \leq f(b)$.

Let us discuss first how to represent Boolean concepts. The most common and widely used representation of a Boolean concept is the *Disjunctive Normal Form* or *DNF*. It is well know that every Boolean concept can be expressed as a DNF formula. A DNF formula is a disjunction of terms, where a term is a conjunction of literals. Respectively, we define the symmetric class of *Conjunctive Normal Form* or *CNF* formulas. A CNF is a conjunction of clauses, where a clause is a

* Supported by the Esprit EC program under project 7141 (ALCOM-II) and the Spanish DGICYT (project PB92-0709).

disjunction of literals. Every Boolean concept can also be represented as a CNF formula.

However, if we restrict a DNF formula, for instance by upper bounding the number of terms by some number k (the class *k-term DNF*) then not every Boolean concept can be represented as a k-term DNF. In other words, we will refer to the class k-term DNF as the subset of Boolean concepts than can be represented as a k-term DNF. We also define the class *k-clause CNF* as the class of concepts that can be represented as a CNF with at most k clauses. Another natural restriction is bounding the size of the terms or clauses. Thus, the class *k-DNF* (*k-CNF*) is the class of Boolean concepts that can be represented as a DNF (CNF) with at most k variables per term (clause). A *read-k DNF* formula is a DNF formula where each variable appears at most k times. Notice that we say that a Boolean concept f belongs to a class $\mathcal{C}$, we are not saying anything about how the concept is actually represented, we only said that this concept can be represented in the class $\mathcal{C}$.

We also consider other class of Boolean concepts. A *decision tree* T is a binary tree such that each internal node has two children, it contains literals in its nodes and each leaf is labeled either with "0" or "1". A decision tree represents a Boolean concept f_T defined as follows. If T is a single leaf with label $b \in \{0,1\}$ then it is the constant concept b. Otherwise, if l_i is the literal in the root of T and T_0 and T_1 are the left and right subtrees respectively, then $f_T(x) = f_{T_0}(x)$ if $l_i(x_i) = 0$ and $f_T(x) = f_{T_1}(x)$ if $l_i(x_i) = 1$. The *depth* of a decision three is the length of its longest path from the root to a leaf. We denote by *k-depth DT* the class of Boolean concepts represented as a decision tree of depth at most k.

We also consider a extension of decision trees, *decision trees with linear operations.* In these trees each node computes a summation (modulo 2) of a subset of the n input variables, and branches according whether the sum is zero or one. This is an extension of the previous decision tree model, since we can still test single variables. Notice that in this extended model of decision tree we can test parity of all variables in a single node while its well known that in the traditional model 2^n nodes are required. We will denote by *k-depth DT* the class of decision trees with linear operations of depth at most k.

Now, we go back to the definitions of the self-duality problem, the main focus of this paper. The dual of a function f denoted by f^d is defined by $f^d(x) = \bar{f}(\bar{x})$. A function is said to be *dual-minor* if $f \leq f^d$, *dual-major* if $f^d \leq f$ and *self-dual* if $f = f^d$, i.e $f(x) \neq f(\bar{x})$ for all $x \in \{0,1\}^n$.

We now formally define the decision problem previously mentioned that we will use through all the paper.

Problem: SD
Input: A DNF formula f.
Question Is f self-dual?

Respectively, we define the problem *MSD* for the case when the input DNF formula is monotone. Through all the paper we will say that *problem SD (MSD) is efficiently solvable for a class of Boolean concepts* $\mathcal{C}$ if there exists an algorithm that on input a formula $f \in \mathcal{C}$ represented as a DNF, the algorithm decides in polynomial time in n and the DNF size of f (its number of terms) whether the formula is self-dual or not. The assumption that the input is given as a DNF is only for unifying purpose of all our results and all previous results on the problem that also assume that the input formula is given as a DNF formula. However, it is easy to see that the results still hold if the input formula is given in any other representation that is polynomial time evaluable, i.e. there is a polynomial time algorithm that for given x and f outputs $f(x)$. We will just refer to problem SD (MSD) meaning that the input is a DNF (monotone DNF) with any restriction.

We review now some of the known results about these problems.

The problem of deciding if a DNF formula is dual-minor is solvable in polynomial time while testing whether a it is dual-major is known to be NP-Complete [2].

Thus, testing dual-minority is solvable in polynomial time while testing dual-majority is NP-Complete, no matter whether the input formula is monotone or not. However, the situation of the more general problem of self-duality is different. We notice that testing self-duality is in fact equivalent to test whether a minor-dual formula is dual-major and therefore the problem is not the same that just testing whether a formula is dual-major. It could be the case that the restriction of f being dual-minor may make the problem easier.

For the monotone case, the complexity of testing self-duality is not known. The best result is a quasi-polynomial time due to Fredman and Khachiyan [5]. More precisely, they showed an $\mathcal{O}(s^{\log s})$ time algorithm for testing whether a monotone DNF formula of at most s terms is self-dual. Thus, problem MSD is not NP-Complete unless $NP \subset DTIME(n^{poly-log(n)})$. However, in the non monotone case the problem SD becomes NP-Complete [2].

Since the polynomial time solvability of MSD is not known and SD is NP-Complete, it is natural to try to efficiently solve the problem for some restricted subclasses of DNF.

For several subclasses of DNF, problem SD is known to be solvable. Among them, we have monotone k-DNF formulas [4], read-once monotone DNF [1], 2-monotonic monotone DNF formulas [11] (this class contains all monotone threshold functions) and any class of monotone formulas that has constant maximum latency [12] (see the reference for a definition of maximum latency).

We complement these results by showing that for the following classes of formulas problem SD is efficiently solvable:

- $\mathcal{O}(\log n)$-depth LDT
- $\mathcal{O}(\log n)$-DNF $\cup$ $\mathcal{O}(\log n)$-CNF (this class contains the classes $\mathcal{O}(\log n)$-term DNF, $\mathcal{O}(\log n)$-clause DNF and $\mathcal{O}(\log n)$-depth DT)
- read-k monotone DNF

We also provide a new characterization of self-dual formulas in terms of its Fourier spectrum from where some of this results are obtained.

None of the known results mentioned earlier imply any of our results. In other words, all the results presented here enlarge the class of formulas where testing self-duality can be done efficiently. As we mentioned before, the problem of testing self-duality is closely related to several other problems in different research areas and therefore it is important to carefully study under which conditions we can solve the problem efficiently.

We start by showing the characterization of self-dual formulas and its applications in the following section. In Section 3 we show that for some other classes of formulas, the problem can be also efficiently solvable by other means. We conclude by discussing our results and some open problems.

2 A characterization of Self-Dual Formulas and its applications

In this section we show a very simple characterization of a self-dual formulas in terms of its Fourier spectrum. The Fourier analysis of Boolean formulas has recently shown up as a powerful tool for proving a variety of results in several fields [8, 10, 9]. To the author's knowledge, this is the first time that the Fourier transform is related to the self-duality problem. We also show how to apply this characterization to efficiently test self-duality for some classes of formulas.

We first recall the definitions of the Fourier transform needed for the characterization. In this section, we will assume that Boolean functions are defined as $f : \{0,1\}^n \mapsto \{-1,+1\}$. Every Boolean function can be uniquely expressed in terms of its Fourier spectrum. That is, $f(x) = \sum_{S \subseteq \{1,\dots,n\}} \hat{f}(S)\chi_S(x)$ where

$$\chi_S(x) = (-1)^{\sum_{i \in S} x_i} = 1 - 2\left(\sum_{i \in S} x_i \bmod 2\right)$$

and the Fourier coefficients $\hat{f}(S)$ calculate the correlation of f and χ_S with respect to the uniform distribution, i.e. $\hat{f}(S) = 2^{-n}\sum_x f(x)\chi_S(x)$. Furthermore, because the vector space of real-valued functions over $\{0,1\}^n$ has dimension 2^n, the set of 2^n parity functions forms an orthonormal basis for this vector space and every function f can be uniquely expressed as a linear combination of parity functions. The *support* of a Boolean function f is the set $\{S|\hat{f}(S) \neq 0\}$, i.e. the set of all sets corresponding to non-zero Fourier coefficients of f.

Now we are ready to state the characterization of self-dual formulas.

Lemma 1. *A Boolean formula $f : \{0,1\}^n \mapsto \{-1,+1\}$ is self-dual if and only if for all S, $S \subseteq \{1,\dots,n\}$ of even cardinality, the corresponding Fourier coefficient, $\hat{f}(S)$, is equal to zero.*

Proof. For the if part, assume that f is self-dual and consider any set S of even cardinality. Clearly, for any vector x, $\chi_S(x) = \chi_S(\bar{x})$ and by definition of self-duality, $f(x) = -f(\bar{x})$. Therefore,

$$
\begin{aligned}
\hat{f}(S) &= 2^{-n} \sum_{x \in \{0,1\}^n} f(x)\chi_S(x) = \\
&= 2^{-n-1} \sum_{x \in \{0,1\}^n} (f(x)\chi_S(x) + f(\bar{x})\chi_S(\bar{x})) = \\
&= 2^{-n-1} \sum_{x \in \{0,1\}^n} \chi_S(x)(f(x) + f(\bar{x})) = 0
\end{aligned}
$$

For the other direction, suppose that for all $S \subseteq \{1, \ldots, n\}$ of even cardinality the value of $\hat{f}(S)$ is equal to zero. We calculate the sum $f(y) + f(\bar{y})$ for any $y \in \{0,1\}^n$. Since for all $S \subseteq \{1, \ldots, n\}$ of odd cardinality, $\chi_S(x) = -\chi_S(\bar{x})$ we can rewrite the sum in the following way:

$$
\begin{aligned}
f(y) + f(\bar{y}) &= \sum_{S:|S| odd} \hat{f}(S)\chi_S(y) + \sum_{S:|S| odd} \hat{f}(S)\chi_S(\bar{y}) = \\
&= \sum_{S:|S| odd} \hat{f}(S)\chi_S(y) - \sum_{S:|S| odd} \hat{f}(S)\chi_S(y) = 0
\end{aligned}
$$

which means that for any y, $f(y) = -f(\bar{y})$, i.e. the function is self-dual. □

The above lemma characterizes self-dual functions in terms of its Fourier coefficients. However, the number of Fourier coefficients in a DNF formula is in general exponentially big and we are not able to use the characterization for testing self-duality of DNF formulas in polynomial time. In fact, we do not expect to be able to use the characterization for testing self-duality for DNF formulas since the problem is NP-Complete. Nevertheless, it is interesting to study classes of formulas that have a "small" support. For one such class, Kushilevitz and Mansour [9] gave an algorithm, which we call KM, that actually finds the small support. We state their result in the following theorem.

Theorem 2. *[9] Let f be a Boolean formula that belongs to the class $\mathcal{O}(\log n)$-depth LDT. Then, on input f, algorithm KM outputs the support of f in polynomial time in n and size of f in its given representation.*

Thus, using previous lemma we are able to check self-duality for that class of formulas, as we show in the following theorem.

Theorem 3. *Problem SD is efficiently solvable for the class of $\mathcal{O}(\log n)$-depth LDT.*

Proof. We run the KM algorithm mentioned in Theorem 2. This algorithm outputs the support of f, which for the case of any $\mathcal{O}(\log n)$-depth decision tree with linear operations contains a polynomial number of sets corresponding to the non zero Fourier coefficients. We check whether one of those sets has even cardinality and answer "no" in case it happens or "yes" otherwise. The correctness follows from Lemma 1. □

3 Other polynomial time solvable classes

In this section we show how we can also solve efficiently problem SD for some other classes of formulas. We start by showing that testing self-duality is equivalent to zero test some other formula when f is known to be minor-dual (recall that this property can be tested in polynomial time).

Lemma 4. *Let f a minor-dual formula. Then, f is self-dual if and only if for all $x \in \{0,1\}^n$, $\bar{f}(x)\bar{f}(\bar{x}) = 0$.*

Proof. Suppose that f is self-dual. Then for all $x \in \{0,1\}^n$, $f(x) = \bar{f}(\bar{x})$ and therefore $\bar{f}(x)\bar{f}(\bar{x}) = 0$. For the other direction, suppose that for all $x \in \{0,1\}^n$ the claim holds but f is not self-dual, i.e there exists an assignment a such that $f(a) \neq \bar{f}(\bar{a})$. Since f is dual-minor, this assignment a has to satisfy $\bar{f}(a) = \bar{f}(\bar{a}) = 1$ and therefore $\bar{f}(a)\bar{f}(\bar{a}) = 1$. □

Thus, if we can zero test the formula $g(x) = \bar{f}(x)\bar{f}(\bar{x})$ (i.e. we can test whether g is logically equivalent to 0) we can check whether f is not self-dual. If g is not constant zero then it has a non empty DNF representation. Furthermore, suppose that the terms in the DNF representation of g have "short" length (they contain "few" variables). Intuitively, this means that a large number of assignments will satisfy some term and therefore, a uniformly chosen random assignment will satisfy g with high probability. Thus we have a randomized algorithm for testing whether g is constant zero (i.e. f is self-dual). This algorithm can be easily made deterministic by using small universal sets, which can be constructed efficiently. More formally, a (n,k)-universal set is a set $X \subseteq \{0,1\}^n$ such that every term of at most k literals is satisfied by at least one assignment in X. In [14], a method for efficiently constructing small (n,k)-universal sets was given. They proved that there exists an algorithm that, given any n and k, $1 \leq k \leq n$, it outputs a (n,k)-universal set of size $\mathcal{O}(k2^{3k}\log n)$ and runs in polynomial time in n and 2^k. We can use this construction together with the previous observation to prove the following.

Lemma 5. *Problem SD is polynomial time solvable for the class of $\mathcal{O}(\log n)$-CNF formulas.*

Proof. First we notice that since f belongs to $\mathcal{O}(\log n)$-CNF then $\bar{f}$ belongs to $\mathcal{O}(\log n)$-DNF since the functions in $\mathcal{O}(\log n)$-CNF are the negation of the functions in $\mathcal{O}(\log n)$-DNF. By Lemma 4, testing if f is self-dual is equivalent to test whether $g = \bar{f}(x)\bar{f}(\bar{x})$ is constant zero or not. Since $\bar{f}$ is a $\mathcal{O}(\log n)$-DNF so it is g. If there exists an assignment $a \in \{0,1\}^n$ such that $g(a) = 1$, then $\bar{f}(a) = \bar{f}(\bar{a}) = 1$ and $f(a) = f(\bar{a}) = 0$. Thus, we can construct a $(n, \log n)$-universal set X in polynomial time [14] and for all assignments in $b \in X$ we check whether $f(b) = f(\bar{b}) = 0$. If g is not zero, then there exist an assignment that satisfy one of its terms of its DNF representation and therefore this assignment has to be in X by definition of X and the fact that g has a representation as a $\mathcal{O}(\log n)$-DNF. Notice that we do not need to construct g, we can obtain the values of g using f and the proof still holds. □

Symmetrically, we can show that if f is a $\mathcal{O}(\log n)$-DNF we can also solve self-duality. The analogous of Lemma 4 will be that f is self-dual if and only if $f(x)f(\bar{x}) = 0$ for any $x \in \{0,1\}$. Thus, the same argument of Lemma 5 proves that we can zero test the class of $\mathcal{O}(\log n)$-DNF. This, together with Lemma 5 proves the following theorem.

Theorem 6. *Problem SD is efficiently solvable for the classes $\mathcal{O}(\log n)$-DNF and $\mathcal{O}(\log n)$-CNF of Boolean formulas.*

This result improves the result for monotone k-DNF formulas [4] since we extend the class to a non monotone one and we allow a non constant number of variables per term. We would like to remark that this result also implies that problem SD is efficiently solvable for the class of $\mathcal{O}(\log n)$-term DNF since this class is properly contained in the class $\mathcal{O}(\log n)$-CNF.

We now show that for another class, namely the class of read-k monotone DNF formulas, we can also test self-duality efficiently. We first need some definitions.

The *sensitive set* $S(x)$ of an assignment x is defined as follows:

$$S(x) = \{y | f(x) \neq f(y), d(x,y) = 1\}$$

where $d(x,y)$ is the Hamming distance of x and y. If $S(x) \neq \emptyset$ and $f(x) = 1$ then we say that x is a *positive sensitive assignment* of f. Let us first give a couple of useful properties of read-k formulas.

Proposition 7. *If x is a positive sensitive assignment of read-k monotone DNF formula f then x simultaneously satisfies at most k terms of f.*

Proof. Let x be a positive sensitive assignment that satisfies $k+1$ terms of a read-k DNF formula f and let $y \in S(x)$ such that y just differs from x in the i-th bit. Since flipping the i-th bit of x from 1 to 0 the formula is not satisfied, x_i must appear in the $k+1$ terms of f that x satisfied. Thus, f is not a read-k formula. □

Proposition 8. *Let f be a read-k monotone DNF formula and let g be a monotone CNF formula. Assume that f is not equivalent to g and there exists an assignment x such that $f(x) = 0$ and $g(x) = 1$. Then, there exists another assignment y such that $x \leq y$, $f(y) = 0$, $g(y) = 1$ and y belongs to the sensitive set of a positive sensitive assignment of f.*

Proof. Let x be an assignment such that $f(x) = 0$ and $g(x) = 1$. We flip 0 bits of x while keeping $f(x) = 0$ until we obtain y such that $x \leq y$ and flipping one bit more of y satisfies f. By monotonicity, g is still satisfied by y and y belongs to the sensitive set of a positive assignment for f. □

In [2], Bioch and Ibaraki have observed that testing self-duality for a monotone DNF formula f is equivalent to check whether a f is equivalent to another monotone CNF formula g that can be easily obtained from f. We can use this observation together with the previous properties about read k monotone DNF formulas to prove the following.

Theorem 9. *The problem MSD is efficiently solvable for the class of read k monotone DNF formulas.*

Proof. We prove that we can test whether a read k monotone DNF formula f is equivalent to some monotone CNF formula g and the theorem follows from the observation of Bioch and Ibaraki [2].

Since f is in DNF and g in CNF we can efficiently check whether there exists an assignment x such that $f(x) = 1$ and $g(x) = 0$. If we find such an assignment we are done. Otherwise, if f and g are still not equivalent, there should be an assignment x such that $f(x) = 0$ and $g(x) = 1$. By Proposition 8 there exists another assignment y that belongs to the sensitive set of a positive sensitive assignment and is still a valid witness of non equivalence between f and g. Moreover, Proposition 7 tell us that all the positive sensitive assignment satisfy at most k terms. Therefore, we check the sensitive sets of all the possible positive sensitive assignments of f. The number of sensitive assignments of f is smaller than n^{k+1} which is polynomial for constant k. Thus, either we find an assignment y which proofs that f is not equivalent to g or, by Proposition 8 it does not exist and f is equivalent to g. □

This result improves the result for read-once monotone DNF that appears in [1] in the terminology of learning theory. The same properties of read-k monotone DNF formulas that we have used here have also been used in the terminology of learning theory in [3].

4 Conclusions

In this paper we have studied the problem of deciding whether a formula is self-dual or not. This problem is known to be related to several other problems in different areas of research (see [2] for further references). The problem is NP-Complete when the input formula is a DNF and its complexity is not known when the input formula is a monotone DNF. Therefore, we have argued that it is important to study this problem for subclasses of DNF formulas in order to determine under which conditions the problem is solvable efficiently. Thus, we have provided polynomial time algorithms for several classes of formulas like $\mathcal{O}(\log n)$-depth LDT, $\mathcal{O}(\log n)$-DNF, $\mathcal{O}(\log n)$-term DNF or read-k monotone DNF. These results greatly enlarge the classes of formulas for which the self-duality problem can be efficiently solved.

It will be very interesting to determine how far can we go on finding polynomial time algorithms for subclasses of DNF formulas. We have shown that restricting the number of terms or the number of variables per term to be logarithmic makes the problem easy. However the problem for a general DNF is NP-Complete and thus it will be very interesting to determine under which restrictions the problem remains NP-Complete and under which conditions becomes polynomial time solvable.

For the case where the input is a monotone DNF the problem is different. The algorithm of Fredman and Khachiyan [5] works in polynomial time for monotone $\mathcal{O}(2^{\sqrt{\log n}})$-term DNF and in quasi polynomial time for unrestricted monotone DNF. We have provide some new classes of monotone DNF formulas where the problem is efficiently solvable and some more were also already known [4, 1, 12]. However, it is still an open problem to clarify the complexity of problem MSD.

5 Acknowledgments

This work was done while the author was at the Department of Computer Science of the Tokyo Institute of Technology. I would like to thank Ricard Gavaldà and Osamu Watanabe for several interesting comments while working on the topic.

References

1. D. Angluin, L. Hellerstein, and M. Karpinski. Learning read-once formulas with queries. *J. ACM*, 40:185–210, 1993.
2. J.C. Bioch and T. Ibaraki. Complexity of identification and dualization of positive Boolean functions. *Information and Computation*, 123:50–63, 1995.
3. C. Domingo. Exact learning of subclasses of cdnf formulas with membership queries. In *Proc. of the 2nd International Conference on Computing and Combinatorics, COCOON*, Lecture Notes in Computer Science, 1090:179–188, Springer-Verlag, 1996.
4. T. Eiter and G. Gottlob. Identifying the minimal transversals of a hypergraph and related problems. Technical Report CD-TR 91/16, Christial Doppler Labor Für Expertensysteme, Technische Universität Wien, 1991.
5. M. Fredman and L. Khachiyan. On the complexity of dualization of monotone disjunctive normal form. Technical Report LCSR-TR-225, Department of Computer Science, Rutgers University, May 1994.
6. H. Garcia-Molina and D. Barbara. How to assign votes in a distributed system. *Journal of the Association for Computing Machinery*, 32:841–860, 1985.
7. T. Ibaraki and T. Kameda. A theory of coteries: Mutual exclusion in distributed systems. *IEEE Trans. on Parallel and Distributed Systems*, pages 779–794, 1993.
8. J. Kahn, G. Kalai, and N. Linial. The influence of variables on Boolean functions. In *Proc. 28th Annu. IEEE Sympos. Found. Comput. Sci.*, pages 68–80. IEEE Computer Society Press, Los Alamitos, CA, 1988.
9. E. Kushilevitz and Y. Mansour. Learning decision trees using the fourier spectrum. *SIAM J. Comput.*, 22(6):1331–1348, 1993. Earlier version appeared in STOC 1991.
10. N. Linial, Y. Mansour, and N. Nisan. Constant depth circuits, Fourier transform and learnability. *Journal of the Association for Computing Machinery*, 40(3):607–620, 1993.

11. K. Makino and T. Ibaraki. A fast and simple algorithm for identifying 2-monotonic positive Boolean functions. Technical Report COMP94-46, IEICE, 1994.
12. K. Makino and T. Ibaraki. The maximum latency and identification of positive Boolean functions. In *Proceedings of ISAAC'94, Algorithms and Computation*, volume 834 of *Lecture Notes in Computer Science*, pages 324–332, 1994.
13. H. Mannila and K.J. Raïhä. Design by example: An application of Armstrong relations. *Journal of Computer and System Sciences*, 22:126–141, 1987.
14. J. Naor and M. Naor. Small-bias, probability spaces: Efficient constructions and applications. *SIAM Journal on Computing*, 22:838–856, 1993.

A Note on Updating Suffix Tree Labels*

Paolo Ferragina[1] and Roberto Grossi[2] and Manuela Montangero[1]

[1] Dipartimento di Informatica, Università di Pisa, Italy
[2] Dipartimento di Sistemi e Informatica, Università di Firenze, Italy

Abstract. We investigate the problem of maintaining the arc labels in the suffix tree data structure [15] when it undergoes string insertions and deletions. In current literature, this problem is solved either by a simple accounting strategy to obtain amortized bounds [10, 18] or by a periodical suffix tree reconstruction to obtain worst-case bounds (according to the global rebuilding technique in [20]). Unfortunately, the former approach is simple and space-efficient at the cost of attaining amortized bounds for the single update; the latter is space-consuming in practice because it needs to keep two extra suffix tree copies. In this paper, we obtain a surprisingly simple real-time algorithm that achieves worst-case bounds and only requires small additional space (i.e., a bi-directional pointer per suffix tree arc). We analyze the problem by introducing a combinatorial coloring problem on the suffix tree arcs.

1 Introduction

Suffix trees [19, 23] and their simple variations, such as compacted tries, are the most important and widely used data structures in many string problems. Their success is mainly due to the ability of encoding all the suffixes of a given string in linear space while allowing the efficient retrieval of a great deal of information about the input string. Weiner [23] and subsequently McCreight [19] were the first ones to use the suffix tree to design an optimal solution to the classical string matching problem, in which we have to determine all the occurrences of a pattern string in the form of substrings of a larger text string. Suffix trees were then employed to solve numerous and fundamental problems that sparked the attention of string matching community in the last two decades (e.g., see [2, 4, 7, 16]). Suffix trees were also applied to some other indexing problems to obtain the *Psuffix tree* [5] and the *Lsuffix tree* [13] (see also [6]).

The aforementioned problems require a *static* suffix tree in the sense that it is not modified after its construction. However, situations occur in which the suffix tree is built on a string that is not fixed a priori and thus can be changed over the time.

In some *text retrieval systems* and some common text editors, such as Gnu Emacs, the incremental text changes are interleaved with on-line pattern queries. Here, it is not reasonable to rebuild the suffix tree after each text update in order

* Supported in part by MURST of Italy. E-mail: ferragin@di.unipi.it, grossi@dsi2.ing.unifi.it

to use the optimal query solutions of [19, 23]. McCreight [19] tried to deal with this dynamic situation by proposing a method for performing some incremental suffix tree changes in response to a text change. However, as the author noted, an update can cost like the whole suffix tree reconstruction in the worst case. For this reason, some more efficient solutions were presented in [9, 14, 21] and most of them used the suffix tree generalization (sometimes called the *generalized* suffix tree [15]) to a dynamic string set.

In *dictionary matching*, a dictionary contains a set of pattern strings that can change over the time by inserting and deleting individual patterns. The user presents some text strings and search for all the occurrences of the dictionary patterns in them. Besides its theoretical importance, dictionary matching has many practical applications ranging from molecular biology [12] to digital libraries [11]. The problem was introduced and solved in [2] (see also [3, 21]). Basically, the technique in [2, 3] inserts a pattern into or deletes it from the dictionary by means of a generalized suffix tree built on the dictionary patterns.

Suffix trees are not only used in string problems but they find an application in *data compression*. In string substitution methods [22, 24, 25], such as the original Ziv and Lempel's scheme, the most time consuming task is string matching because we have to determine the text parts that occur earlier (called blocks) and then replace them by shorter references to their previous occurrences. In statistical data compression, such as the PPM methods [8], we have to support the operation of finding a *context*, which consists of some (maybe long) strings. Suffix trees can help in these tasks; however, when compressing large input data, they make considerable use of computational and space resources. For this reason, the compression operates only on a text part (called *window*) which is then shifted rightward thus making the indexing problem dynamic and requiring the update of suffix trees [10].

At this point, it is clear that updating suffix trees is a common problem in several applications, so that proper solutions have been devised so far. In this paper, we examine the problem of maintaining the arc labels in a (generalized) suffix tree built on a dynamic string set. Fiala and Green [10] and later Larsson [18] studied this problem and both gave amortized solutions. Here, we propose a surprisingly simple algorithm that obtains worst-case bounds and requires small additional space.

2 The Suffix Tree Definition

Given a string $X[1,m]$, we use $X[i,j]$ to denote its substring made up of its characters in positions $i, i+1, \ldots, j$ (where $1 \leq i \leq j \leq m$) and call $X[1,i]$ a prefix and $X[j,m]$ a suffix. We assume that its last character $X[m] = \$$ is an endmarker that guarantees that any two suffixes cannot be each other's prefix.

The *suffix tree* [19, 23] for $X[1,m]$, denoted by ST_X, is a compacted trie that stores X's suffixes in its leaves: Each arc is *labeled by a pointer triple* (X,i,j) that represents the substring $X[i,j]$. Any two sibling arcs have labels whose corresponding strings begin with different characters. Each internal node v has at

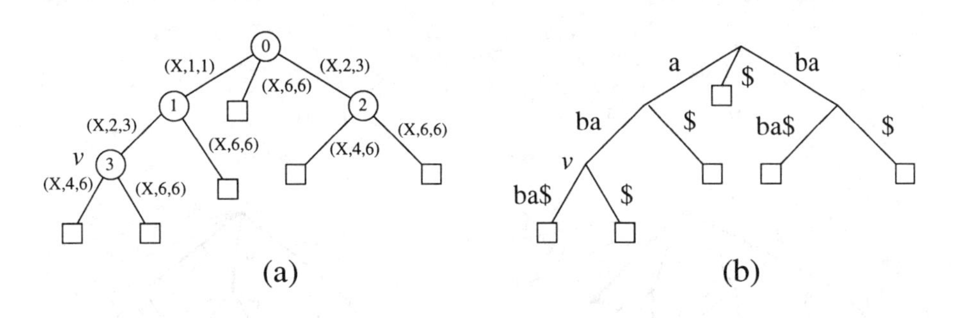

Fig. 1. (a) The suffix tree for string X = 'ababa\$'. We have: $W(v)$ = 'aba' and v stores its length 3. For our convenience, we illustrate in (b) the suffix tree showed in (a) by explicitly writing down the string $X[i,j]$ represented by triple (X,i,j).

least two children and corresponds to the string $W(v)$ obtained by concatenating the strings represented by the arc labels found along the downward path leading to v. For this reason, v stores the integer $|W(v)|$. An example of a suffix tree is shown in Figure 1.

It is worth noting that labeling the arcs by pointer triples is crucial to reduce the suffix tree space occupancy from $O(m^2)$ to $O(m)$. We use the following property for retrieving the string $W(v)$ by simply accessing v's parent $p(v)$:

Property 1 If (X,i,j) labels an arc $(p(v),v)$ and ℓ is the integer $|W(p(v))|$ stored in $p(v)$, then $W(v) = X[i-\ell,j]$.

This property allows us to prove the following result:

Lemma 1. *For any two suffix tree arcs, such that one is the other's ancestor, we can label the shallowest of the two arcs by using the other arc's label in* $O(1)$ *time.*

The suffix tree can be built incrementally in $O(m)$ time for a bounded alphabet [19]. Given $i > 1$ (the base case $i = 1$ is trivial), let us assume that we have built the compacted trie for the suffixes $X[j,m]$ inductively (with $j = 1,2,\ldots,i-1$) and we want to install $X[i,m]$. Let lp_i be $X[i,m]$'s longest prefix that occurs in this compacted trie: lp_i is a prefix of $W(u')$ and $W(p(u'))$ is a proper prefix of lp_i, for a (unique) node u' in the compacted trie. The insertion of $X[i,m]$ requires the efficient retrieval of lp_i and u' to create the leaf storing $X[i,m]$ (see [19]). If $W(u') = lp_i$, then we insert this leaf as a child of u'; otherwise, we split arc $(p(u'),u')$ by inserting a new node u, such that $W(u) = lp_i$, and by installing the new leaf as u's child. When $i = m$, the resulting compacted trie is the suffix tree for X.

We can generalize the suffix tree to a string set $\Delta = \{X_1,\ldots,X_k\}$ and obtain the so-called *generalized suffix tree* [15], denoted by ST_Δ, which is the

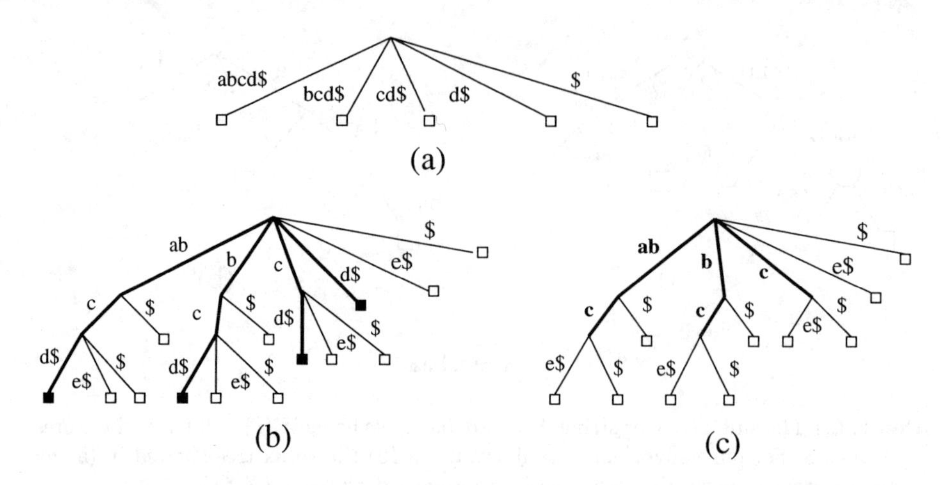

Fig. 2. (a) The suffix tree for the string $X = abcd\$$. (b) The suffix tree after the insertion of strings $ab\$$, $abc\$$, $abce\$$. The black leaves denote X's suffixes and the black paths spell out them. These paths are labeled by $\Theta(|X|^2)$ triples referring to X. In (c), string X is deleted and the $\Theta(|X|^2)$ bold labels need to be kept consistent.

compacted trie obtained by "superimposing" the suffix trees ST_X incrementally for all $X \in \Delta$. Two arcs are partially superimposed whenever their labels have a common prefix. Since ST_Δ is isomorphic to the suffix tree built on the string $D = X_1\$ \cdots \$X_k\$$ (where endmarker \$ does not match itself), Property 1 and Lemma 1 still hold. We can use the algorithms in [15] to modify string set Δ as follows:

(1) We can insert a new string X into Δ by updating ST_Δ in $O(|X|)$ time (for a bounded alphabet). We transform suffix tree ST_Δ into suffix tree $ST_{\Delta \cup \{X\}}$ by appending string $X\$$ to D (i.e., $D := DX\$$) and by installing a leaf for each suffix $X[i,m]\$$ ($1 \leq i \leq m$) as explained above. See Figure 2.b.

(2) We can delete a string X from Δ in $O(|X|)$ time (for a bounded alphabet). Specifically, we delete all the leaves that correspond to the suffixes of $X\$$ (they are only $|X|$ in number, see Figure 2.c). We also have to delete some of their parent nodes if they become *unary* (i.e., with only a single child).

3 The Suffix Tree Label Updating Problem

From the previous discussion, it follows that we can identify *three* basic operations that constitute the key steps in updating a suffix tree:

- Ins-Leaf-Node(f,u) : We insert leaf f into the suffix tree as a child of node u. We have to choose the proper label for the arc (u, f).
- Ins-Leaf-Arc(f,e) : We split arc $e = (p(u), u)$ by inserting a new node v and install leaf f as v's child. We have to choose the proper label for the three new arcs: $(p(u), v)$, (v, u) and (v, f).

– DEL-LEAF(f) : We remove leaf f from the suffix tree. If its parent $p(f)$ becomes unary, we remove $p(f)$ and install its child, say u, as $p(p(f))$'s child. We have to choose the proper label for the arc $(p(p(f)), u)$.

Although we can use the three basic primitives above to modify the suffix tree topology, we have to deal with the problem of keeping the arc labels consistent. We indeed say that a label (X, i, j) is *consistent* if and only if it refers to a string X *currently* in Δ. We are now ready to state:

Problem 2 (Label Update Problem). Given a (generalized) suffix tree ST_Δ, we want to keep its arc labels *consistent* under string insertions and deletions in Δ (i.e., operations INS-LEAF-NODE, INS-LEAF-ARC and DEL-LEAF).

Inserting a new string into Δ and keeping the suffix tree labels consistent is not much of a problem. On the contrary, each time we remove a string X from Δ, we run into the problem of having to retrieve all the arcs previously labeled by X's substrings and *relabeling* these arcs by consistent labels. We describe below three possible approaches to solve Problem 2:

(1) A straightforward approach consists of keeping, for each string $X \in \Delta$, the list of the suffix tree arcs labeled by X's substrings. Each time a string X is deleted from Δ, we have to delete $|X|$ leaves and some of their parents (if they become unary, see above). We then relabel all the arcs in X's list consistently, that is, we relabel each arc involved, in constant time, by using Lemma 1 on one of its descending arcs. This can be very expensive in the worst case because X's list can grow arbitrarily up to $\Theta(|X|^2)$ size after that a few string insertions break the arcs leading to the leaves storing X's suffixes. See Figure 2 for an example.

(2) Fiala and Green's method [10] allows us to keep the arc labels consistent *on-line* in $O(1)$ amortized time per updated leaf. They show that we can assign some credits to the arcs and to the three basic update operations above in order to guarantee that each arc has at least one credit available each time we have to traverse it when relabeling some arcs. This solution is simple and practical to implement but does not make us sure to achieve $O(1)$ worst-case bounds per updated leaf (see [18] for an alternative amortized solution).

(3) A worst-case solution can be obtained by using Overmars' *global rebuilding technique* [20]. We delete a string only *logically* and keep two suffix tree *carbon copies* to limit the garbage space: One copy becomes active when the total size of both the suffix tree and the string set is halved, the other when the total size is doubled. This approach introduces some non consistent labels but guarantees that their number is not too large by switching to the proper carbon copy according to the string set size. This avoids the time consuming task of updating the non consistent labels. Unfortunately, this technique wastes space and its implementation is not straightforward because of the schedule of the update operations on the carbon copies. Although its asymptotical performance is optimal in the worst case, it is unreasonable in practical applications and does not guarantee that the labels are updated in real time.

We now describe an optimal worst-case and space efficient method to update suffix tree labels under string insertions and deletions. Our solution turns out to be very simple and real-time, and this is fundamental in McCreight's suffix tree update [19] and data compression [10] algorithms. We only add an extra bi-directional pointer per arc label, so that no more than one arc has to be relabeled after inserting or deleting a leaf.

4 A Dynamic Arc Coloring Problem

We abstract the problem of updating the arc labels in Problem 2 by introducing a combinatorial problem on its arcs. We use the following terminology: given a tree, all the arcs leading to its leaves are called *external* while the remaining arcs are called *internal*.

Problem 3 (Arc Coloring Problem). Given a rooted tree T of n nodes (all of them are non unary) and a color set $C = \{1, 2, \ldots, |C|\}$ (where $|C| \leq n$), let us assume that T's arcs are colored by C's colors according to the following conditions:

1. For every color $c \in C$, the number of *internal* arcs colored c does not exceed the number of *external* arcs colored c.
2. For every *internal* arc colored $c \in C$, there is at least one descending *external* arc colored c.

We wish to maintain the above two conditions under the following tree update operations:

- INS-LEAF-NODE(f,u,c) : We insert leaf f as a child of node u and color arc (u, f) by c. Node u has at least two children before the operation.
- INS-LEAF-ARC(f,e,c) : We split arc $e = (p(v), v)$ by inserting a new node w and install new leaf f as w's child. We color (w, f) by c.
- DEL-LEAF(f) : We remove leaf f from the suffix tree. If $p(f)$ becomes unary (and v is its only child), we remove $p(f)$ and install v as $p(p(f))$'s child.

We can satisfy Condition (1) alone by counting the number of arcs of the same color. Whenever an arc needs its recoloring, we choose a color among the ones for which the number of internal arcs is strictly smaller than the number of (same colored) external ones. Conversely, Condition (2) makes Problem 3 nontrivial. We point out that Problem 3 translates into Problem 2 smoothly. Specifically, we set T as the suffix tree ST_Δ and we assign each color to a different string in Δ. Color c is assigned to an arc if its label refers to the string colored c. Condition (1) makes us sure that each string deletion needs to relabel a linear number of arcs in ST_Δ. Condition (2) makes us sure that *an arc recoloring can be translated into a suffix tree arc relabeling* by Lemma 1. INS-LEAF-NODE, INS-LEAF-ARC and DEL-LEAF operations on T immediately relate to the three basic primitives operations on ST_Δ. Given this simple transformation, we can prove the following result:

Theorem 4. *Problem 3 can be solved by taking $O(1)$ worst-case time per operation* INS-LEAF-NODE, INS-LEAF-ARC *and* DEL-LEAF.

Theorem 4 states that for each tree update operation we need to change color to $O(1)$ arcs. Therefore, we can use the transformation above and recolor (i.e., relabel) a suffix tree arc in $O(1)$ time by Lemma 1. As a result, INS-LEAF-NODE, INS-LEAF-ARC and DEL-LEAF take $O(1)$ time each in the suffix tree and thus we can state:

Theorem 5. *The suffix tree labels can be maintained consistent in Problem 2 under the insertion and deletion of an m-length string in $O(m)$ worst-case time.*

4.1 The Binary Tree Case

We begin by examining the simple case of a binary tree T without unary nodes. Since each node has either two or zero children, we cannot use the INS-LEAF-NODE operation because it requires a node with at least two children. Hence, inserting a new leaf into T always involves an arc splitting, while deleting a leaf from T always causes its parent to get unary and, thus, requires its removal. Consequently, we only have to implement INS-LEAF-ARC and DEL-LEAF in Problem 3 and satisfy Conditions (1) and (2). Specifically, we augment the suffix tree by linking pairs of arcs according to the following invariant:

- Each external arc can be linked to *no more* than one of its ancestor (internal) arcs and, in this case, both of them have the same color.
- Each internal arc colored c is linked to exactly one of its descending external arcs colored c, for any color c.

The invariant above implies Conditions (1) and (2). Furthermore, it guarantees that some arcs are linked in pairs, formed by an internal and an external arc. We assume that each link is bi-directional and point out that the only possible nonlinked arcs are external.

While operation INS-LEAF-ARC can be easily implemented to maintain the invariant, operation DEL-LEAF can require to recolor the internal arc linked to the external arc removed. Our algorithm takes advantage of the *binary* tree topology, which makes us sure that an external arc removal determines also the removal of an internal arc. This in turn makes one color available and we use it to recolor another internal arc (i.e., the one linked to the the external one removed). We implement the two basic primitives as follows:

INS-LEAF-ARC(f,e,c): We create a node w and split arc $e = (u, v)$ into two arcs $e_1 = (u, w)$ and $e_2 = (w, v)$. We then create arc $e_3 = (w, f)$ colored c. We are left with the problem of choosing e_1's and e_2's colors in order to preserve the invariant above. Our strategy is to assign e_3's color to e_1 and e's color to e_2. We also link (internal arc) e_1 and (external arc) e_3 together. We then replace e by e_2 in its linked list.

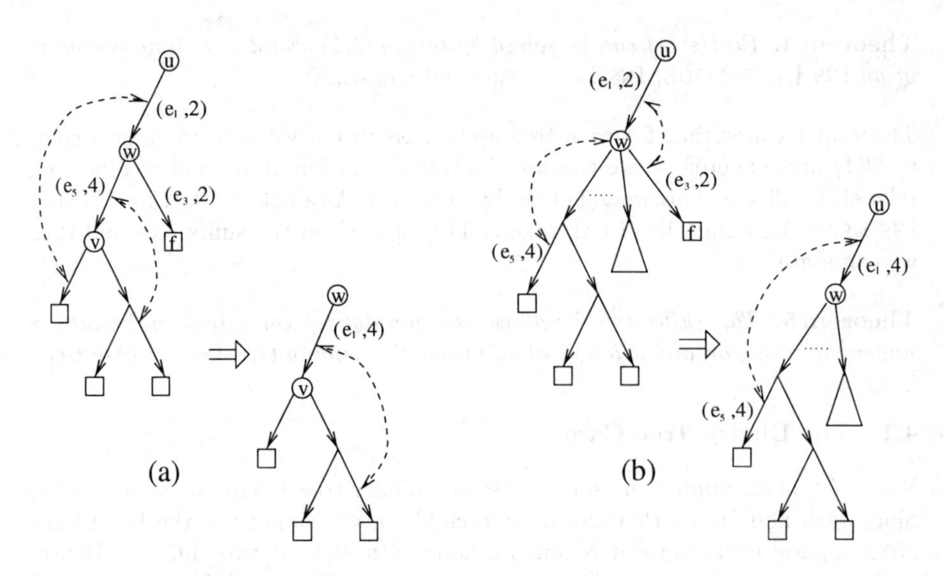

Fig. 3. DEL-LEAF in a binary tree (a) and in a general tree (b). Pair (e', c) indicates that arc e' has color c. Dashed (double) arrows represent the bi-directional links between arc pairs.

DEL-LEAF(f): We delete leaf f and its parent w (since it becomes unary). We then merge arcs $e_1 = (u, w)$ and $e_2 = (w, v)$ together into arc $e = (u, v)$ and remove arc $e_3 = (w, f)$. We have to describe how to color e and maintain the invariant. Three cases follow according to arc e_3:

Case 1 : The external arc e_3 is not linked to any other internal arc. In this case, the removal of arc e_3 from T cannot affect any internal arcs. We therefore assign e_2's color to e and replace e_2 by e in its linked list. The external arc linked to e_1 is no longer linked. See Figure 3.a.

Case 2 : The external arc e_3 is linked to arc e_1 (which is incident to w). Since e_1 is removed too, this case is similar to Case (1) and thus we go on in an identical way.

Case 3 : The external arc e_3 is linked to one of its (ancestor) arcs, say $e_4 \neq e_1$. We assign e_1's color to e_4 and replace e_1 by e_4 in its linked list. We then assign e_2's color to (merged arc) e and replace e_2 by e in its linked list.

The correctness of our approach readily follows from the invariant, which can be proved to hold inductively by the above case analysis. Since each case takes $O(1)$ time to be managed, we obtain:

Lemma 6. *Given a binary tree T, Problem 3 can be solved by taking $O(1)$ worst-case time per operation* INS-LEAF-NODE, INS-LEAF-ARC *and* DEL-LEAF.

4.2 The General Case

In the case of a general tree T without unary nodes, a node can have more than two children and the approach shown in Section 4.1 is no longer applicable. We might think to transform an arbitrary tree into a binary tree by implementing the children list of a node by means of a *balanced* binary tree. Although this is a well-known and simple approach, it is not clear how to color the arcs created with this transformation and how to preserve Conditions (1) and (2) in $O(1)$ time under the tree rebalancing of the children lists. We propose a solution that works for an arbitrary tree T without using any transformation.

As previously mentioned, when we delete an external arc in the binary tree case, we surely remove an internal arc and reuse its color if necessary. On the contrary, we run into a difficulty for an arbitrary tree because an external arc removal does not necessarily imply the removal of any internal arcs. For example, let us examine the *pathological* case in which we delete a leaf f whose parent w has at least three children. When we delete f, we do not need to delete w and thus we cannot go on as in Section 4.1. In particular, if the removed external arc (w, f) is linked to an internal (ancestor) arc, we have to recolor the internal arc but we do not have any colors available. We deal with this more demanding situation by introducing another invariant:

- Each external arc is either: (i) linked to one of its ancestor (internal) arcs; (ii) linked to one of its ancestor nodes; or (iii) it is not linked at all. In case (i), the two linked arcs have the same color; in case (ii), we say that the node is the *target* of the external arc.
- Each internal arc colored c is linked to exactly one of its descending external arcs colored c, for any color c.
- If an internal node has $d > 2$ children, then it is the target of $d - 2$ external arcs (which form its *target list*).

This invariant is stronger than the one introduced for the binary case and thus implies Conditions (1) and (2). We assume that each link is bi-directional. The invariant makes us sure that the external arcs in the target list of a node can be used to assign colors when dealing with the pathological case. We implement the three basic primitives as follows:

INS-LEAF-NODE(f,u,c): We create the external arc $e = (u, f)$ and color it c. Since node u has now more than two children, we link the (external) arc e to the target node u. This preserves the invariant since u has one more child and so its target list size increases by one.

INS-LEAF-ARC(f,e,c): We proceed like the binary tree case by creating a new node w which is used to split arc $e = (u, v)$. Specifically, we assign the same color c to both $e_1 = (u, w)$ and $e_3 = (w, f)$, and link them together. It is worth noting that w is not a target node because it has two children (i.e., v and f). Moreover, if e is linked to an arc or a target node, we replace e by $e_2 = (w, v)$ in this link.

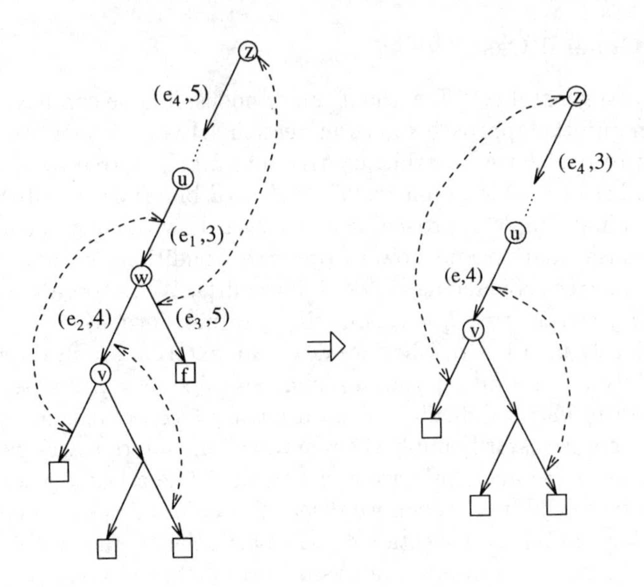

Fig. 4. Case 1.3 for DEL-LEAF in a general tree.

DEL-LEAF(f): Let w be f's parent and e_3 be arc (w, f). We have to deal properly with the target nodes, so that we distinguish among three main cases depending on the number of w's children.

Case 1 : w has two children: one is leaf f and the other is a node v (we let $e_2 = (w, v)$ and $e_1 = (u, w)$, where u is w's parent). By the invariant, w is not the target of any external arcs and so we go on as in the binary tree case, except that we have to deal with a stronger invariant. We change T's topology by removing arcs e_1, e_2 and e_3 and by inserting arc $e = (u, v)$. We then assign e_2's color to e. Moreover, if e_2 is linked to an arc or to a target node, we replace e_2 by e in this link. We still have to deal with the case where e_3 is linked to another arc or to a target node. We distinguish among the following three subcases:

1.1 If e_3 is linked to e_1 or it is not linked to any other arc, then we proceed as in the binary tree case.
1.2 If e_3 is linked to an ancestor arc $e_4 \neq e_1$, we assign e_1's color to e_4; and we make e_1's linked external arc point to e_4 (as in the binary tree case).
1.3 If e_3 is linked to an ancestor target node z (here, $z \neq w$ because w cannot be a target node), we replace e_3 in z's target list by the external arc linked to e_1. Since we remove e_1, its linked external arc is no longer linked to any other arc or target node and thus we can use it to replace e_3 (see Figure 4).

Case 2 : w has more than two children: one of them is leaf f and the external arc $e_3 = (w, f)$ is in w's target list. We remove arc e_3 from w's target list.

(We preserve the invariant because w has one less child and its target list size decreases by one.)

Case 3 : w has more than two children: one of them is leaf f and the external arc $e_3 = (w, f)$ is *not* in w's target list. We remove an arbitrary arc, say e_5, from w's target list in constant time and use it to manage the arc or the target node possibly linked to e_3. (We preserve the invariant because the number of w's children and w's target list size decrease by one.) We distinguish among three subcases on e_3:

3.1 If e_3 is linked to an ancestor arc, say e_4 (maybe, $e_1 = e_4$), we use e_5's color to recolor (internal) arc e_4 and we link e_4 and e_5 together. See Figure 3.b, where $e_1 = e_4$.

3.2 If e_3 is linked to an ancestor target node z (with $z \neq w$), then we make e_5's new target node be z and replace e_3 by e_5 in z's target list.

3.3 If e_3 is *not* linked to any ancestor arc or node, then we are done. (We already removed arc e_5 from w's target list to decrease its size by one.) It is worth noting that e_5 now is not linked to anything else.

The correctness of our approach readily follows from the invariant and our case analysis above. Since each case takes $O(1)$ time to be managed, we can implement INS-LEAF-NODE, INS-LEAF-ARC and DEL-LEAF in constant time thus proving Theorem 4. At this point, we can use the transformation described in Section 3 and implement suffix tree operations: INS-LEAF-NODE, INS-LEAF-ARC and DEL-LEAF in constant time, thus proving our main Theorem 5.

References

1. AHO, A. V., AND CORASICK, M. J. Efficient string matching: an aid to bibliographic search. *Communication of the ACM 18* (1975), 333–340.
2. AMIR, A., AND FARACH, M. Adaptive dictionary matching. In *IEEE Symposium on Foundations of Computer Science* (1991), pp. 760–766.
3. AMIR, A., FARACH, M., IDURY, R. M., POUTRÉ, H. L., AND SCHÄFFER, A. A. Improved dictionary matching. *Information and Computation 119* (1995), 258–282.
4. APOSTOLICO, A., AND PREPARATA, F. Optimal off-line detection of repetitions in a string. *Theoretical Computer Science 22* (1983), 297–315.
5. BAKER, B. S. A theory of parameterized pattern matching: Algorithms and applications. In *ACM Symposium on Theory of Computing* (1993), pp. 71–80.
6. BRESLAUER, D. The suffix tree of a tree and minimizing sequential transducers. In *Combinatorial Pattern Matching* (1996).
7. CHANG, W. I., AND LAWLER, E. L. Sublinear approximate string matching and biological applications. *Algorithmica 12* (1994), 327–344.
8. CLEARY, J. G., TEEHAN, W. J., AND WITTEN, I. H. Unbounded length contexts for PPM. In *IEEE Data Compression Conference* (1995), pp. 52–61.
9. FERRAGINA, P., AND GROSSI, R. Optimal on-line search and sublinear time update in string matching. In *IEEE Symposium on Foundations of Computer Science*, 604–612, 1995. Also *SIAM Journal on Computing* (to appear).

10. Fiala, E. R., and Green, D. H. Data compression with finite window. *Communication of the ACM 32*, 4 (1989), 490–505.
11. Fox, A. E., et al. (eds.) Special Issue on "Digital Libraries" *Comm. ACM*, 38:4 (1995).
12. Frenkel, K. A. The human genome project and informatics. *Communication of the ACM 34* (1991), 41-51.
13. Giancarlo, R. A generalization of the suffix tree to square matrices, with applications. *SIAM Journal on Computing 24* (1995), 520–562.
14. Gu, M., Farach, M., and Beigel, R. An efficient algorithm for dynamic text indexing. In *ACM-SIAM Symposium on Discrete Algorithms* (1994), pp. 697–704.
15. Gusfield, D., Landau, G. M., and Schieber, B. An efficient algorithm for all pairs suffix-prefix problem. *Information Processing Letters 41* (1992), 181–185.
16. Kosaraju, S.R. Efficient tree pattern matching. In *IEEE Foundations of Computer Science* (1989), 178–183.
17. Landau, G. M., and Vishkin, U. Fast parallel and serial approximate string matching. *Journal of Algorithms 10* (1989), 157–169.
18. Larsson, N. J. Extended application of suffix trees to data compression. In *IEEE Data Compression Conference* (1996).
19. McCreight, E. M. A space-economical suffix tree construction algorithm. *Journal of the ACM 23*, 2 (1976), 262–272.
20. Overmars, M. H. *The design of Dynamic Data Structures.* Springer-Verlag Lecture Notes in Computer Science #156, 1983.
21. Sahinalp S. C. and Vishkin U. Efficient approximate and dynamic matching of patterns using a labeling paradigm. In *Proc. of IEEE Symposium on Foundations of Computer Science*, 1996.
22. Storer, J., and Szymanski, T. Data compression via textual substitution. *Journal of the ACM 29*, 4 (1982), 928–951.
23. Weiner, P. Linear pattern matching algorithm. In *IEEE Symp. on Switching and Automata Theory* (1973), pp. 1-11.
24. Ziv, J., and Lempel, A. A universal algorithm for sequential data compression. *IEEE Trans. Info. Theory 23*, 3 (1977), 337–343.
25. Ziv, J., and Lempel, A. Compression of individual sequences via variable-rate coding. *IEEE Trans. Info. Theory 24*, 5 (1978), 530–536.

Relaxed Balanced Red-Black Trees

S. Hanke[1], Th. Ottmann[2], and E. Soisalon-Soininen[3]

[1] Institut für Informatik, Universität Freiburg, Am Flughafen 17, D-79110 Freiburg
e-mail: hanke@informatik.uni-freiburg.de

[2] Institut für Informatik, Universität Freiburg, Am Flughafen 17, D-79110 Freiburg
e-mail: ottmann@informatik.uni-freiburg.de

[3] Laboratory of Information Processing Science, Helsinki University of Technology, Otakaari 1 A, FIN-02150 Espoo, Finland. e-mail: ess@cs.hut.fi

Abstract. *Relaxed* balancing means that, in a dictionary stored as a balanced tree, the necessary rebalancing after updates may be delayed. This is in contrast to *strict* balancing meaning that rebalancing is performed immediately after the update. Relaxed balancing is important for efficiency in highly dynamic applications where updates can occur in bursts. The rebalancing tasks can be performed gradually after all urgent updates, allowing the concurrent use of the dictionary even though the underlying tree structure is not completely in balance. In this paper we propose a new scheme of how to make known rebalancing techniques relaxed in an efficient way. The idea is applied to the red-black trees, but can be applied to any class of balanced trees. The key idea is to accumulate insertions and deletions such that they can be settled in arbitrary order using the same rebalancing operations as for standard balanced search trees. As a result it can be shown that the number of needed rebalancing operations known from the strict balancing scheme carry over to relaxed balancing.

1 Introduction

A *dictionary* is a scheme for storing a set of data such that the operations *search*, *insert*, and *delete* can be carried out efficiently. Standard implementations of dictionaries using balanced search trees like red-black trees, AVL-trees, half-balanced trees, and others presuppose that each update operation is followed by a sequence of rebalancing steps which restore the respective balance condition. Maintaining the balance conditions assures that the trees cannot degenerate into linear lists and search and update operations can be performed in a number of steps which is always logarithmic in the number of keys stored in a tree.

In a concurrent environment, however, uncoupling the updating (insertion and deletion) from the rebalancing transformations may increase the possible amount of concurrency and speed up updates considerably. This leads to the notion of *relaxed balance.* Instead of requiring that the balance condition is restored immediately after each update operation the actual rebalancing transformations can be delayed arbitrarily and interleaved freely with search and update operations.

Relaxed balancing was first suggested in [8] for red-black trees. The first actual solution, presented by Kessels [10], is for relaxed balancing in AVL-trees [1] where the allowed updates are only insertions. Nurmi et al. [14] extend the work of Kessels to the general case in which deletions are also allowed. In [11] the solution of [14] is analyzed, and it is shown that for each update operation in a tree with maximum size n, $O(\log n)$ new rebalancing operations are needed. Relaxed balanced B-trees are introduced in [14] and further analyzed in [12]. In [13] Nurmi and Soisalon-Soininen propose a relaxed version of red-black trees which they call a *chromatic tree*. Boyar and Larsen [5] analyze the proposal of [13] and show that after a minor modification the number of rebalancing operations per update is $O(\log(n+i))$, if i insertions are performed on a tree which initially contains n leaves. Boyar et al. [3] prove for a slightly modified set of rebalancing operations that only an amortized constant amount of rebalancing is necessary after an update in a chromatic tree. In [21] Soisalon-Soininen and Widmayer propose a relaxed version of AVL-trees which fulfills, despite the local nature of its operations, some global properties. For example, they show that in their solution no rotation will be performed if the underlying search tree happens to fulfill the AVL tree balance condition before any rebalancing has been done.

Except for the recent paper [21], all previous solutions are not wholly based on the standard balancing transformations but require a large number of different new transformations.

In this paper we propose a new technique of how to make known rebalancing algorithms relaxed in an efficient way. We show that essentially the same set of rebalancing transformations as used in the strict case can also be used for the relaxed case, and that the number of needed rebalancing operations known from the strict balancing scheme carry over to relaxed balancing.

In order to illustrate the key ideas and to clarify the ideas underlying our solution as much as possible, we restrict ourselves to the case of red-black trees. But we emphasize that the ideas of marking items for deletions, allowing trees to grow randomly below the balanced part, and to settle accumulated update and rebalancing requests in top-down manner using the standard rebalancing operations carry over to many other classes of search trees as well. The aim of our proposal is to extend the constant-linkage-cost update algorithm for red-black trees [19] in such a way that updates and local structural changes are uncoupled and may be arbitrarily delayed and interleaved. A key idea in our solution is the assumption that the deletion of a key in a tree leads to a *removal request* only; the actual removal of a leaf is considered to be a part of the structural change to restore the balance condition. In this way we put completely aside the problem of deletion which has complicated the previous solutions of relaxed balancing.

2 Red-black trees

The trees in this paper are leaf-oriented binary search trees, which are full binary trees (each node has either two or no children). The nullary nodes of a tree are called the external nodes or leaves while the other nodes are said to be internal nodes. We assume that the keys (chosen from a totally ordered universe) are stored in the leaves of the binary tree. The internal nodes contain routers, which guide the search from the root to a leaf.

For simplicity, we do not distinguish between search trees with stored items (in left-to-right order at the leaves) and their underlying graph structure.

A binary search tree is *balanced* if its height is bounded by $O(\log N)$, where N is the number of keys stored in the tree.

In the following we consider the variant of red-black trees proposed by Sarnak and Tarjan [19]. The nodes of the tree are coloured red or black, respectively. As the balance condition it is required that

1. each search path from the root to a leaf consists of the same number of black nodes,
2. each red node (except for the root) has a black parent, and
3. all leaves are black.

We give now a short review of the update algorithms for strictly balanced red-black trees. Then we will explain in Section 3 how to realize relaxed balancing by using the same rebalancing transformations.

The rebalancing transformations of red-black trees need only a constant number of structural changes (at most two rotations or a rotation plus a double rotation) to restore the balance condition after an update operation.

2.1 Insertions

In order to insert a new key x into a strictly balanced red-black tree we first locate its position among the leaves and replace the leaf by an internal red node with two black leaves. The two leaves now store the old key (where the search ended) and the new key x.

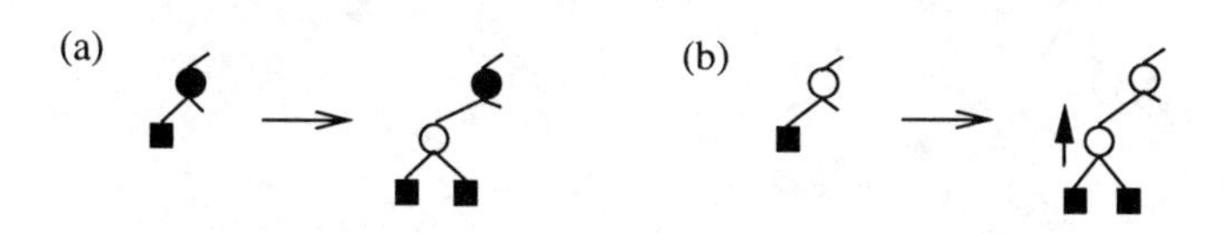

Fig. 1. Insertion of a new item and call of the rebalancing procedure *up-in* (denoted by "↑"). Filled nodes denote black nodes and nodes without fill denote red nodes.

If the parent of the new internal node p is red (as well as p itself) then the resulting tree is no longer a red-black tree. In order to correct this and to restore the balance condition we call the rebalancing procedure *up-in* for the new inner node p, cf. Figure 1.

Whenever the rebalancing procedure *up-in* is called for a node p then p is a red node. Note that if p's parent is red as well then the grandparent of p must be black (if it exists). The task of the rebalancing procedure is to achieve that p

obtains a black parent while the number of black nodes on any search path from the root to a leaf is not changed. For this the *up-in* procedure flips the colours of some nodes in the immediate vicinity above p and

1. either performs a structural change (at most one rotation or double rotation) involving a few nodes occuring in the immediate vicinity above p in order to restore the balance condition and stops, cf. Figure 2a–d,
2. or (exclusively) calls itself recursively for p's grandparent and performs no structural change at all, cf. Figure 2e.

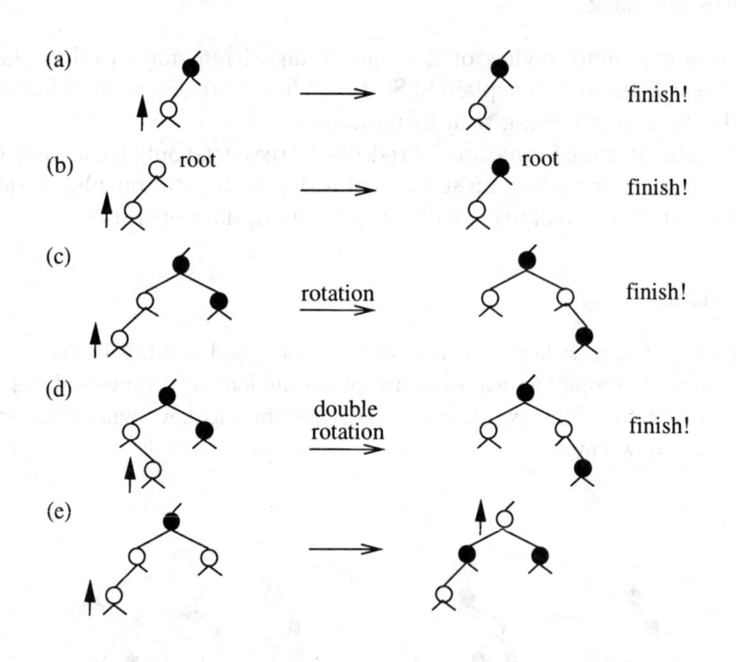

Fig. 2. Local structural changes by the rebalancing procedure *up-in*

For a node p with an up-in request Case 2 above applies if and only if p's parent and the sibling of p's parent are both red.

2.2 Deletions

In order to delete a key from a strictly balanced red-black tree, we first locate the leaf where the key is stored and then remove the leaf together with its parent. (Note that the balance condition implies that the remaining sibling of the leaf is either a leaf as well or a red node which has two leaves as its children.) If the removed parent was black then the red-black tree structure is now violated. It can be restored easily if the remaining node is red (change its colour). Otherwise, the removal leads to the call of the rebalancing procedure *up-out* for the remaining leaf, cf. Figure 3.

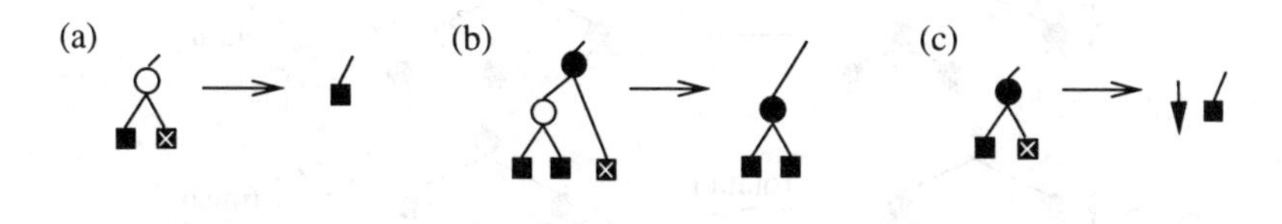

Fig. 3. Deletion of an item (marked by "×") and call of the rebalancing procedure *up-out* (denoted by "↓")

Whenever the rebalancing procedure *up-out* is called for a node p then p is a black node and each search path to a leaf of the subtree with root p has one black node too few. The task of the procedure *up-out* is to increase the black height of the subtree rooted at p by one. In order to achieve this the *up-out* procedure changes the colours of some nodes in the immediate vicinity beside and above p and

1. either performs a structural change (at most two rotations or a rotation plus a double rotation) involving a few nodes occuring in the immediate vicinity beside and above p in order to restore the balance condition and stops, cf. Figure 4a–d,
2. or (exclusively) calls itself recursively for p's parent and performs no structural change at all, cf. Figure 4e.

For a node p with an up-out request Case 2 above applies if and only if p's parent, p's sibling, and both children of p's sibling are all black.

3 Relaxed balancing

In this section we show how to realize relaxed balancing by using the same rebalancing transformations as in the case of strict balancing.

We first uncouple the real insert and delete operations (i.e. searching of a leaf and the insertion or removal of an item, respectively) from the rebalancing procedures *up-in* and *up-out*. Instead of calling the rebalancing procedure immediately after the update operation we only deposit an *up-in request* "↑" or an *up-out request* "↓", respectively, at the corresponding node as shown in Section 2 and do nothing.

In relaxed balancing we allow to interrupt the restructuring procedure after each single restructuring step. We also allow to carry out several of single restructuring steps concurrently at different locations in the tree. The only requirement will be that no two of them interfere at the same location. In order to

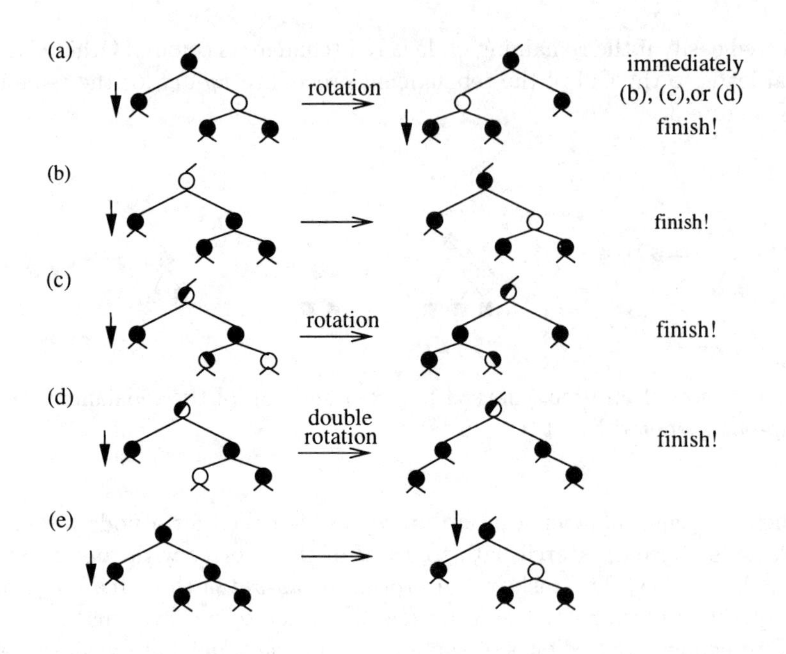

Fig. 4. Local structural changes by the rebalancing procedure *up-out*. Half filled nodes denote nodes that are either black or red.

make this more precise we first specify what we mean by a one-step restructuring at a node p with an up-in or an up-out request: To handle such a request means either to settle it by a local structural change and halt, as shown in Figure 2a–d (Figure 4a–d), or to shift the request to the grandparent (parent) and handle it at a later time, as shown in Figure 2e (Figure 4e).

Now we define a relaxed balanced red-black tree as follows.

A *relaxed balanced red-black tree* is a full binary tree with red and black nodes where the red nodes may have up-in requests, the black nodes may have up-out requests, and as the relaxed condition it is required that

1. For each search path from the root to a leaf, the sum of the number of black nodes plus the number of up-out requests is the same,
2. each red node (except for the root) has either a black parent or an up-in request, and
3. all leaves are black.

The idea here is that each one-step rebalancing operation keeps the tree as a relaxed balanced red-black tree, and gradually transforms it towards a red-black tree. As we will see in the case of accumulated insertions the relaxed balance condition can be guaranteed very easily. But in the case of accumulated deletions we run into many problems if we allow that arbitrary leaves can be removed together with their (possibly black) parents. (Recall that the rebalancing procedure *up-out* increases the black height of a subtree rooted at a black node with an up-out request only by one.)

This leads to the simple but crucial idea of our balancing scheme. Instead of removing a leaf together with its parent immediately after locating it among the leaves we only deposit a *removal request* "×" at that leaf. That is, the leaf is not physically removed but only marked to be deleted later after eventually conflicting up-in or up-out requests have bubbled up in the tree. The actual removal is considered to be part of the rebalancing transformation. To handle such a removal request is only allowed if thereby the relaxed balance condition is not violated. Therefore, we may have to handle or bubble up conflicting requests first, before we can actually remove the leaf with the removal request.

It should now be obvious how to settle a removal request. If the removal request applies to a leaf which has a red parent or a red sibling it can easily be settled (remove the leaf together with its parent and, if necessary, colour the sibling black). Otherwise, the removal of the leaf leads to an up-out request. This is as depicted in Figure 3.

3.1 Interleaving updates

As we have seen, a single insertion replaces a leaf by a red node with two leaves as its children and may lead to an up-in request for the new red node. Now observe that we can do the same also in the case when an (accumulated) insertion falls into a leaf which has a red parent with an up-in request as the result of a previous insertion. In this way, a sequence of insertions may lead to a growth below the red-black tree as in the case of natural (unbalanced) binary search trees. Each of the newly created internal nodes with a red parent has an up-in request attached to it.

If an insertion falls into a leaf which has a removal request and/or an up-out request, then if the leaf has a removal request abandon the removal request and (re-)insert the key at that leaf. Otherwise, if the leaf has an up-out request remove the up-out request and replace the leaf by an internal black node with two leaves.

Several deletions can be accumulated by depositing a removal request at a leaf for each single deletion. Only if a deletion falls into a leaf which has a red parent with an up-in request as the result of a previous insertion then we delete the leaf immediately and abandon the up-in request.

3.2 Concurrent handling of rebalancing transformations

First we observe that several removal requests can be settled concurrently in an arbitrary order as long as it is assured that no two of them interfere at the same nodes. So, a removal request for a node which itself or the parent of which has an up-out request—as the result of a previously settled removal request—cannot be handled before the up-out request has been settled or bubbled up in the tree. This assumption assures that two removal requests appended to both leaves of a black node are always handled correctly.

Analogously, if the sibling of a leaf with a removal request has an up-in request then the up-in request must be settled or bubbled up in the tree before the removal request can be handled.

Now observe that several up-in and up-out requests can also be settled concurrently in an arbitrary order as long as it is assured that no two of them interfere at the same nodes. If we assure that a request is never shifted into the (large enough) surrounding area of another request, it is always possible to carry out the one-step rebalancing operations correctly. If a transformation shifts a request to an ancestor node occurring in the surrounding area of another request, this transformation cannot be carried out before the other request has been settled or bubbled up in the tree. That means, the conflict of two requests can simply be avoided by handling requests in top-down manner if they would meet in the same area, otherwise, in arbitrary order. There is always a request that can be carried out, since the topmost of those requests can be settled without any interference of the other ones.

In this way we get a simple algorithm for relaxed balancing by expanding the standard balancing technique. Now we show how to tune this algorithm for relaxed balanced red-black trees by choosing the surrounding area as small as possible. As result we will obtain the rebalancing transformations for chromatic trees proposed by Boyar et al. [3].

Let the *vicinity* of an up-in or an up-out request of a node p denote the set of neighbouring nodes the colours of which have to be changed or which will obtain new subtrees by the transformation explained in Section 2 that settles the request of p. (Note that in accordance with our definition nodes that must be considered in order to decide which kind of transformation applies, but which are not affected by the transformation itself, do not lie in the vicinity of p's request.) A request of a node p *is in conflict with* a request of a node q if q lies in the vicinity of p's request.

Obviously, the rebalancing transformations for red-black trees have the following properties:

Fact 1 *The vicinity of an up-in or an up-out request of a node p never contains nodes of the subtree rooted at p.*

Fact 2 *For each node q occuring in the vicinity of an up-in request of a node p, $depth(q) < depth(p)$. Furthermore, if q is black then q must be p's grandparent.*

Our aim is to show that for each given relaxed balanced red-black tree which contains up-in or up-out requests there is always a request that is not in conflict with other requests so that at least one rebalancing transformation can be carried out. We want to show that if a request of a node p is in conflict with other requests then at least one of these other requests is not in conflict with the request of p.

From Fact 1 and Fact 2 it follows

Fact 3 *If an up-in or up-out request of a node p is in conflict with a an up-in or up-out request of a node q and $depth(q) < depth(p)$ then the q's request is not in conflict with the request of p.*

Fact 4 *If a request of a node p is in conflict with a request of a node q and depth(q) = depth(p) then p has an up-out request. Furthermore, if q has an up-in request then q's request is not in conflict with the request of p.*

Fact 5 *Assume that a request of a node p is in conflict with a a request of a node q, depth(q) > depth(p), and there is no other request in p's vicinity with smaller depth than q. If q has an up-in request, then the request of q is not in conflict with any other requests.*

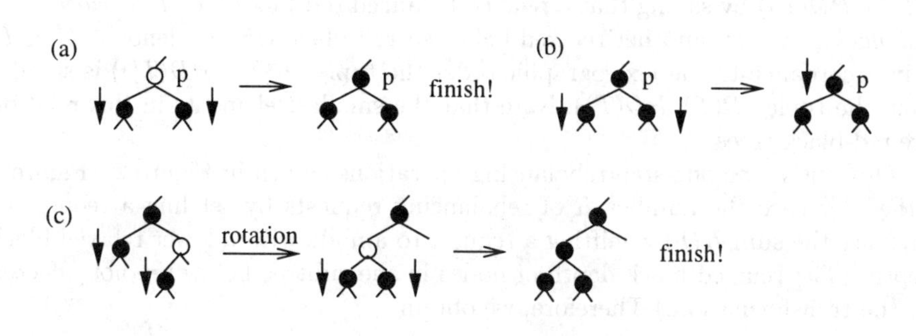

Fig. 5. Additional transformations.

Observe that two up-out requests of sibling nodes are in conflict with each other meaning that we cannot settle one of the requests by using the restructuring *up-out* without any interference of the other one. However, this problem can be solved by using the simple transformation shown in Figure 5a,b. If the parent p of those nodes is red then we colour p black and remove the up-out requests from both children of p. Otherwise, we replace the up-out requests of p's children by an up-out request for p (recursive shift). As the result of this transformation we have abandoned at least one up-out request by using only colour changes.

This additional transformation can be combined analogously with the rotation shown in Figure 4a as the operations shown in Figure 4b–d. If a node p with an up-out request has a red sibling the child of which has an up-out request, the transformation shown in Figure 5c settles both up-out requests.

Lemma 1. *Given a relaxed balanced red-black tree which does not satisfy the balance condition for red-black trees, the tree has at least one node at which one of the rebalancing operations shown in Figure 2—Figure 5 can be carried out while preserving the relaxed balanced red-black tree structure.*

Next we observe that any valid sequence of one-step rebalancing operations which is long enough, will ultimately transform a relaxed balanced red-black tree into a strictly balanced red-black tree again.

Let p be a node of a relaxed balanced red-black tree T. By $\varphi(p)$ we denote the parent node of p in T. The *relaxed black depth* of p, $rbd(p)$ is defined by

$$rbd(p) = \begin{cases} rbd(\varphi(p)) & \text{if } p \text{ is red} \\ rbd(\varphi(p)) + 1 & \text{if } p \text{ is black and has no up-out request} \\ rbd(\varphi(p)) + 2 & \text{if } p \text{ is black and has an up-out request} \end{cases}$$

where we set $rbd(\varphi(\text{root})) := 0$. Let $R(T)$ denote the number of rebalancing requests in T (i.e. the number of up-in, up-out, and removal requests), and let $RD(T)$ denote the sum of all relaxed black depths of nodes in T which have an up-in or an up-out request.

We characterize the balance of a relaxed balanced red-black tree by the tuple $(R(T), RD(T))$ by saying that a relaxed balanced red-black tree T' is *closer to a red-black tree* than another relaxed balanced red-black tree T, denoted $T' < T$, if in accordance to the lexicographic order the tuple $(R(T'), RD(T'))$ is smaller than the tuple $(R(T), RD(T))$. Note that the smallest elements in this relation are red-black trees.

Obviously, the one-step rebalancing operations shown in Figure 2—Figure 5 either decrease the number R of rebalancing requests by settling a request or decrease the sum RD by shifting a request to a node with a lower relaxed black depth. (The relaxed black depth of nodes in the subtree below are not affected by the transformation.) Therefore, we obtain

Lemma 2. *Let T be a relaxed balanced red-black tree with at least one rebalancing request, and let T' be the relaxed balanced red-black tree that has been obtained from T by applying one of the one-step rebalancing transformations shown in Figure 2—Figure 5. Then $T' < T$.*

We get easily an upper bound on the necessary one-step rebalancing operations. One update operation causes at most one rebalancing request. Such a rebalancing request is—in accordance to the relaxed black depth—bubbled up in the tree by using only colour flips until it is finally removed by using a constant number of colour flips and pointer changes. This leads to the main theorem of this paper:

Theorem 3. *Let T be a red-black tree and assume that insertions, deletions and rebalancing transformations (possibly interspersed) are applied to T. Then the number of structural changes (pointer changes) needed to restore the balance condition is $O(i + d)$ where i is the number of insertions and d the number of deletions. The number of colour changes is $O(\log(n + i))$ where n is the size of T.*

We also state the following obvious but important property for relaxed balanced red-black trees and their update scheme.

Theorem 4. *Let T be a red-black tree and assume that insertions, deletions but no rebalancing transformations are applied to T such that a tree T' is obtained. Assume further that T' has no leaves with (unsettled) removal requests and that the sequence of updates has not changed the shape of the original tree. Then T' does not contain any unsettled up-out or up-in requests.*

A possible sequence of those updates is a sequence of insertions of keys $k_1, \ldots, k_r$ which is followed by a sequence of deletions of $k_1, \ldots, k_r$ (or vice versa). Note, that the relaxed balancing scheme proposed in [12] generates even for such a sequence of updates a tree which is full of balance conflicts.

The expanded set of one-step rebalancing operations we have obtained by tuning the basic algorithm for relaxed balancing is essentially the same set of rebalancing operations proposed by Boyar et al. [3], if we assume that the weight of a node in [3] is always $w \leq 2$, and, whenever a rebalancing transformation is carried out, the nodes with a lower depth in the vicinity of the handled request always have $w \leq 1$. It should be clear how to adapt this transformations to the general case $w \geq 0$ proposed in [3] by allowing the accumulation of several up-out requests at single nodes and to move up-out requests from one node to another even if they are not handled by such a transformation.

4 Conclusion

Our aim was to keep the relaxed balancing scheme conceptually as simple as possible. For this reason we have restricted our presentation to the class of red-black trees.

The basic idea of our relaxed balancing scheme (accumulation of insertions and deletions and settling them in an arbitrary order, using the same rebalancing operations as for standard balanced search trees) is applicable to other classes of search trees as well. The idea is simply to let the tree grow below the original leaves as a random binary search tree and to mark an original leaf, if it has to be deleted. Then this set of unsettled updates can be handled just as a sequence of standard updates which are followed immediately by rebalancing transformations. All what is necessary is to assure that the postponed updates are settled top-down and no two restructurings interfere at the same nodes. Of course, this simple scheme has also opened a lot of room for tuning the individual update algorithms.

References

1. Adel'son-Vels'kii, G.M, Landis, E.M.: "An algorithm for the organisation of information"; Soviet Math. Dokl., 3 (1962), 1259–1262.
2. Bayer, R., McCreight, E.: "Organization and maintenance of large ordered indexes"; Acta Informatica, 1 (1972), 173–189.
3. Boyar, J., Fagerberg, R., Larsen, K.: "Amortization Results for Chromatic Search Trees, with an Application to Priority Queues"; 4th International Workshop on Algorithms and Data Structures, Lecture Notes in Computer Science, 955 (1995), 270–281.
4. Boyar, J., Fagerberg, R., Larsen, K.: "Chromatic priority queues"; Preprint 15, Department of Mathematics and Computer Science, Odense University (1994).

5. Boyar, J., Larsen, K.: "Efficient rebalancing of chromatic search trees"; Journal of Computer and System Sciences, 49 (1994), 667–682.
6. Ellis, C.S.: "Concurrent search in AVL-trees"; IEEE Transactions on Computers, C-29 (1980), 811–817.
7. Ellis, C.S.: "Concurrent search and insertions in 2–3 trees"; Acta Informatica, 14 (1980), 63–86.
8. Guibas, L.J., Sedgewick, R.: "A dichromatic framework for balanced trees"; Proc. 19th IEEE Symposium on Foundations of Computer Science, (1978), 8–21.
9. Keller, A., Wiederhold, G.: "Concurrent use of B-trees with variable-length entries"; SIGMOD Record, 17, 2 (1988), 89–90.
10. Kessels, J.L.W.: "On-the-fly optimization of data structures"; Comm. ACM, 26 (1983), 895–901.
11. Larsen, K.: "AVL trees with relaxed balance"; Proc. 8th International Parallel Processing Symposium, IEEE Computer Society Press, (1994), 888–893.
12. Larsen, K., Fagerberg, R.: "B-trees with relaxed balance"; Proc. 9th International Parallel Processing Symposium, IEEE Computer Society Press, (1995), 196–202.
13. Nurmi, O., Soisalon-Soininen, E.: "Chromatic binary search trees: A structure for concurrent rebalancing"; Acta Informatica 33 (1996), 547–557.
14. Nurmi, O., Soisalon-Soininen, E., Wood, D.: "Concurrency control in database structures with relaxed balance"; Proc. 6th ACM Symposium on Principles of Database Systems, (1987), 170–176.
15. Olivié,H.J.: "A new class of balanced search trees: half-balanced binary search trees"; R.A.I.R.O. Theoretical Informatics, 16, 1 (1982), 51–71.
16. Ottmann, Th., Soisalon-Soininen, E.: "Relaxed Balancing Made Simple"; Techn. Report 71, Institut für Informatik, Universität Freiburg, Germany (1995), also appeared as electronic version, anonymous FTP ftp.informatik.uni-freiburg.de, in /documents/reports/report71/, also http://hyperg.informatik.uni-freiburg.de/Report71.
17. Ottmann, Th., Wood, D.: "Updating binary trees with constant linkage cost"; International Journal of Foundations of Computer Science, 3, (1992), 479–501.
18. Sagiv, Y.: "Concurrent operations on B*-trees with overtaking"; Journal of Computer and System Sciences, 33, 2 (1986), 275–296.
19. Sarnak, N., Tarjan, R.E.: "Planar point location using persistent search trees"; Comm. ACM, 29 (1986), 669–679.
20. Shasha, D., Goodman, N.: "Concurrent search structure algorithms"; ACM Transaction on Database Systems, 13, (1988), 53–90.
21. Soisalon-Soininen, E., Widmayer, P.: "Relaxed balancing in search trees"; Advances in Algorithms, Languages, and Complexity: Essays in Honor of Ronald V. Book (eds. D.-Z.Du and K.-I. Ko), Kluwer Academic Publishers, Dordrecht (1997). To appear.

The Algorithmic Complexity of Chemical Threshold Testing

Peter Damaschke

FernUniversität, Theoretische Informatik II
58084 Hagen, Germany
Peter.Damaschke@fernuni-hagen.de

Abstract. Let us be given a set of water samples where possibly some are contaminated with a chemical substance. We wish to find these "defective" samples and their concentrations. Assume that there is an indicator available that can only detect whether the concentration is at least a known threshold, but we may split, merge, and dilute different samples. This problem is a generalized version of the well-known group testing problem and is also related to learning Boolean threshold functions.
First we consider the problem of approximating the concentration in a single sample by a threshold indicator only. We present a practical and efficient strategy that achieves a close approximation using a small number of tests and auxiliary operations. This strategy is formulated as a pebble moving game on a suitable graph.
Then we give principal asymptotic complexity results for finding, in a set of samples, those samples where the concentration exceeds a given limit. The results for the number of tests are asymptotically optimal, furthermore we propose a way for the efficient use of the other resources.

1 Introduction

In the well-known group testing problem, we are given a set of n elements, where each of them is either defective or good. We wish to identify all defectives. For this, the following kind of tests is available: We may choose any subset X. The result of testing X is positive if X contains at least one defective, and negative otherwise. A typical application of this problem is chemical or medical testing. Assume that we have a set of n samples of (say) water. A sample is defective if it is contaminated with a certain (say toxic) substance. Further assume that there is a test available which indicates the presence of this substance. Then we can draw subsamples from subsets X of samples, merge them and test these mixtures. There are also applications more related to computer science, such as memory circuit testing; see [7].

It is a simple matter to show that $\Theta(r \log \frac{n}{r})$ tests are required to identify r defectives out of n elements. A lot of work has been done to bound the factor hidden in Θ, to give tight exact bounds in special cases and for related problems, and to estimate the algorithmic complexity of the resulting test strategies. We mention [1] [3] [5] [6] [7] [8] [10] [11] [13] as some of the recent papers.

Reviewing the chemical application of group testing, we observe the following points: (a) One is not interested in the concentrations of the substance in the samples, we only wish to classify them into defective and good ones. (b) It is assumed that we can draw arbitrary many subsamples from each given sample, as many as required by the test strategy. (c) The test indicates even the smallest trace of the substance, regardless of its concentration.

Particularly assumption (c) may become problematic in practice if we have few defectives with small concentrations, since the efficient group testing strategies use large subsets X. It seems more appropriate to assume an indicator which gives a positive answer if and only if the concentration is at least a known fixed threshold. On the other hand, (b) is not so unrealistic, since the known strategies access to each sample only logarithmically often. Furthermore one should be interested in the approximate concentrations of the substance in the contaminated samples, in contrast to (a). In the following we specify this problem which includes classical group testing as a special case.

So let us be given a set of samples (of water). Each of them is characterized by the concentration of the questionable substance in it. We can draw parts of samples, dilute them by water, merge them, and test them by our threshold indicator. As a further natural assumption, there is a minimum quantity of water, called a unit, we can deal with by our equipment. That means, we can only handle integer multiples of a unit.

More formally, each of the n samples is characterized by a non-negative real c, the concentration. As the measure for the concentration we simply choose the threshold of our indicator. That means, the indicator detects the substance iff $c \geq 1$. There may be only a restricted quantity of each sample available. Note that, in any sample, c is previously unknown, and we can gather informations about c only by tests. Additionally, we have arbitrarily many units of pure water, i.e. with concentration $c = 0$. For simplifying speech, this reservoir is also considered to be a sample, although with known concentration.

We can perform the following basic operations:

- Draw an arbitrary integer number of units from a sample.
- Merge k units of samples with concentrations $c_1, \ldots, c_k$. This results in k new units, each with the average concentration $\sum_{i=1}^{k} c_i/k$.
- Test a unit with concentration c by the indicator. The reply is positive iff $c \geq 1$, otherwise negative. We assume that a test neither destroys the unit nor affects the concentration, thus any tested unit remains unchanged and can be used for further operations.

Note that merging also includes the case of dilution by pure water. These operations reflect what can be really done with liquid samples, the only restriction is that we can handle only discrete units of liquid. Especially note that a merging step is irreversible, i.e. we cannot later dissolve a mixture into its former components.

Our goal is, roughly speaking, to determine the samples with highest concentrations and to approximate these concentrations up to a prescribed accuracy, using a possibly small number of tests and of other resources, such as the number of draw and merge operations, and the number of units involved in the whole test process.

Before we discuss this problem, let us see in which sense it generalizes classical group testing. For any fixed limit L, a sample is called an L-defective if $c \geq L$. Assume that $L \geq n$, and that any sample fulfills either $c < 1$ or $c \geq L$. (That means, a sample is either clean, up to the sensitivity of our indicator, or it is intensively contaminated.) Then a group test on subset X is obviously equivalent to the following procedure: Merge one unit from each sample of X, and test this mixture. In this view, the assumptions of group testing look particularly artificial which motivates our study of the more realistic thresholding problem with arbitrary concentrations.

The paper is organized as follows. In Section 2 we show how to approximate a concentration in a single sample (even in a single unit of a sample) with fair accuracy, using a moderate number of tests, merge operations, and water units. Our favorite strategy, which is the main technical contribution, can be nicely described as a pebble game on certain graphs. Pebble games (chip games) are an extremely useful tool for complexity-theoretic questions (see e.g. [12]) and for the analysis of search strategies (as e.g. in [2]). In fact, the game we will define reminds to the pebble game used for proving time vs. space results, however our rules are somewhat different.

In Section 3 we give optimal asymptotic bounds (up to constant factors) for the complexity of finding all L-defectives in a set of n samples. This will be called generalized group testing. It is a typical situation that much less than n samples are contaminated, so we are interested in bounds depending on the total contamination of the given set of samples, too, and we will take this into account. The proofs of the results concerning the test number include quite simple tree-like strategies and counting arguments. We also consider the amount of other resources.

Section 4 points out a relationship to learning positive threshold functions and lists some questions that remain open in our paper.

Throughout the paper, log means $\log_2$.

2 Approximating a Concentration

In this section, let us be given a single sample where we wish to determine the concentration c up to a prescribed accuracy, using a minimum worst-case number of tests by a threshold indicator. We also wish to minimize the number of necessary draw and merge operations. Note that c is equal to the dilution factor where the indicator changes its answer. So we equivalently search for this unknown dilution factor. In the following we study the most unfortunate case that we have only one unit of our sample at hand, and we look what is still possible then. Results for an arbitrary sample size can be easily obtained as a byproduct of this analysis.

First we show that $\lfloor c \rfloor$ can be determined in $O(\log c)$ steps. That means, if we want to enclose c into an interval of length 1 then we succeeed in a logarithmic number of steps. Later we consider the problem of estimating c with higher precision, and we study the amount of water for dilution.

For convenience of denotation we describe a unit with concentration rc simply by the dilution ratio r. Hence the originally given unit is described by 1, and a test is positive if $r \geq 1/c$. A collection of q units each with concentration rc is abbreviated by $q \times r$. Further we write a collection of units with several concentrations formally as the sum of these terms; the value of the sum must always be 1. Merging of units means simply to form the average.

Proposition 1. *We can determine $\lfloor c \rfloor$ using $O(\log c)$ tests, draw and merge operations.*

Proof. The strategy is essentially a simple binary search. The important thing is that we can submit the suitable queries to the indicator.

Initially test the original sample 1. If the result is negative then we know that $\lfloor c \rfloor = 0$. So assume that the result is positive. Dilute the sample by one unit of water. This yields $2 \times \frac{1}{2}$.

In the first stage, always test a unit with smallest concentration. In the positive case, dilute it by one unit of water, such that it is replaced by two units of half concentration. Repeat this until a negative answer is obtained. This results in a sequence of units $\frac{1}{2^1} + \frac{1}{2^2} + \ldots + \frac{1}{2^{n-1}} + 2 \times \frac{1}{2^n}$ where $2^{n-1} \leq c < 2^n$. If $n = 1$ then $\lfloor c \rfloor = 1$ is clear. If $n > 1$ then merge the last three units. This yields $\frac{1}{2^1} + \frac{1}{2^2} + \ldots + \frac{1}{2^{n-2}} + 3 \times \frac{1}{2^{n-2}3}$. This completes the first stage.

Now suppose that the following invariant is true: For some non-negative integers n and k, with k odd, we have the situation $\frac{1}{2^1} + \frac{1}{2^2} + \ldots + \frac{1}{2^n} + k \times \frac{1}{2^n k}$, and we know $(k-1)2^n \leq c < (k+1)2^n$.

After the first stage this is true with $k = 3$ and $n := n - 2$. Test the last unit. If the answer is negative then we know $(k-1)2^n \leq c < k2^n$ which is equivalent to $((2k-1)-1)2^{n-1} \leq c < ((2k-1)+1)2^{n-1}$. Merging the last $k+1$ units gives $(k+1) \times \frac{1}{2^{n-1}(k+1)}$. Stretching them by water to $2k-1$ units yields $(2k-1) \times \frac{1}{2^{n-1}(2k-1)}$. Together with the previous samples this recovers our invariant. Similarly, if the test gives a positive answer then we know $k2^n \leq c < (k+1)2^n$ which is equivalent to $((2k+1)-1)2^{n-1} \leq c < ((2k+1)+1)2^{n-1}$. Now merge the last $k+1$ units and stretch them to $2k+1$ units.

The process terminates when $n = 0$ is reached, and there remains the sequence $k \times \frac{1}{k}$ with odd k. A final test decides whether $\lfloor c \rfloor = k - 1$ or k. The $O(\log c)$ bound is obvious. □

Although the $O(\log c)$ result is asymptotically optimal, there is a serious drawback making the strategy practically useless in this form, namely, $\Theta(c)$ units of water are required which is wasteful in the case of large c. Unfortunately, this is optimal for many integer c which we can easily see: Note that the average of a set of positive ratios cannot contain new prime factors in the denominator, provided that we represent every ratio by relatively prime numerator and denominator.

For c prime we easily conclude that c units are necessary for generating $1/c$ by merging operations. A similarly bad assertion holds if c contains a large prime factor. On the other hand, there is no proper reason to demand integer values for the ends of our output intervals containing c. We will show by a different strategy that output intervals of length 1 and shorter can be achieved by a logarithmic amount of tests and units, and by a polylogarithmic number of simpler merge operations.

Less importantly, for small c we would like to have finer output intervals than integer steps. So the problem arises which ratios r of a concentration can be actually merged within moderate complexity bounds, particularly for small $1/r$.

Next we provide the announced well-behaved strategy. In the proof of Proposition 1 we first searched for the powers of 2 with the unknown c in between, and then always halved the actual search interval with respect to c. In our new strategy, we will always halve the search interval on a scale linear in $1/c$ (and thus linear in r rather than in c.) If a unit for the next merging step is not at hand then we will generate it from previous units by repeating possibly few of the last steps. By this we lose the unimportant property that all occuring r are inverse integers, and we lose the great simplicity of instantaneous descriptions of the search process, but we gain efficiency in all resources.

Theorem 2. *There exists a strategy for estimating an unknown concentration* $c \geq 1$ *in one unit of a sample, with the following properties: It uses* t *tests,* t *units of water, and* $t^2/4$ *merge steps, each incorporating only two units. It outputs an interval of length at most* $c^2/(2^t - c)$ *including* c. *We can stop the search after any test number* t *with* $\log c < t \leq c/2$, *since* $\log c$ *is already determined after* $\lfloor \log c \rfloor + 1$ *tests.*

Proof. The case $c < 1$ can be detected by one test, so assume $c \geq 1$.

For a fixed unknown c define a sequence r_i approximating $r = 1/c$ in the following inductive way: Let $r_0 = 0$, $r_1 = 1$, and call r_0 the mate of r_1. If $r_0, \ldots, r_i$ are already defined and r_j $(j < i)$ is the mate of r_i then let $r_{i+1} = (r_j + r_i)/2$. Note that $r \in (u_i, v_i]$ where one of u_i, v_i is r_{i+1} and the other endpoint is r_j or r_i. This endpoint becomes the mate of r_{i+1}. Let d be the largest index such that r_0 is still the mate of r_d.

Note that $2^{d-1} \leq c < 2^d$. Trivially we have $|u_t - v_t| = 1/2^t$ which yields $|1/u_t - 1/v_t| = 1/2^t u_t v_t$. Furthermore we have $v_t \geq 1/c$ and $u_t \geq 1/c - 1/2^t$ which is positive for $t \geq d$. Together this implies $|1/u_t - 1/v_t| \leq c^2/(2^t - c)$.

A basic step in our search strategy will consist of mixing two units of r_{i+1} from one unit of r_i and its mate, and the mate of each newly created member r_{i+1} is chosen due to the test result for r_{i+1}. Clearly, we have to test each new member of the sequence exactly once. Since every merge step is concerned with only two units, and every new member of the sequence can be tested immediately after production, all we have to study is the number of units and merge steps for producing a sequence of the desired length t.

We may conveniently describe the process as producing and pebbling a certain acyclic directed graph. Define a graph with nodes r_i and the following edges:

A directed edge joins r_i and r_{i+1}, and if r_j is the mate of r_i then a directed edge goes from r_j to r_{i+1}. So every node except r_0 and r_1 has exactly two predecessors.

We consider the following pebble game. Initially we have a potentially infinite number of pebbles on r_0 and one pebble on r_1. All other nodes are empty. A node containing at least one pebble is called occupied. Whenever both predecessors of a node are occupied, we are permitted to move one pebble from each predecessor to the considered node. Our goal is to reach r_t in a possibly small number of steps, moving a possibly small total number of pebbles.

The equivalence of this game to our chemical test problem is evident: The value of each node r_i is the average of the values in its predecessors, the edges of the graph depend on the test results at the nodes (and thus only on c), each pebble is a unit, r_0 contains the water reservoir and r_1 the given sample, each move is a merge step, and the last node we reach by a pebble and its mate correspond to the ends of the output interval. Note that our graph is infinite, but only the initial part up to the last reached node is known in any moment. We could equivalently consider a two-person game on a growing finite graph where the player moves the pebbles and creates new nodes, and an adversary fixes the mate of every new node, and thus the edges. We have to prove that, for any t, we can deliver a pebble to r_t within the asserted complexity bounds.

Our strategy works simply as follows. In any move, choose the node with maximum index which is empty but both predecessors are occupied, and move a pebble from each predecessor to this node. The game stops if we arrive at the desired t or no further move is possible.

In the following we list some properties of this strategy. For our analysis we partition the set of nodes into consecutive levels, by the following rule: If r_i is the mate of r_{i+1} then both nodes belong to different levels (that is, r_{i+1} opens a new level), otherwise they belong to the same level. Particularly, r_0 and $r_1, \ldots, r_d$ are the first two levels.

Claim 1. (trivial) At any time, every node except r_0 contains at most two pebbles. $\diamond$

Claim 2. If both r_i and its mate r_j are occupied then a further move with destination r_k $(k > j)$ will be made.

If r_{i+1} is empty then we have a move to r_{i+1} or to a later node. If r_{i+1} is occupied then the claim follows by an inductive argument, since one of the occupied nodes r_j and r_i is the mate of r_{i+1}. $\diamond$

Claim 3. At any time and in any level (except r_0), we have at most one node with two pebbles, and all later nodes in this level are empty.

The claim is true in the initial state of the game. We show that that it remains valid after each move. Consider a move from r_i and its mate r_j to r_{i+1}.

First let r_{i+1} be in the same level as r_i. If there is an occupied node right from r_{i+1} on the same level then, by Claim 2, the actual move would have a destination right from r_{i+1}, a contradiction. (Note that r_j is occupied and is the mate of all nodes of the level.) Hence all nodes right from r_{i+1} on the same level are empty. If r_i contains two pebbles then one of them is moved to r_{i+1}. A previous node in the level cannot contain two pebbles, since r_i is occupied, and the claim was true before the move. So r_{i+1} is the only node with two pebbles in this level, after the move.

Now let r_{i+1} be the first node in its level. If some node in this level was occupied immediately before the considered move then, by Claim 2 again, the actual move has a destination right from r_{i+1}, a contradiction. (Note that r_i is occupied and is the mate of all nodes in the next level.) Hence the claim remains true also in this case. $\diamond$

Claim 4. If r_t is the rightmost occupied node then $r_1, \ldots, r_t$ together contain at most t pebbles.

The claim is true in the initial state with $t = 1$, and it remains true after any move not affecting r_0. We have to prove that it remains true also whenever a new pebble from r_0 has been moved.

It suffices to give, after such a move, an injection from the set of nodes with two pebbles into the set of empty nodes up to r_t. Consider any node r_j with two pebbles. If r_j is not the last node of its level then we assign it to some later node in the same level which is empty, due to Claim 3.

Let r_j be the last node of its level and $j > d + 1$. Then we assign it to some node of the next level. This next level must be completely empty: If it contains some occupied node r_i then this node was already occupied before the previous move, since r_0 cannot send a pebble behind r_{d+1}. Also r_j was occupied, for the same reason. Since r_j is the mate of r_i, by Claim 2 the previous move would have a destination right from r_i, a contradiction.

If $r_j = r_d$ then we assign it to the node r_1 which remains always empty. It remains the case that $r_j = r_{d+1}$, the third level contains only this node, and the fourth level contains an occupied node. Then the previous step must have moved a pebble from r_0 and r_d to r_{d+1}, otherwise r_{d+1} would have been already occupied, and the strategy would have chosen another move, not from r_0. Since r_d was occupied before the move, Claim 3 implies that either r_d was the only node in the first level containing two pebbles before the move, or there was no such node. Hence after the move there is no node with two pebbles in the first level, and we are free to assign r_{d+1} to r_1.

Note that the assignment defined above is in fact injective. $\diamond$

Claim 5. (trivial) As long as one of $r_1, \ldots, r_d$ is occupied, we always have another move available. $\diamond$

Claim 6. If $r_1, \ldots, r_d$ are all empty then the further nodes contain more than 2^{d-1} pebbles.

For the proof we describe the contents of $r_1, \ldots r_d$ by a word of symbols 0,1,2, called the occupation word, indicating the number of pebbles on each node. Initially the occupation word is 10^{d-1}. Note that in the first d moves, this becomes 01^{d-1}, and two pebbles are brought into r_{d+1}. In the following moves, the occupation word is changed only if a pebble from r_d is requested by later nodes. This can be easily seen: Our strategy always chooses the rightmost free node with occupied predecessors as the next destination, and the function assigning the mate to each node is monotone. Hence, as long as there exists such a triple of nodes within the subgraph after r_d, the nodes up to r_d will not be affected. Furthermore r_d is the only node among $r_1, \ldots r_d$ which can immediately send pebbles to nodes behind r_d, hence a request cannot access to the other nodes of the second level.

What happens in case of a request? If the last symbol of the occupation word is 1 then it switches to 0. If it is 0 then all later nodes must wait until a pebble is delivered to r_d. Assume that the occupation word has the form $w10^k$. Then our strategy will change it in the next k moves to $w01^{k-1}2$, and then it serves the actual request, thus transforming the occupation word into $w01^k$. Note that, between requests, the occupation word always consists only of symbols 0,1. This is clear by the above observation, using induction on the number of requests (or by Claim 3). Moreover, if we consider the occupation word as the binary expansion of an integer, then serving a request corresponds to decrementing this integer by 1. The initial occupation word 01^{d-1} was the binary expansion of $2^{d-1} - 1$. Each request brings at least one new pebble into the region after r_d. Together with the first two pebbles these are more than 2^{d-1}. $\diamond$

Claim 7. For any $t \leq c/2$ we can deliver a pebble to r_t.

If $r_1, \ldots, r_d$ never becomes empty then Claim 5 implies that we can continue the game up to infinity. Since every move carries some pebbles from left to right, but the number of pebbles on any node is bounded due to Claim 1, after a finite number of steps we will occupy the next node in the sequence. Thus we reach any node.

If $r_1, \ldots, r_d$ becomes empty then Claims 4 and 6 together imply that we have already reached node $2^{d-1} + 1 > c/2$. $\diamond$

Claim 8. If we abort the game at r_t then we have moved a total of at most t pebbles and performed at most $t^2/4$ moves.

The first assertion rephrases Claim 4. Any move increases the index of one pebble by at least 1 and the index of a second pebble by at least 2. Using Claim 1, the final sum of indices of all t pebbles can be easily bounded by $3t^2/4$. Thus the number of moves is bounded by $t^2/4$. $\diamond$

The theorem follows now directly from the last claims. $\square$

It seems that some of the above estimations can be further improved.

Corollary 3. *An interval of length at most 1, including an unknown concentration* $c > 16$ *in one unit of a sample can be found by about* $2 \log c$ *tests, using approximately* $2 \log c$ *units of water, and* $\log^2 c$ *merge steps.* $\square$

This improves the first unsatisfactory attempt of Proposition 1. The supposition $c > 16$ comes from the condition $t \approx 2\log c \leq c/2$. Any additional test improves the accuracy by a factor 1/2, but requires $O(\log c)$ further merge steps. For smaller c, a direct inspection shows that small output intervals are also quickly reachable in the same way. We omit this separate consideration. Finally, note that the strategy is also computationally efficient due to its very simple rule.

3 Asymptotic Results for Generalized Group Testing

Now we wish to find all L-defectives in a set of n samples. It is evident that we cannot expect a strategy for arbitrary real $L \geq 1$, since we must merge suitable units from the given samples for performing tests whether $c \geq L$. But fortunately, appropriate values L occur frequently enough: A positive ratio $r \leq 1$ is called (u, v)-mergeable if we can merge a unit with r times the original concentration from one unit of a sample, using at most u units of water and v merging steps. Theorem 2 already guarantees the existence of $(t, t^2/4)$-mergeable ratios r such that the gaps between neighbored $1/r$ values are rather small. Moreover, there is a lot of other possible merging strategies. From these observations and concrete experiments with small u, v we have evidence that the set of (u, v)-mergeable ratios is quite "dense". Here we cannot go deeper into this matter which deserves a separate study.

In the following we make some general assumptions. Let n be a power of 2, otherwise we can add $O(n)$ dummy samples with concentration 0. We only consider such limits L where m/L is (u, v)-mergeable for some m being a power of 2, and for u, v not too large. Clearly, any $m/2^i L$ is $(u + i, v + i)$-mergeable then.

Let s be the "pruned sum" of all concentrations c_i of all given samples, that is $s = \sum_{i=1}^{n} \min\{c, L\}$. Note that $s \leq nL$. Since we are interested in instances where much less than n samples are contaminated, we will always assume even $s < nL/2$, without further mentioning this explicitly. (If the number of L-defectives is $\Omega(n)$ then we will need anyhow $\Omega(n)$ tests which is the uninteresting case.)

In order to straight out the presentation, we consider in the next theorem only the number of tests. We presume that L is appropriately chosen such that, for some power m of 2 with $m/L = \Omega(1)$, the ratio m/L is mergeable at low cost. By the above discussion, this is not a severe restriction, rather there is a rich scale of suitable values L.

Theorem 4. *Let L be a limit and m a power of 2 such that $m/L = \Omega(1)$ is mergeable. Then all L-defectives out of n samples can be found optimally by $O(\frac{s}{L}\log\frac{nL}{s})$ tests if $L \geq n$ or $s \geq n$, and by $O(\frac{n}{L} + \frac{s}{L}\log(L+1))$ tests if $L, s < n$.*

Proof. First we argue that the bounds are optimal. Consider instances with $r = s/L$ samples each with $c = L$, and the remaining $n - r$ samples with $c = 0$. (Assume that r is integer.) There are $\binom{n}{r}$ possible outputs merely of this type, so we get the information theoretic lower bound of $\Omega(\log \binom{n}{r})$. Standard calculations show that this is $\Omega(r \log \frac{n}{r}) = \Omega(\frac{s}{L} \log \frac{nL}{s})$. If $n > s$ then this is also $\Omega(\frac{s}{L} \log(L+1))$. (In case $L = 1$ note that we have generally supposed $s < nL/2$, hence $nL/s > 2$.) Furthermore, if $rL \log(L+1) = \Omega(n)$ then the last bound is also $\Omega(n/L)$. It remains to establish the $\Omega(n/L)$ bound for smaller r. Since we are proving a lower bound, we may assume that all L-defectives except one are already known "for free", and only the last L-defective must be found among the remaining $n-r+1 = \Omega(n)$ samples. Observe that any test giving a negative answer can exclude at most L samples (namely those sharing at least $1/L$ of the tested mixture). In the worst case we get only negative answers, so we must execute $\Omega(n/L)$ tests.

The upper bound is proved by a straightforward search strategy. We split the set of samples recursively into sets of half size, thus obtaining a complete binary tree of subsets. Each node of the tree represents a subset where the number b of elements is a power of 2. Trivially, for each of these subsets we can form an "average sample". In order to test whether the sum of concentrations in a subset is at least L ($L \geq b$) we must dilute this average sample down to a factor b/L. Due to our assumption, b/L is in fact mergeable for every $b \leq m$. So we can perform these kind of tests on all subsets of at most m elements. For simplifying speech, we identify the diluted average samples with the corresponding tree nodes.

First consider the case $L \geq n$. Since $m = \Omega(L)$ this means $m = \Omega(n)$. We may even assume $m \geq n$, since we do not take care to constant factors.

The strategy is a simple top-down traversal. Test the root by the indicator. If the test is negative then we are sure that there is no L-defective, and we can stop. In the positive case continue the procedure recursively on the two subtrees. Eventually, the "positive" leaves are the L-defectives.

There exist $O(s/L)$ paths of nodes which are tested with positive result. This is clear since the lowest "positive" nodes represent disjoint subsets, each with sum of concentrations at least L, and descendants of "negative" nodes are not further tested. Now cut the tree into two parts containing the nodes with distance at most $\log \frac{s}{L}$ from the root, and the nodes with larger distance, respectively. The part near the root contains $O(s/L)$ nodes. As said above, the latter part includes $O(s/L)$ tested paths, each of length $O(\log n - \log \frac{s}{L})$. Altogether this gives the asserted bound for the number of nodes.

Next let be $L < n$ but $s \geq n$. In this case, split the set of samples into n/m sets of size m. Since $m = \Omega(L)$ was supposed, these are $\Theta(n/L)$ sets of size $\Theta(L)$. Apply the above tree strategy on each subset.

Let s_i be defined as s, but restricted to the ith of these subsets. The previous partial result implies for the ith subset a bound of $O(\frac{s_i}{L} + \frac{s_i}{L} \log \frac{L^2}{s_i})$ tests if $s_i < L^2/4$. But if $s_i \geq L^2/4$ then this bound also holds, since $O(L)$ is a trivial bound and $O(L) = O(s_i/L)$. Hence the sum of all these terms is $O(\frac{s}{L} + \frac{1}{L} \sum_i s_i \log \frac{L^2}{s_i})$. This can also be written as $O(\frac{s}{L} + \frac{s}{L} \log L^2 + \frac{1}{L} \sum_i s_i \log \frac{1}{s_i})$. For any fixed number k of summands, the last sum is maximized if all $s_i = s/k$, hence it is maximized at all for $k = n/L$. This gives a total bound of $O(\frac{s}{L} \log \frac{nL}{s})$, as asserted.

It remains the case $L, s < n$. Then we apply the same strategy. The difference is that now only $O(s/L)$ subsets are candidates for containing L-defectives. We find them by $O(n/L)$ tests at the roots of the trees. Then we must further proceed only with the $O(s/L)$ candidate sets. So the maximum k in the above estimation becomes s/L which implies the asserted upper bound. □

Corollary 5. *For L and m as above, all L-defectives out of n samples can be found by $O(\frac{n}{L} + \frac{s}{L}\log n)$ tests.* □

Proposition 6. *Provided that m/L is (u,v)-mergeable, the above strategy needs 2 units from each sample, and for each test $O(u + \log m)$ units of water, and $O(v + \log m)$ merge steps.*

Proof. Instead of merging complete units just at the time when needed for tests, we form our subset tree in advance in a bottom-up fashion. First draw two units from each sample, that is, for each leaf of the tree. If we have already two units for each of two sibling nodes then we merge one from both siblings, and forward the two new units to the father node. So we can inductively continue this process up to the root, and finally any node keeps one unit containing the average of samples of the represented subset. So we only need two units from each sample to generate one unit for each subset occuring in the tree.

Only immediately before each test we dilute the average sample in the considered node down to b/L where b is the size of the subset. Since m/L is (u,v)-mergeable, this can be done using $O(u + \log m)$ units and $O(v + \log m)$ merge steps. □

4 Conclusions

There is an interesting connection between generalized group testing and learning positive Boolean threshold functions by membership queries. For the definition of threshold functions and the learning model we refer to [4] [9] and presume that the reader is familiar with that. Note that the strategy of Section 3 uses only tests of the following type: Check whether the sum of concentrations in a chosen subset is at least L. But this is exactly the type of queries allowed in learning positive threshold functions. The best known learning algorithms for this function class, due to [4] [9], need either $O(nm)$ queries and $O(n^2m + nm^2)$ computation time, or $O(nm^2)$ queries and time, where m denotes the number of maximal false and minimal true vectors of variables. (These algorithms actually work on a more general function class including the positive threshold functions.) In the light of group testing, it is a natural question whether sparse threshold functions where only r of the n weights are positive and the others are 0 can be learned faster. A first result is given by the following

Proposition 7. *A positive threshold function in n variables with r positive weights and m extremal false and true vectors can be learned by $O(rm(r+\log n))$ queries.*

Due to lack of space, we omit the (not very deep) proof which combines the result of [4] and a binary search technique. We ask for an improved result.

Other interesting problems are suggested by the previous sections. We mention some of them:

Improve the complexity bounds for approximating a single concentration.

Which is the complexity of finding approximately r samples with the highest concentrations, for prescribed r? (A natural idea is to run the strategy of Section 3 with several values of L.)

Give tight bounds for generalized group testing (not only up to a constant factor) and competitive strategies, such as known for plain group testing.

Which complexity results do we get if the tests destroy the involved units?

Characterize the (u, v)-mergeable ratios r, (a) if we allow arbitrary merge steps, and (b) if only merge steps with at most 2 (or another constant of) units are permitted.

References

1. I.Althöfer, E.Triesch: Edge search in graphs and hypergraphs of bounded rank, *Discrete Math.* 115 (1993), 1-9
2. J.A.Aslam, A.Dhagat: Searching in the presence of linearly bounded errors, *23th ACM STOC* 1991, 486-493
3. A.Bar-Noy, F.K.Hwang, I.Kessler, S.Kutten: A new competitive algorithm for group testing, *Discrete Applied Math.* 52 (1994)
4. E.Boros, P.L.Hammer, T.Ibaraki, K.Kawakami: Polynomial time recognition of 2-monotonic positive Boolean functions given by an oracle, *SIAM J. Computing*, to appear
5. P.Damaschke: A tight upper bound for group testing in graphs, *Discrete Applied Math.* 48 (1994), 101-109
6. P.Damaschke: A parallel algorithm for nearly optimal edge search, *Info. Proc. Letters* 56 (1995), 233-236
7. P.Damaschke: Searching for faulty leaves in binary trees, in: M.Nagl (ed.), *Graph-Theoretic Concepts in Computer Science – 21st Int. Workshop WG'95*, Aachen 1995, *LNCS* 1017 (Springer), 265-274
8. D.Z.Du, F.K.Hwang: Competitive group testing, *Discrete Applied Math.* 45 (1993)
9. K.Makino, T.Ibaraki: A fast and simple algorithm for identifying 2-monotonic positive Boolean functions, *Proc. 6th ISAAC'95*, *LNCS* 1004, Springer 1995, 291-300
10. D.Z.Du, H.Park: On competitive group testing, *SIAM J. Comp.* 23 (1994), 1019-1025
11. D.Z.Du, G.L.Xue, S.Z.Sun, S.W.Cheng: Modifications of competitive group testing, *SIAM J. Comp.* 23 (1994), 82-96
12. R.Reischuk: *Einführung in die Komplexitätstheorie* (in German), Teubner, Stuttgart 1990
13. E.Triesch: A group testing problem for hypergraphs of bounded rank, 1995, submitted

A Meticulous Analysis of Mergesort Programs

Jyrki Katajainen*[1] and Jesper Larsson Träff[2]

[1] Department of Computer Science, University of Copenhagen, Universitetsparken 1, DK-2100 Copenhagen East, Denmark, Email: jyrki@diku.dk
[2] Max-Planck-Institut für Informatik, Im Stadtwald, D-66123 Saarbrücken, Germany, Email: traff@mpi-sb.mpg.de

Abstract. The efficiency of *mergesort programs* is analysed under a simple unit-cost model. In our analysis the time performance of the sorting programs includes the costs of key comparisons, element moves and address calculations. The goal is to establish the best possible time-bound relative to the model when sorting n integers. By the well-known information-theoretic argument $n \log_2 n - O(n)$ is a lower bound for the integer-sorting problem in our framework. New implementations for two-way and four-way bottom-up mergesort are given, the *worst-case* complexities of which are shown to be bounded by $5.5n \log_2 n + O(n)$ and $3.25n \log_2 n + O(n)$, respectively. The theoretical findings are backed up with a series of experiments which show the practical relevance of our analysis when implementing library routines for internal-memory computations.

1 Introduction

Given a sequence of n elements, each consisting of a key drawn from a totally ordered universe $\mathcal{U}$ and some associated information, the *sorting problem* is to output the elements in ascending order according to their keys. We examine methods for solving this problem in the internal memory of a computer under the following assumptions:

1. The input is given in an array and the output should also be produced in an array, which may be the original array if the space is tight.
2. The key universe $\mathcal{U}$ is the set of integers.
3. There is no information associated with the elements. Hence, we do not make any distinction between the elements and their keys.
4. Comparisons and moves are the only operations allowed for the elements.
5. Comparisons, moves, and basic arithmetic and logical operations on integers take constant time.
6. An integer can be stored in one memory location and there are $O(n)$ locations available in total.

From assumption 4 it follows that any sorting method must use at least $\Omega(n \log_2 n)$ time (see, e.g., [11, Section 5.3]). However, without this assumption n integers can be sorted in $O(n \log_2 \log_2 n)$ time [1], or even in $O(n)$ time if integers are small (see, e.g., [11, pp. 99–102]).

* Supported partially by the Danish Natural Science Research Council under contract No. 9400952 (project "Computational Algorithmics").

Mergesort is as important in the history of sorting as sorting in the history of computing. A detailed description of bottom-up mergesort, together with a timing analysis, appeared in a report by Goldstine and von Neumann [6] as early as 1948. Today numerous variants of the basic method are known, for instance, top-down mergesort (see, e.g., [17, pp. 165–166]), queue mergesort [7], in-place mergesort (see, e.g., [8]), natural mergesort (see, e.g., [11, pp. 159–163]), as well as other adaptive versions of mergesort (see [5, 14] and the references in these surveys). The development in this paper is based on bottom-up mergesort, or straight mergesort as it was called by Knuth [11, pp. 163–165].

As much as this paper is about sorting, it is about the timing analysis of programs. In contrast to *asymptotical analysis* (or big-oh analysis), the goal in *meticulous analysis* (or little-oh analysis) is to analyse also the constant factors in the running time, especially the constant in the leading term of the function expressing the running time.

To facilitate the meticulous analysis of programs, a model has to be defined that assigns a cost for all primitive operations of the underlying computer. The running time of a program is then simply the sum of the costs of the primitive operations executed on a particular input. Various cost models have been proposed for this purpose. In his early books [10, 11] Knuth used the MIX model where the cost of a MIX assembly language instruction equals the number of memory references that will be made by that instruction, including the reference to the instruction itself. That is, the cost is one or two for most instructions, except that the cost of a multiplication was defined to be ten and that of a division twelve. In his later books [12, 13] Knuth has used a simpler memory-reference model in which the cost of a memory reference is one, whereas the cost of the operations that do not refer to memory is zero.

Knuth analysed many sorting programs in his classic book on sorting and searching [11]. His conclusions were that quicksort is the fastest method for internal sorting but, if a good worst-case behaviour is important, one should use heapsort since it does not require any extra space for its operation.

In this paper we complement Knuth's results by analysing also *multiway mergesort*, which was earlier considered to be good only for external sorting. In the full paper we furthermore analyse a new implementation of the in-place mergesort algorithm developed by Katajainen et al. [8] and the adaptive mergesort variant proposed by van Gelder (as cited in [5]). Our conclusions are different from those of Knuth. In our cost model multiway mergesort is the fastest method for integer sorting, in-place mergesort is the fastest in-place sorting method, and adaptive mergesort is competitive with bottom-up mergesort together with the advantage of being adaptive. An explanation of these results is that the inner loop of mergesort is only slightly more costly than that of quicksort but the outer loop of multiway mergesort is executed less frequently.

The rest of the paper is organized as follows. In Section 2 we introduce a subset of the C programming language, called *pure C*, and an associated cost model. In Section 3 the *worst-case* performance of multiway bottom-up mergesort is analysed under this cost model. The theoretical analysis is backed up by a series of experiments in Section 4. The results are summarized in Section 5 (see Table 2).

2 A C cost model

The model of computation used throughout this paper is a Random Access Machine (RAM) which consists of a *program*, *memory* and a collection of *registers*. A register and a memory location can store an integer. The actual computations are carried out in the registers. In the beginning of the computation the input is stored in memory, and the output should also be produced there.

The machine executes programs written in *pure C* which is a subset of the C language [9]. All the primitive operations of pure C have their counterparts in an assembly language for a present-day RISC processor [15, Appendix A.10]. The execution of each primitive operation is assumed to take one unit of time. The reader is encouraged to compare this assumption with the actual costs of C operations measured by Bentley et al. [3] on a variety of computers.

The data manipulated by pure C programs are *integers* (`int`), *constants*, and *pointers* (`int*`). If a variable is defined to be of type `int`, we assume that the value of this variable is stored in one of the registers, that is, the actual type is **`register int`**. Also pointer variables, whose content indicates a location in memory, are kept in registers, i.e., their type is **`register int*`**.

A pure C program is a sequence of possibly labelled statements. Let `x`, `y`, `z` be not necessarily distinct integer variables, `c` and `d` constants, `p` and `q` pointer variables, and ℓ a label of some statement in the program under execution. The primitive statements of pure C are listed below.

1. *Load statement* "`x = *p;`" loads the integer stored at the memory location pointed to by `p` into register `x`.
2. *Store statement* "`*p = y;`" stores the integer from register `y` at the memory location pointed to by `p`.
3. *Move statement* "`x = y;`" copies the integer from register `y` to register `x`. Also the form "`x = c;`" is possible.
4. *Arithmetic statement* "`x = y` $\oplus$ `z;`" stores the sum, difference, product or quotient of the integers in registers `y` and `z` into register `x` depending on $\oplus \in$ `{+,-,*,/}`. Also the forms "`x = c` $\oplus$ `z;`" and "`x = y` $\oplus$ `d;`" are possible.
5. *Branch statement* "`if (x` $\lhd$ `y) goto` ℓ`;`" branches to the statement with label ℓ, where $\lhd \in \{<,>,=,\leq,\geq,\neq\}$. Also the forms "`if (c` $\lhd$ `y) goto` ℓ`;`" and "`if (x` $\lhd$ `d) goto` ℓ`;`" are possible.
6. *Jump statement* "`goto` ℓ`;`" branches unconditionally to the statement with label ℓ.
7. *Empty statement* "`;`" does nothing.

Thus pure C statements are simply normal C statements involving at most three addresses. It is easy to translate the C control structures into pure C.

In the basic model the cost of all pure C primitives is assumed to be the same. In reality, computers are more complicated since the actual running time of a program depends on pipelining of the instructions and caching of the data. Therefore, the cost given by the model can only be treated as an approximation of the exact running time. It might be possible to get a more accurate estimate of the running time by assigning a *weight* to every primitive operation, although this still ignores the context in which an operation is being executed.

Knuth has used a simple variant of the weighted pure C model in his recent books [12, 13] when comparing the practical efficiency of different programs. In his *memory-reference model* the cost of every load and store statement is one, whereas the cost of all other primitives involving only registers is zero. The classical goodness measures used in the sorting literature are the number of *key comparisons* and the number of *element moves*. In the following we study only the pure C cost of various mergesort programs, but from these programs the number of memory references, key comparisons and element moves carried out can be readily calculated (cf. Table 2).

In order to make our programs more readable, we will later on use the array syntax "`a[i]`" instead of the pointer syntax "`*(a+i)`". According to our cost model, the cost of the load statement "`x = a[i]`" as well as the store statement "`a[i] = y`" is two. However, if the array is accessed sequentially, it is possible to do a more efficient translation into pure C. The programs to be presented contain very few statements that could be executed in parallel. In our program descriptions independent statements are written on the same line, in order to show where the execution might benefit from parallelism in the hardware (pipelining, instruction parallelism).

3 The worst-case performance of mergesort programs

Let `a` be an array of n elements to be sorted. Further, assume that `b` is another array of the same size. We say that the first element of a subarray of `a` or `b` is its *head* and the last element its *tail*. *Multiway bottom-up mergesort* sorts the elements in passes. Initially, each element of `a` is thought to form a sorted subarray of size one. In each *pass* the subarrays are grouped together such that each group consists of, say, m consecutive subarrays (except the last one which might be smaller), after which the subarrays in every group are m-way merged from `a` to `b`. This way the number of sorted subarrays is reduced from n to $\lceil n/m \rceil$. Then the rôle of `a` and `b` is switched and the same process is repeated until only one subarray remains, containing all the elements in sorted order. The heart of the construction is the merge function, which repeatedly moves the smallest of the heads of (at most) m shrinking subarrays to the output zone in `b`. In the following we present some improvements over textbook implementations, and analyse the performance of these improved implementations.

3.1 Two-way bottom-up mergesort

In two-way bottom-up mergesort, or briefly *two-way mergesort*, two subarrays are merged at a time. Sedgewick [17, pp. 173–174] pointed out that it is advantageous to reverse the order of elements in every second subarray. This way the maximum of the tails of the subarrays will function as a sentinel element saving one pointer test. However, this method will not necessarily retain the order of equal elements, i.e., the resulting sorting method is no longer stable. We present a different optimization which preserves stability and is even more efficient than the reversal method.

In a normal textbook program (see, e.g., [2, p. 63]) a merge of two subarrays is accomplished in a loop where the smallest of the two heads is moved into the output zone, the involved indices are updated accordingly, and at the end of each iteration it is tested whether either of the subarrays is exhausted. However, during one iteration

```
void MERGE(a[], h1, t1, h2, t2, b[], h3, t3) {
        i = h1; j = h2; m = h3;
        u = a[i]; v = a[j];
        if (a[t1] > a[t2]) goto test1;
        goto test2;
first:  b[m] = u;
        i = i + 1; m = m + 1;
        u = a[i];
test1:  if (u ≤ v) goto first;
        b[m] = v;
        j = j + 1; m = m + 1;
        v = a[j];
        if (j ≤ t2) goto test1;
        for (; i ≤ t1; i++, m++) b[m] = a[i];
        return;
second: b[m] = v;
        j = j + 1; m = m + 1;
        v = a[j];
test2:  if (u > v) goto second;
        b[m] = u;
        i = i + 1; m = m + 1;
        u = a[i];
        if (i ≤ t1) goto test2;
        for (; j ≤ t2; j++, m++) b[m] = a[j];
        return; }
```

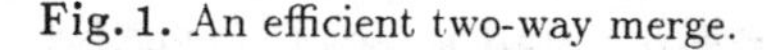

Fig. 1. An efficient two-way merge.

only the position of one of the heads, not both, is updated. Therefore, one of these tests is superfluous, and can be avoided by more careful programming. Another way of speeding up the program is to check prior to the loop which of the tails is smaller and then write a separate code for the two possible cases. After this it is no longer necessary to test whether the end of the subarray with the larger tail is reached. Moreover, we apply Sedgewick's rule [16, p. 853] that no inner loop should ever end with a jump statement. These ideas are implemented in Fig. 1 as function `MERGE`.

In `MERGE` the value of `a[j]` (`a[i]`) is read before the test whether `j` (`i`) is out of the range or not. However, an element is never used if it is from neither of the subarrays. If required, the reference outside array **a** can be avoided by sorting the first $n-1$ elements and thereafter inserting the last element into its proper location. By binary search this requires only $O(\log_2 n)$ comparisons, after which at most n element moves are to be done.

Depending on the relative order of the tails, there are two symmetric cases. Let us consider the case where the tail of the first subarray is larger than that of the second subarray. In the first inner loop of `MERGE`, a comparison is performed to decide from which subarray an element should be moved to the output zone and, if this element comes from the second subarray, a test is performed at the end to see if that subarray is exhausted or not. Thus, the cost of one iteration is five or six, provided that pointers are used instead of array cursors. In the second `for`-loop the cost of copying the elements to the output zone is five per element.

In each of the $\lceil \log_2 n \rceil$ merging passes there is at most one merge where the second subarray is shorter than the first subarray or where the second subarray is missing altogether. The overall cost caused by these special merges is proportional to $\sum_{d=0}^{\lceil \log_2 n \rceil} 2^d$, which is $O(n)$. In a normal case the subarrays being merged are of the same size. When considering these normal merges only, the extra test is necessary for at most half the elements in each merging pass. Furthermore, the number of normal merges is clearly never more than $n - 1$. Therefore, the overall cost caused by the normal merges is bounded by $5.5n \log_2 n + O(n)$. To sum up, the *worst-case* performance of bottom-up mergesort is $5.5n \log_2 n + O(n)$.

3.2 Four-way bottom-up mergesort

Four-way bottom-up mergesort, or simply *four-way mergesort*, merges four, instead of two, subarrays in each merge. This reduces the number of passes from $\lceil \log_2 n \rceil$ to $\lceil \log_4 n \rceil$, where n denotes the size of the input as earlier. In the implementation of four-way merge it is important to find the minimum of the four heads fast. For example, the heads could be kept in a priority queue from which the smallest element is always removed, and after each removal a new element (if any) from the same subarray is inserted into the priority queue. Practical experiments, e.g., those carried out by Katajainen et al. [8], indicate that the best data structure for this purpose is an unordered list, even though with this structure the number of element comparisons will increase from about $n \log_2 n$ to $1.5n \log_2 n$. In the present paper we show that a faster implementation is obtained if a program state is used to remember the relative order of the four heads, instead of using a data structure. The state is changed accordingly when a head is updated.

As in two-way mergesort, in each merging pass there is at most one special merge where the number of merged subarrays is less than four or, if four, the last subarray is shorter than the others. The cost of these special merges is clearly proportional to $\sum_{d=0}^{\lceil \log_4 n \rceil} 4^d$, which is $O(n)$. Therefore, we concentrate on the normal merges where the subarrays are of the same size.

A normal merge goes through *four phases*: in Phase i, $i \in \{0, 1, 2, 3\}$, the end of i subarrays has been reached, that is $4 - i$ of the subarrays being merged still have elements left. Phase 3 reduces to copying in which the cost of the inner loop is five per element. Phase 2 is a two-way merge, the inner loop of which costs at most six per element. Let us now consider how a three-way merge, i.e., Phase 1 can be accomplished efficiently.

In a three-way merge there are *three cases* depending on which of the tails of the subarrays being merged is smallest. All these cases can be handled in a similar fashion. Therefore, we describe here only the case where the tail of the third subarray is smaller than the other two tails. The program fragment taking care of this case is given in Fig. 2 as a block diagram. The blocks with consecutive numbers should be placed physically after each other in the final program.

Let `u`, `v`, and `x` denote the heads of the first, second, and third subarray, respectively. In the program two invariants are maintained:

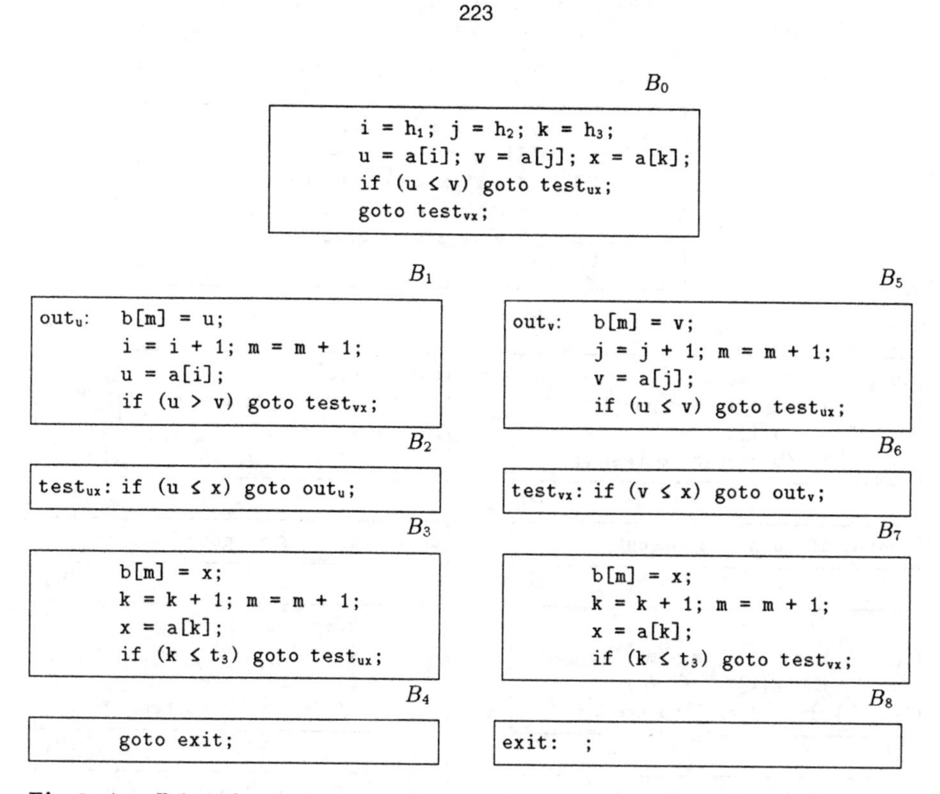

Fig. 2. An efficient three-way merge when the last subarray has the smallest tail.

1. prior to the execution of block B_2, $u \le v$, and
2. prior to the execution of block B_6, $u > v$.

So the outcome of the tests in blocks B_2 and B_6 determines which of the elements u, v or x is to be moved to the output zone. After the test control switches to block B_1, B_3, B_5, or B_7. In each of these blocks five statements are executed before a new test in B_2 or B_6. Therefore, the cost of the inner loop is six per element moved to the output zone.

Let us finally consider Phase 0 where four subarrays are being merged. Now there are *four cases* depending on which of the subarrays has the smallest tail. Due to symmetry, we study only the case where the last subarray is exhausted first. A block diagram for this particular four-way merge is given in Fig. 3. The block numbering indicates again the order of the blocks in the final program.

Let u, v, x, and y denote the heads of the four subarrays. In the program fragment of Fig. 3 the following invariants are maintained:

1. prior to B_2, $u \le v$ and $x \le y$,
2. prior to B_4 and B_6, $u \le v$ and $x > y$,
3. prior to B_9, $u > v$ and $x \le y$, and
4. prior to B_{11} and B_{13}, $u > v$ and $x > y$.

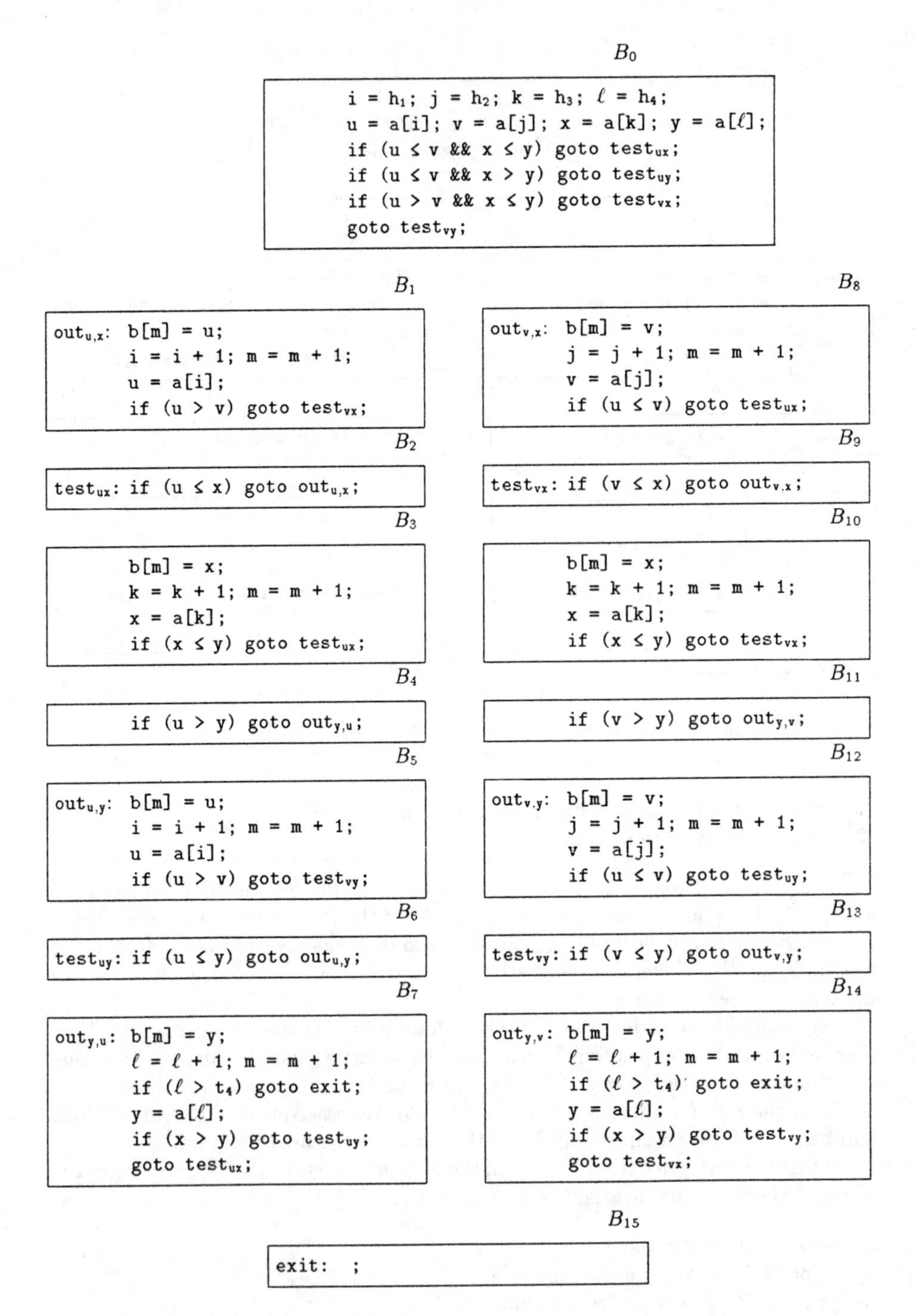

Fig. 3. An efficient four-way merge when the last subarray has the smallest tail.

After a test in B_2, B_4, B_6, B_9, B_{11}, or B_{13} control switches to B_1, B_3, B_5, B_7, B_8, B_{10}, B_{12}, or B_{14}, where an element is moved to the output zone and another test is carried out so that the above invariants hold before the next iteration starts. Each time an element is taken from the last subarray, i.e., when the control is in B_7 or B_{14}, six or seven statements must be executed in addition to the test. If an element is taken from any of the other subarrays, only five additional statements are necessary before starting the next iteration.

To sum up the above calculations, the cost of a normal four-way merge is bounded by $3 \cdot 6 \cdot 4^d + 8 \cdot 4^d + O(1)$ if the size of the merged subarrays is 4^d. In the dth merging pass, $d \in \{0, 1, \ldots, \lceil \log_4 n \rceil\}$, at most $n/4^{d+1}$ normal merges are carried out so their total cost is $6.5n + O(n/4^d)$. Therefore, the cost caused by the normal merges over all passes is $3.25n \log_2 n + O(n)$. Since the cost of special merges is only $O(n)$, the *worst-case* performance of four-way mergesort is $3.25n \log_2 n + O(n)$.

Fig. 2 and 3 describe the four-way mergesort program only in part. Our actual C implementation is about 450 lines of pure C code, corresponding to four variants of Fig. 3, three variants of Fig. 2, and the two-way merge of Fig. 1. One could go a step further and write a program for eight-way mergesort. Back-on-the-envelope calculations show that the worst-case complexity of eight-way mergesort is below $3n \log_2 n + O(n)$. Hence, the performance of eight-way mergesort is — at least in theory — superior to that of four-way mergesort, but the size of the resulting program may invalidate the simple cost model (if the program no longer fits in the program cache). This issue is discussed in the full paper.

4 Experimental results

To test the theoretical predictions we have implemented the mergesort programs developed in the previous sections. For the sake of comparison we have also implemented versions of quicksort and heapsort whose meticulous analyses are given in the full paper. The sorting programs implemented include

1. textbook two-way bottom-up mergesort which avoids unnecessary copying (for the merge routine, see, e.g., [2, p. 63]);
2. efficient two-way mergesort based on the merge function given in Fig. 1; This implementation uses arrays, and according to the simple cost model performs seven or eight instructions in its inner loop.
3. two-way mergesort optimized as above, but using pointers; According to our cost model this version performs five or six instructions in its inner loop.
4. efficient four-way mergesort as developed in Section 3.2, but using pointers;
5. iterative implementation of quicksort taken from the book of Sedgewick [17, pp. 118 and 122];
6. bottom-up heapsort as described by Carlsson [4], but realizing multiplications and divisions by 2 with shifts.

All mergesort programs have the same functionality: they are called with an input array and a work array, and return a pointer to the sorted output, which is either of the two arrays of the call. The running time is measured as the time to execute the entire call to the sorting function. Hence, the time spent in the allocation of

size	mergesort (textbook)	two-way (improved)	two-way (pointers)	four-way (pointers)	quicksort (iterative)	heapsort
100 000						
time (ms)	360	280	240	150	250	440
mems	6 533 608	3 799 996	3 799 996	2 366 613	3 449 301	3 720 253
comps	1 566 804	1 666 803	1 666 803	1 817 456	2 051 474	1 699 474
moves	1 700 000	1 700 000	1 700 000	900 000	1 078 344	1 910 113
200 000						
time (ms)	760	580	510	310	530	1 010
mems	13 866 344	7 999 996	7 999 996	4 733 339	7 268 981	7 839 933
comps	3 333 172	3 533 171	3 533 171	3 829 517	4 381 170	3 598 572
moves	3 600 000	3 600 000	3 600 000	1 800 000	2 250 168	4 019 950
500 000						
time (ms)	1 980	1 510	1 340	840	1 480	3 030
mems	36 716 236	20 999 996	20 999 996	12 833 142	20 536 818	20 900 453
comps	8 858 118	9 358 117	9 358 117	10 102 406	13 082 640	9 646 895
moves	9 500 000	9 500 000	9 500 000	5 000 000	5 857 320	10 700 208
1 000 000						
time (ms)	4 150	3 190	2 830	1 680	3 020	6 850
mems	77 435 094	43 999 996	43 999 996	25 666 483	41 918 488	43 803 693
comps	18 717 547	19 717 546	19 717 546	21 212 307	26 534 802	20 294 732
moves	20 000 000	20 000 000	20 000 000	10 000 000	12 190 528	22 401 825
2 000 000						
time (ms)	8 520	6 510	5 690	3 550	6 550	15 440
mems	162 865 458	91 999 996	91 999 996	55 333 870	88 233 067	91 605 995
comps	39 432 729	41 432 728	41 432 728	44 471 255	5 648 4052	42 589 483
moves	42 000 000	42 000 000	42 000 000	22 000 000	25 363 742	46 802 976
5 000 000						
time (ms)	23 440	18 120	16 100	9 810	16 780	43 460
mems	444 005 872	249 999 996	249 999 996	148 332 746	234 561 702	242 367 609
comps	107 002 936	112 002 935	112 002 935	119 544 746	152 193 888	113 150 419
moves	115 000 000	115 000 000	115 000 000	60 000 000	66 405 096	123 683 773

Table 1. Results for integer sorting. "mems" denotes the number of memory references, "comps" the number of key comparisons, and "moves" the number of element moves.

work space is excluded. Quicksort and heapsort are implemented similarly, in order to make the comparisons as fair as possible.

In Table 1 we give the CPU time spent by the various programs as measured by the system call `clock()`. Times are in milliseconds. We also give the memory reference counts, measured as proposed by Knuth in [13, p. 464-465], the number of key comparisons performed, and the number of element moves. Experiments were carried out on a Sun4 SPARCstation 4 with 85 MHz clock frequency and 64 MBytes of internal memory. We used the GNU C-compiler `gcc` with optimization level `O4`. The running times are averages of several runs with integers drawn randomly from the interval $\{0, 1, \ldots, n-1\}$ by using the C library function `random()`. The sorting programs have all been run with the same input.

The running times follow the analysis according to the simple cost model quite well, whereas counting only memory references is definitely too simple. For instance, the two-way mergesort program based on Fig. 1 makes the same number of memory references as the version doing sequential access with pointers. The simple cost model predicts respective costs $7.5n \log_2 n$ and $5.5n \log_2 n$, which corresponds reasonably well to the measured running times. Quicksort is marginally worse than the pointer version of two-way mergesort, despite its better best-case behaviour. Also in this case, simply counting memory references does not give an accurate prediction of the relative behaviour of the two programs; quicksort is seen to make slightly fewer memory references than mergesort. Four-way mergesort is clearly the best of the sorting programs tested. Compared to two-way mergesort it is about a factor of 1.54 faster, which is not too far from the predicted 5.5/3.25 (the lower-order term is slightly bigger for four-way mergesort compared to that for two-way mergesort).

Compiler optimization gave surprising improvements to all sorting programs, about a factor of 2 to 3. However, the relative quality of the programs was invariant to these optimizations. It is also worth noting that compiler optimizations could *not* improve the textbook mergesort program to perform better than the improved two-way mergesort program based on Fig. 1, even for the latter compiled without optimization.

5 Summary

The theoretical findings of this paper (and the full version) are summarized in Table 2. We plan to carry out a more thorough experimental evaluation of all the sorting algorithms listed there (including the interplay with the compiler).

Acknowledgements

The first author would like to thank Alistair Moffat, Lee Naish, and Tomi Pasanen for their helpful comments.

References

1. A. Andersson, T. Hagerup, S. Nilsson, and R. Raman, Sorting in linear time?, in *Proceedings of the 27th Annual ACM Symposium on the Theory of Computing*, ACM Press, New York, N.Y., 1995, pp. 427–436.
2. S. Baase, *Computer Algorithms: Introduction to Design and Analysis*, 2nd Edition, Addison-Wesley Publishing Company, Reading, Mass., 1988.
3. J. L. Bentley, B. W. Kernighan, and C. J. van Wyk, An elementary C cost model, *UNIX Review* **9** (1991) 38–48.
4. S. Carlsson, Average-case results on Heapsort, *BIT* **27** (1987) 2–17.
5. V. Estivill-Castro and D. Wood, A survey of adaptive sorting algorithms, *ACM Computing Surveys* **24** (1992) 441–476.
6. H. H. Goldstine and J. von Neumann, Planning and coding of problems for an electronic computing instrument, Part II, Volume 2, reprinted in *John von Neumann Collected Works*, Volume V: *Design of Computers, Theory of Automata and Numerical Analysis*, Pergamon Press, Oxford, England, 1963, pp. 152–214.

C program	key comparisons	element moves[1]	pure C primitives[2]	memory references[3]
two-way mergesort				
worst case	$n \log_2 n$	$n \log_2 n$	$5.5n \log_2 n$	$2\ell n \log_2 n$
four-way mergesort				
worst case	$n \log_2 n$	$0.5n \log_2 n$	$3.25n \log_2 n$	$\ell n \log_2 n$
in-place mergesort[5]				
worst case	$n \log_2 n$	$n \log_2 n$	$3.75n \log_2 n$	$2\ell n \log_2 n$
adaptive mergesort[5]				
worst case	$n \log_2 n$	$1.5n \log_2 n$	$8n \log_2 n$	$3\ell n \log_2 n$
randomized quicksort[5]				
best case[4]	$n \log_2 n$	$\Theta(n)$	$3n \log_2 n$	$kn \log_2 n$
standard heapsort[5]				
best case[4]	$n \log_2 n$	$0.5n \log_2 n$	$6n \log_2 n$	$(0.5k + \ell)n \log_2 n$
bottom-up heapsort[5]				
best case	$n \log_2 n$	$n \log_2 n$	$11n \log_2 n$	$(k + 2\ell)n \log_2 n$

[1] A move of an element from one location to another in memory.
[2] When sorting integers.
[3] When the size of a key is k words and that of an element ℓ words.
[4] Assuming that all keys are distinct.
[5] Analysed in the full paper.

Table 2. The behaviour of various sorting programs when sorting n elements, each consisting of a key and some information associated with this key. The low-order terms in the quantities are omitted.

7. M. J. Golin and R. Sedgewick, Queue-mergesort, *Information Processing Letters* **48** (1993) 253–259.
8. J. Katajainen, T. Pasanen, and J. Teuhola, Practical in-place mergesort, *Nordic Journal of Computing*, **3** (1996) 27–40.
9. B. W. Kernighan and D. M. Ritchie, *The C Programming Language*, 2nd Edition, Prentice-Hall, Englewood Cliffs, N.J., 1988.
10. D. E. Knuth, *The Art of Computer Programming*, Volume 1: *Fundamental Algorithms*, Addison-Wesley Publishing Company, Reading, Mass., 1968.
11. D. E. Knuth, *The Art of Computer Programming*, Volume 3: *Sorting and Searching*, Addison-Wesley Publishing Company, Reading, Mass., 1973.
12. D. E. Knuth, *Axioms and Hulls*, Lecture Notes in Computer Science **606**, Springer-Verlag, Berlin/Heidelberg, Germany, 1992.
13. D. E. Knuth, *The Stanford GraphBase: A Platform for Combinatorial Computing*, Addison-Wesley Publishing Company, Reading, Mass., 1993.
14. A. M. Moffat and O. Petersson, An overview of adaptive sorting, *The Australian Computer Journal* **24** (1992) 70–77.
15. D. A. Patterson and J. L. Hennessy, *Computer Organization & Design: The Hardware/Software Interface*, Morgan Kaufmann Publishers, San Francisco, Calif., 1994.
16. R. Sedgewick, Implementing Quicksort programs, *Communications of the ACM* **21** (1978) 847–857. Corrigendum *ibidem* **23** (79) 368.
17. R. Sedgewick, *Algorithms*, 2nd Edition, Addison-Wesley Publishing Company, Reading, Mass., 1988.

BSP-Like External-Memory Computation

Jop F. Sibeyn* Michael Kaufmann†

Abstract

In this paper we present a paradigm for solving external-memory problems, and illustrate it by algorithms for matrix multiplication, sorting and list ranking. Our paradigm is based on the use of BSP algorithms. The correspondence is almost perfect, and especially the notion of x-optimality carries over to algorithms designed according to our paradigm.

The advantages of the approach are similar to the advantages of BSP algorithms for parallel computing: scalability, portability, predictability. The performance measure here is the total work, not only the number of I/O operations as in previous approaches. So the predicted performances are more useful for practical applications.

1 Introduction

Sequential Computation. The von Neumann model (RAM model) is strongly established, and the availability of a generally accepted model has allowed for a tremendous progress in the theory of algorithms. The basic assumption of this model is that all stored data can be accessed in unit-time. This is not an absolute truth, not even in the current practice: data in cache can be accessed at least ten times faster than data that are not in cache.

In the future, with further increasing sizes of data, this assumption will break down even further (see [2] for a detailed description). Yet, most programmers accept the RAM model without discussion, and may try to exploit some cache-features in a final optimization. Even right now we can encounter sequential problems that should not be dealt with on the basis of the von Neumann model. On a typical work-station, accessing a datum in main memory is several thousand times faster than accessing a datum that is stored on the hard-disc. The effect of this cannot be neglected, and thus special algorithms are necessary for solving problems that are so large, that their data do not fit into the main memory: *external-memory algorithms*. There is only a limited number of researchers who have intensively considered this field [1, 4, 3, 10].

Parallel Computation and the BSP Model. The situation in parallel computation is different: the lack of a unifying model may be viewed as one of the main reasons why progress in this domain has been so much slower than in sequential computation. Here numerous models are around, and a lot of energy is still invested in model discussions,

*Max-Planck-Institut für Informatik, Im Stadtwald, 66123 Saarbrücken, Germany. E-mail: jopsi@mpi-sb.mpg.de.

†Wilhelm-Schickard-Institut, Universität Tübingen, Sand 13, 72076 Tübingen, Germany. E-mail: mk@informatik.uni-tuebingen.de

and the wheel has been reinvented several times. There exists an enormous gap between most theoretical models and the practice of parallel machines.

In 1990 Valiant has proposed the BSP (Bulk Synchronous Parallel) model to bridge this gap and to provide the community with a possible unifying model. In a recent paper [9], McColl gives a detailed overview of the current state of BSP. He states that the great strength of the model lies in its

- scalability,
- portability,
- predictability.

Each of these features is clearly of crucial importance. Of course, the model also has its draw-backs, most importantly, some programming flexibility is sacrificed for uniformity, which may lead to less efficient algorithms for a particular parallel machine or problem. Indeed the BSP model has become quite popular in the past few years (see the proceedings of Euro-Par '96, LNCS 1123, to find at least ten papers dealing with theoretical and practical aspects of the model and its modifications).

In the BSP model, the performance of a parallel computer is characterized by only three parameters:

p: the number of processors;

g: the total number of local operations performed by all processors in one second, divided by the total number of words delivered by the communication network in one second in a situation of continuous traffic.

l: the number of time steps required for performing a barrier synchronization;

In the model, a parallel computation is subdivided in supersteps, at the end of which a barrier synchronization and a routing is performed. Hereafter all request for data that where posted during a preceding superstep are fulfilled. We consider the cost of a superstep. If w is the maximum number of internal operations in a superstep, and h is the maximum number of words sent or received by any processing unit, *PU*, then the number of time steps $T_{\text{superstep}}$ to perform it, is given by [9]

$$T_{\text{superstep}}(w, h) = w + h \cdot g + l. \tag{1}$$

The parameter l can be interpreted to take the start-up latency into account, and may depend on the diameter of the network. g takes care of the throughput of the network: the larger it is, the weaker the network. For a true PRAM, $l = g = 0$. More details on the BSP model are given in [15, 8, 9].

This Paper. We introduce a paradigm that may help to find good external-memory algorithms as the result of a guided search through algorithms in the already much further developed field of parallel algorithms. We start with some examples, that set the direction for this search. Then we consider the limitations of an earlier proposal.

In Section 4, we propose the paradigm itself. We prove that the quality measure of BSP algorithms caries over to algorithms designed according to our paradigm. This feature fundamentally distinguishes our approach from anything before: parallel algorithms do not only serve as a source of inspiration, but by well-established analytical means the most suitable candidate can be selected. This we consider to be our main result. The paper is concluded with some examples, illustrating how to handle the paradigm.

2 Guiding Examples

2.1 Structured Versus Random Access

Consider the following, almost identical programs:

```
    Program A;
var row: array [0 .. N − 1] of int ;
function rand: select random number from {0, 1, ..., N − 1};
(   for i := 0 to N − 1 do row[i] := rand ) .

    Program B;
var row: array [0 .. N − 1] of int ;
function rand: select random number from {0, 1, ..., N − 1};
(   for i := 0 to N − 1 do row[rand] := i ) .
```

The only difference is that in Program A a random number is assigned to consecutive positions of the array, and that in Program B consecutive numbers are attributed to random positions of the array.

The computational effort of both programs is the same, but the memory is accessed differently. We have implemented them in C on a SPARC10 workstation running under Solaris with a 32MB main memory (of which about 22MB are freely available) and a 80MB swap space. The resulting time consumptions are given in Table 1. Program C will be described in Section 2.2.

N	T_A/N	T_B/N	T_C/N
$2 \cdot 10^6$	2.13	3.58	3.54
$5 \cdot 10^6$	2.19	4.54	3.78
$6 \cdot 10^6$	2.26	1071	3.86
$8 \cdot 10^6$	2.30	3525	3.93
$16 \cdot 10^6$	2.40	12563	4.11

Table 1: The time consumptions in seconds for the programs A, B and C per million entries of the array.

The time consumptions for Program A and B increase linearly until $N = 5 \cdot 10^6$ with very small differences. For $N > 5 \cdot 10^6$, the array does not fit into the main memory. Program A is almost not effected but for Program B, in which the memory access is chaotic, such that almost every step means a page fault, the time consumption explodes.

In the remainder of the text M denotes the memory-size, N the problem-size, and B the page-size. All numbers should be specified in terms of bytes or in terms of integers (four bytes). We refer to bytes.

Program A models the access pattern of a well-written program operating on an array or a matrix, for which we may assume that the elements of the list are handled in order. If $N > M$, then the number of page faults is only $pf_A = (N - M)/B$. Program B models the access pattern of a program operating on a list or a graph. In this case, the access pattern cannot be structured by the programmer, and will in general be chaotic. Thus, for $N > M$, the expected number of page faults is given by $pf_B = (N - M/B) \cdot (1 - M/N)$. For large N, the difference with pf_A approximates B. Thus, with typical page-sizes around 8KB, pf_B may be a several thousand times larger than pf_A. As for trivial programs, like Program A and B the time is determined by how fast the data can be loaded, this results in a similar factor between the time consumptions.

2.2 Blocked Random Access

Program A and B are extreme cases. Program C has an intermixed structure:

```
    Program C;
var row: array [1 .. N] of int ;
function rand: select random number from {0, 1, ..., m − 1};
(   for i := 0 to N/m − 1 do
        for j := 1 to m do row[i · m + rand] := j ) .
```

Again the accessed row position is chosen randomly, but in this case at any time only from a block of size m. In our implementation we chose $m = 10^6$.

The time consumptions for various values of N are given in Table 1. In Table 1, we see for $N < M$ that Program C has approximately the same behavior as Program B. On the other hand, for $N > M$, Program B behaves dramatically different from Program C: just as Program A, it slows down only by 15%. This is not surprising either: as long as $m < M$, the number of page faults in Program C is approximately the same as in Program A. Namely, if this condition holds, then all pages that are relevant to a block of size m are loaded into the memory once, then accessed for a while, and finally replaced.

3 PRAM Algorithms?

In Section 4, we present a paradigm which allows us to perform external-memory algorithms with essentially the memory-access pattern of Program C. This reduces the time consumption from that of Program B to that of Program C. The paradigm is based on the observation that under certain conditions simulating parallel algorithms leads to good external-memory algorithms.

Already Chiang e.a. [3] proposed to simulate parallel algorithms for external-memory computation. They suggested to simulate PRAM algorithms. In such a sequential simulation, the data have to be paged-in for every communication step. With some examples we illustrate that this may be problematic, and does not easily lead to good external-memory algorithms.

Example 1 *Consider multiplying two $n \times n$ matrices in parallel. The standard PRAM algorithm, see for example [6], uses $n^3 / \log n$ PUs and solves the problem in $\mathcal{O}(\log n)$ time. Simulating this sequentially would require $\mathcal{O}(n^3 / \log n)$ memory, which is excessive (considering that n is very large).*

Taking some more care, we could simulate a PRAM algorithm with $P = 3 \cdot n^2/M$ PUs. Assuming that $P \leq n$, an obvious PRAM algorithm would be to let every PU multiply $n/P = M/(3 \cdot n)$ rows of A with all bundles of n/P columns of B. The choice of P is inspired by the fact, that now all data of a PU, plus the computed results just fit into the main memory. Simulating this requires that all elements of B are paged-in $P = \Theta(n^2/M)$ times.

Alternatively, we could multiply all $\sqrt{M/3} \times \sqrt{M/3}$ submatrices of A and B, and add the results together. This algorithm needs to page all data into the main memory only $\mathcal{O}(n/\sqrt{M})$ times, which is the square root of the above.

In this extremely simple case the choice of the right algorithm is clear, but only through the back-door: knowing the optimal external-memory algorithm, we see that indeed this is also a good PRAM algorithm.

In Example 1, the choice of the best external-memory algorithm is not obvious, but at least it goes back on an optimal strategy for multiplying matrices on PRAMs (by recursive two-division). The situation may be even worse. Our following example shows that in some cases the best external-memory algorithm is not (a modification of) an optimal PRAM algorithm with a reduced number of PUs.

Example 2 *Consider sorting N numbers. Suppose that $M < N \leq M^{3/2}/2$ (the example can be generalized for larger N). A good sorting strategy is to apply column-sort [7] or a variant thereof. This requires that all data are paged-in only a constant number of times. As a PRAM algorithm, column-sort is not very interesting: it does not lead to a $\mathcal{O}(\log N)$ algorithm.*

A comparably good external-memory algorithm, is to sort all subsets of size M, and then to apply an N/M-way merge. As a PRAM algorithm this does not make sense, this is a typical sequential approach.

A much better PRAM algorithm is based on repeated two-way merging. On a PRAM, each of the two best algorithm has work $\mathcal{O}(N \cdot \log N)$, and time $\mathcal{O}(\log N)$ [5]. As an external-memory algorithm, it requires that all data are paged in at least $\mathcal{O}((\log(N/M))$ times.

In general, for deriving external-memory algorithms, one must bring along a considerable understanding of the problem, to choose from the several possible PRAM algorithms the best one, and sometimes even look elsewhere. This is far from a mechanical process. PRAM algorithms might serve as a source of inspiration, but only in a very loose sense. The fundamental problem is, that algorithms that are indistinguishable as PRAM algorithms, from the point of view of their complexity, may perform very differently when simulated sequentially. The principle cause is, that in PRAM algorithms communication is for free, and that thus PRAM algorithms may comprise a large number of communication steps. In other words, the model has been oversimplified, missing certain essential characteristics.

4 Paradigm

4.1 BSP-Like Algorithms

The good performance of Program C motivates us. Each of the operations on a chunk of size m can be viewed as the operations of a PU on its internal memory. In our case, we perceive Program C as the simulation of a virtual parallel machine with $P = N/m$ PUs. Of course, so far these PUs operate isolatedly, but the communication in a parallel machine can be modeled by writing 'messages' into a $P \times P$ matrix. Our paradigm is a formalization of these observations:

Paradigm 1 *Divide the memory space into P blocks of suitable size. Then develop a BSP-like algorithm for a virtual machine with P processors and execute it sequentially.*

More concrete we state as follows:

Consider a problem $\mathcal{P}$ of size $N > M$. Solve $\mathcal{P}$ as follows:

1. *Set $m = M/3$.*

2. *Suitably divide the memory space of size N into $P = \lceil N/m \rceil$ blocks $B_0, \ldots, B_{P-1}$, of size m at most.*

3. *Design a* BSP-like algorithm *for $\mathcal{P}$, for a virtual machine with P PUs.*

4. *Execute the BSP-like algorithm on the sequential computer.*

We still have to specify Step 3 and Step 4:

Definition 1 *A BSP-like algorithm, is an algorithm that proceeds in discrete* supersteps, *and satisfies the following conditions:*

- *In Superstep s, $s \geq 1$, PU_i, $0 \leq i < P$, operates only on the data in the block B_i and on the messages $Mes(j, i, s)$, $0 \leq j \leq P - 1$.*
- *In Superstep s, $s \geq 1$, PU_i, $0 \leq i < P$, generates messages $Mes(i, j, s + 1)$ to be 'send' to PU_j, $0 \leq j \leq P - 1$. The size of $Mes(i, j, s + 1)$ is at most m/P.*
- *The initial messages, $Mes(i, j, 1)$, $0 \leq i, j < P$ are void.*

Lemma 1 *If a BSP-like algorithm is executed on a parallel computer, then no PU ever has to store more than M data at the same time.*

Proof: In any Superstep s, PU_i must store at most the data in block B_i, plus the data in the messages $Mes(j, i, s)$ and $Mes(i, j, s + 1)$, $0 \leq j < P$. In total, these are at most $m + 2 \cdot P \cdot (m/P) = 3 \cdot m = M$ data. □

Lemma 1 implies that BSP-like algorithms are indeed suitable for execution on a sequential computer with a main-memory of size at least M:

Program SEQUENTIAL_EXECUTION
for $s \geq 1$ **do**
  for $0 \leq i < P$ **do**
    Page-in B_i and $Mes(j, i, s)$, $0 \leq j < P$.
    Perform the operations of PU_i in Superstep s.
    Page-out B_i and $Mes(i, j, s + 1)$, $0 \leq j < P$.

In the last step, for $s \geq 2$, $Mes(i, j, s + 1)$ may overwrite $Mes(i, j, s - 1)$ to save storage. This gives

Lemma 2 *A BSP-like algorithm can be executed on a sequential computer with main-memory size M and storage capacity $3 \cdot N$, with $3 \cdot M/B$ paging operations per superstep.*

If one is willing to program at the level of the operating system, then such explicit context switches as occur in SEQUENTIAL_EXECUTION might be performed several times faster than just leaving the paging to the standard pager.

4.2 Relation with BSP-Algorithms

An external-memory algorithm that work in accordance with the paradigm is called BSP-like, because, with a suitable definition of the parameters, it is a direct simulation of a parallel BSP algorithm. As we already implicitly assumed before: *In our cost estimates, we only consider paging-in operations.*

Corresponding to the parameters (p, g, l), we have (P, G, L):

$$
\begin{aligned}
P &= \lceil 3 \cdot N/M \rceil, \\
G &= \frac{\#\{\text{internal operations per second}\}}{\#\{\text{words that can be paged-in per second}\}}, \\
L &= M \cdot G/3i, \text{ the expression for synchronization / start-up time.}
\end{aligned}
$$

Notice that G is *not* a big number: it takes many steps to page-in a whole page of size B, but a page contains many words as well. For the computer plus hard-disc on our desk, $G \simeq 2 \cdot 10^6/4 \cdot 10^5 = 5$. The definition of G is completely general, and does not presuppose that there is only one hard-disc.

With these definitions, the correspondence with the BSP model is almost perfect. Let W_s be the total number of operations performed during Superstep s, and let

$$H_s = \sum_{0 \leq i,j < P} \mathit{Size}\{\mathit{Mes}(j,i,s)\},$$

then we can express the cost of a superstep as concisely as in (1):

Theorem 1 *The number of time steps $T_{superstep}$, to perform Superstep s, is given by*

$$T_{superstep}(W_s, H_s) = W_s + H_s \cdot G + P \cdot L.$$

Proof: Paging-in all data takes $\sum_i (B_i + \sum_j \mathit{Size}\{\mathit{Mes}(j,i,s)\}) \cdot G = P \cdot M/3 \cdot G + H_s \cdot G$. Performing the operations themselves takes W_s steps. □

Let S be the number of performed supersteps. Then we get the following expression for the total number of time steps T_{ext}, for solving the problem:

$$T_{\text{ext}}(P,G,L) = \sum_{s=1}^{S} (W_s + H_s \cdot G + P \cdot L). \tag{2}$$

The first term is due to computation, the last two to 'communication'. Considerations that apply to the BSP model can be repeated here. It also leads to our main theorem, which shows that the choice of the parameters is correct:

Theorem 2 *Consider an external-memory problem of size N, with parameters (P,G,L). Suppose, that for parameters $(p,g,l) = (P,G,L)$ there is a BSP algorithm in which no PU ever stores more than N/p data, running in $T_{par}(p,g,l)$ time steps. Let $T_{ext}(N,M)$ be the number of time steps required for the corresponding BSP-like algorithm. Then,*

$$T_{ext}(N,M) \leq p \cdot T_{par}(p,g,l).$$

Proof: Let S be the number of performed supersteps, then we know from (1), that

$$T_{\text{par}}(p,g,l) = \sum_{s=1}^{S} (w_s + h_s \cdot g + l).$$

Here w_s is the maximal work any PU has to perform in Superstep s, and likewise is h_s the maximum number of packets any PU has to send or receive at the end of Superstep s. Thus, for the corresponding BSP-like algorithm, $W_s \leq p \cdot w_s$ and $H_s \leq p \cdot h_s$. Combining these facts with (2), finishes the proof. □

4.3 Quality Measure

From the theory of BSP algorithms, we also adopt the following quality measure:

Definition 2 *An external-memory algorithm is x-optimal if the total number of time steps for its execution T_{ext} satisfies $T_{ext} \leq x \cdot T_{seq}$. Here T_{seq} gives the minimum number of steps for solving the problem on a sequential computer with an infinite memory. It is said to be asymptotically x-optimal, if $T_{ext} \leq (x + o(1)) \cdot T_{seq}$ for $M \to \infty$.*

Note that T_{seq} corresponds to the work that an algorithm has to do. In earlier papers [1, 3], the number of paging operations was considered as unique quality measure for external-memory algorithms. In many cases this may be adequate, but generally, this is only half the story. There are external-memory problems, for which asymptotical one-optimality is achievable. Not considering the work of such an algorithm, would amount to not considering its very essence.

Example 3 *Multiplication of two $n \times n$ matrices is a problem for which asymptotical one-optimality can be achieved (if we forget about the sub-cubic algorithms). As we have seen in Example 1, this problem can be solved such that each number is paged-in only $\mathcal{O}(n/\sqrt{M})$ times. Thus, the total paging takes $\mathcal{O}(n^3/\sqrt{M} \cdot G)$ time steps. It can be checked easily, that the number of computation steps equals that of the trivial row-times-column algorithm, which requires $\Omega(n^3)$ steps. Thus, the algorithm is $1 + \mathcal{O}(G/\sqrt{M})$ optimal. With constant G, this converges to 1 for $M \to \infty$.*

All this was known, but we have not seen before, that the importance of considering also the work of an external-memory algorithm was exposed so clearly. One of the strong points of our BSP-like paradigm, is that *both* important cost factors are 'automatically' taken into account. This gives more useful predictions of the performance.

In fact, the BSP-like paradigm seems even more suited for the design and analysis of external-memory algorithms, than the BSP model itself for parallel algorithms:

- The BSP model does not take the possibility of exploiting locality into account. On the communication time on a parallel computer, locality issues may be decisive.
- Start-up time, the effect that sending small packets is relatively expensive, is only very partially represented in the BSP model.
- The computation is assumed to proceed in rounds.

For BSP-like algorithms running on a sequential computer, the first two points do not carry over. The hard-disc behaves like a completely connected network: the cost of a communication pattern is solely determined by the amount of data to transfer plus the costs of a barrier synchronization. The third point remains. It is fundamental to any BSP-like approach, and though this might be a limitation, we believe that operating in rounds is a natural scheme that supports the understanding.

4.4 Typical Parameter Values

Values of M, N. Presently $M \simeq 10^7$, and $N \leq 10^{12}$.

Values of P, G, L. From the estimates for M and N, it follows that $P \leq 1000$. For a very fast computer, with a very slow hard-disc we may have $G = 100$, but mostly we will have $G < 10$. If there are several hard-discs, then G may even lie around 1. L has a rather extreme value: $L \simeq M$, this is far larger than the value of l in most parallel computers. Thus, our system can be compared to a parallel system that communicates through powerful dial-up links.

Value of W_s/H_s. For hard problems with little 'locality', W_s/H_s is a constant: after a constant number of internal operations a new argument is necessary. This situation occurs for problems like list ranking.

In other problems, e.g. those that have good mesh algorithms, there is much less need for communication. More formally, algorithms for d-dimensional meshes, $d \geq 1$, a constant, can be simulated with, $W_s/H_s = \Theta(M^{1/d})$. This implies that x-work-optimal algorithms for a finite dimensional mesh lead to an asymptotically x optimal external-memory algorithm. In another guise, this idea was already exploited in [13].

5 Designing BSP-Like Algorithms

One can try to design BSP-like algorithms from scratch. Obviously, minimization of the number of supersteps must be one of the principal goal, observing that L is much larger than G. Doing this, the BSP-like paradigm provides a framework which allows for exact cost prediction. This is valuable in its own right, but in addition, we provide an 'algorithms machinery' for generating algorithms that work according to the paradigm. Only the latter gives our paradigm its full right of existence.

5.1 Inheritance of Quality

We restate the goal of external-memory computation to be conform with our BSP-like framework:

Goal 1 *For a given problem of size N, that must be solved on a computer with memory size M, the goal is to come with an algorithm that is x-optimal for the minimum x.*

Theorem 3 *Consider an external-memory problem of size N, with parameters (P, G, L). Suppose, that for parameters $(p, g, l) = (P, G, L)$ there is an x-optimal BSP-algorithm for solving this problem on a parallel computer. If no PU ever stores more than N/p data, then the corresponding BSP-like algorithm is also x-optimal.*

Proof: Let $T_{\text{par}}(p, g, l)$ be the time for a BSP algorithm, running on a parallel computer with parameters (p, g, l). x-optimality is defined by $T_{\text{par}}(p, g, l) \leq x \cdot T_{\text{seq}}/p$. Multiplying left and right side with p, the theorem follows by combining with Theorem 2 and the definition of x-optimality for BSP-like algorithms. □

At this point we could conclude by supplying some more references to work on BSP algorithms: for many important problems extensive research has been performed on algorithms that give good, sometimes optimal, performance for large ranges of the parameters (p, g, l). All these results carry over immediately!

5.2 Limitations

Algorithms that Copy Data. In Theorem 2 and 3, we explicitly assume that in the BSP algorithm the PUs never store more than N/p data. This condition is necessary:

Example 4 *Consider computing $A \cdot B$, where A and B are $n \times n$ matrices. Initially each PU holds n^2/p entries of A and B. Now PU_i, $0 \leq i < p$, copies its entries of A*

to the PUs with indices $(i + k \cdot \sqrt{p}) \bmod p$, *for all* $0 \leq k < \sqrt{p}$; *and all its B entries to the PUs with indices* $(i + k) \bmod p$, $0 \leq k < \sqrt{p}$. *Hereafter, all products can be computed without further communication. In order to compute the sums, one more round of communication is enough. The work in each PU is optimal, each PU exchanges at most* $\mathcal{O}(n^2/\sqrt{p})$ *data, and the number of routing rounds is two.*

If there is enough storage capacity in each PU, if l is very large, and if g is moderate, then the algorithm of the example may be competitive. However, the corresponding BSP-like algorithm does not make sense as an external-memory algorithm.

Geometrically Decreasing Problem Sizes. Theorem 3 holds generally, but there are good external-memory algorithms that are not found by looking for the best BSP algorithm with parameters (P, G, L). In this sense our approach shares the weaknesses of the PRAM approach (see Example 2). This is not due to the BSP-like paradigm itself, but rather to our definition of L.

The sketched problem arises for algorithms with geometrically decreasing problem sizes. Generally, if during the algorithm the relevant set of data varies in size, then the number of 'PUs' can be varied accordingly. This is perfectly consistent with our BSP-like paradigm, and in (2), one only has to replace P by P_s, the number of PUs in Superstep. Such an algorithm can even be viewed as the sequential simulation of a BSP algorithm: it is easy to modify SEQUENTIAL_EXECUTION such that if in a BSP algorithm the problem size in Superstep s is N_s, that then the number of simulated PUs equals $P_s = \lceil 3 \cdot N_s/M \rceil$.

Example 5 *Consider ranking a list of length* $N > M$. *Because in all algorithms, the work of a PU is only a small factor larger than the total amount it has to route, the* routing volume, *we may concentrate on the number of supersteps and the routing volume. What is the best BSP algorithm for* $(p, g, l) = (P, G, L)$?

We compare the standard algorithms pointer jumping and independent-set removal (better algorithms and details are given in [12]). Pointer jumping can be performed in $\log N$ *supersteps, with a volume of* $2 \cdot \log N \cdot M$. *Independent-set removal requires approximately* $12 \cdot \log(N/M)$ *supersteps, and routing volume* $12 \cdot M$. *So, the costs of these algorithms,* $T_{par,\,poj}$ *and* $T_{par,\,isr}$, *respectively, are given by*

$$\begin{aligned} T_{par,\,poj} &\simeq 2 \cdot \log N \cdot (M/3) \cdot g + \log N \cdot l, \\ T_{par,\,isr} &\simeq 12 \cdot (M/3) \cdot g + 12 \cdot \log(N/M) \cdot l. \end{aligned}$$

It depends on N and M, but especially of l/g, *which algorithm is the better. In our case,* $l = (M/3) \cdot g$. *For this* l, *pointer jumping is better iff* $3 \cdot \log N < 12 + 12 \cdot \log(N/M)$. *For* $N = 10^9$ *and* $M = 10^6$, *this happens. As a BSP algorithm, for these* (p, g, l), *pointer jumping will indeed be better than independent-set removal. However, the costs of the corresponding BSP-like algorithms (neglecting the work) are given by*

$$\begin{aligned} T_{ext,\,poj} &\simeq 3 \cdot \log N \cdot N \cdot G \simeq 90 \cdot N \cdot G, \\ T_{ext,\,isr} &\simeq \sum_s (3 \cdot (0.75^s \cdot N) \cdot G + 6 \cdot (0.75^s \cdot P) \cdot L) \leq 36 \cdot N \cdot G. \end{aligned}$$

Thus, as an external-memory algorithm, independent-set-removal is clearly faster.

The problem is that Theorem 3 is not tight, it gives only a one-sided guarantee. In a BSP algorithm there is real a difference between reducing the load in the PUs and reducing the number of PUs, while in a BSP-like algorithm, if L is adapted, this is more or less the same. Therefore do not think that (P, G, L) can be defined such that Theorem 3 becomes tight: the correspondence between the problems is not one-to-one.

6 The Paradigm at Work

In the remainder we demonstrate the potential of the paradigm on some basic problems.

Matrix Multiplication. We consider the multiplication of two $n \times n$ matrices on a sequential computer with $n < M < n^2$. In Example 3, we have seen that the third algorithm from Example 1 is asymptotically one-optimal. Here we show that it is distinguished from the second algorithm in Example 1, by looking for the best BSP matrix-multiplication algorithm for the instance with $(p, g, l) = (3 \cdot n^2/M, G, M \cdot G/3)$.

For the work in the BSP algorithm, we only consider multiplications. $T_{\text{par, stripe}}$ and $T_{\text{par, square}}$ denote the time for the algorithms in Example 1. $T_{\text{par, square}}$ is smaller:

$$T_{\text{par, stripe}} = \sum_{s=1}^{n^2 \cdot 3/M} M^2/(9 \cdot n) + M/3 \cdot g + l = n \cdot M/3 + n^2 \cdot g + n^2 \cdot 3/M \cdot l,$$

$$T_{\text{par, square}} = \sum_{s=1}^{n \cdot \sqrt{3/M}} ((M/3)^{3/2} + M/3 \cdot g + l) = nM/3 + ng\sqrt{M/3} + nl\sqrt{3/M}.$$

Sorting. Next, we consider sorting N numbers. What is the best BSP algorithm for $(p, g, l) = (3 \cdot N/M, G, M \cdot G/3)$? We only compare the first and the last algorithm from Example 2: column-sort [7] and merge sort [5]. The time consumptions of these algorithms are denoted $T_{\text{par, column}}$ and $T_{\text{par, merge}}$, respectively. $T_{\text{seq, sort}}(N)$ is the time for sorting N numbers with the best sequential sorting algorithm. For $T_{\text{par, merge}}$ we only give a lower estimate. A modification of an algorithm in [11] gives a BSP algorithm for column-sort consumption:

$$T_{\text{par, column}} = (1 + o(1)) \cdot T_{\text{seq, sort}}(M/3) + M \cdot g + 3 \cdot l = (1 + o(1)) \cdot T_{\text{seq}}(M/3),$$

$$T_{\text{par, merge}} \geq T_{\text{seq, sort}}(M/3) + \Omega(M \cdot g) + \Omega(\log(M/N) \cdot l).$$

$T_{\text{par, merge}}$ is larger. Again, the better external-memory algorithm can be found by just analyzing BSP performances.

Transitive Closure. Now we turn to problems for which we did not know an external-memory algorithm so far. The first problem is that of computing transitive closure of a general graph, and the related problems (LU decomposition, all pairs shortest paths, etc.), which sequentially are solved in $\mathcal{O}(n^3)$ time for a graph with n nodes. There is a BSP algorithm [8] with $T_{\text{par, closure}} = n^3/p + n^2/\sqrt{p} \cdot g + \sqrt{p} \cdot l$. This requires only $\mathcal{O}(n^2/p)$ memory per PU, so the condition of Theorem 2 is satisfied:

Theorem 4 *There is an external-memory transitive closure algorithm running in*

$$T_{\textit{ext, closure}}(N, M, G) = \mathcal{O}(N^{3/2} + N^{3/2} \cdot G/M^{1/2}).$$

Fast Fourier Transform. We consider the fast Fourier transform, FFT, for a vector of length N. The sequential complexity of this problem is $\mathcal{O}(N \cdot \log N)$.

On a parallel computer with $p \leq \sqrt{N}$, there is a BSP algorithm, recently described in [14] with $T_{\text{par, FFT}} = N \cdot \log N/p + N/p \cdot g + l$. Similarly to Theorem 4, this implies

Theorem 5 *If $M^2 \geq N$, then there is an external-memory algorithm for the FFT problem with running time $T_{\textit{ext, FFT}} = \mathcal{O}(N \cdot \log N + N \cdot G)$.*

For transitive closure and for FFT, the time for paging data is asymptotically negligible. Both algorithms are asymptotically $\mathcal{O}(1)$-optimal.

7 Conclusion

We proposed the use of BSP-like algorithms as a general approach in the design and analysis of external-memory algorithms. With our paradigm, the work of an algorithm can be modeled in an integrated way. Earlier approaches mostly neglected the work, which is not correct in general. The quality of an BSP-like algorithm is expressed by its x-optimality. We provided kind of a machine for generating good external-memory algorithms, by proving the intimate link between BSP algorithms for parallel computers, and BSP-like algorithms: x-optimal BSP algorithms, give rise to x-optimal BSP-like algorithms. In many, but not all, cases the optimal external-memory algorithm can be identified by searching for the optimal BSP algorithm. We demonstrated this for several specific problems. Given the fact that there is far more research activity in the field of BSP algorithms (see [9] for references), it is to be expected that there are many more external-memory algorithms to be discovered in this way.

References

[1] Aggarwal, A., J.S. Vitter, 'The Input/Output Complexity of Sorting and Related Problems,' *Communications of the ACM*, 31(9), pp. 1116–1127, 1988.

[2] Bilardi, G., F.P. Preparata, 'Horizons of Parallel Computation,' *Journal of Parallel and Distributed Computing*, 27, pp. 172-182, 1995.

[3] Chiang, Y-J, M.T. Goodrich, E.F. Grove, R. Tamassia, D.E. Vengroff, J.S. Vitter, 'External-Memory Graph Algorithms,' *Proc. 6th Symposium on Discrete Algorithms*, pp. 139–149, ACM-SIAM, 1995.

[4] Cormen, T.H., *Virtual Memory for Data Parallel Computing*, Ph. D. Thesis, Department of Electrical Engineering and Computer Science, Massachusetts Institute of Technology, 1992.

[5] Cole, C., 'Parallel Merge Sort,' *SIAM Journal of Computing*, 17(4), pp. 770–785, 1988.

[6] JáJá, J., *An Introduction to Parallel Algorithms*, Addison-Wesley, 1992.

[7] Leighton, F.T., 'Tight Bounds on the Complexity of Parallel Sorting,' *IEEE Transactions on Computers, C-34(4)*, pp. 344–354, 1985.

[8] McColl, W.F., 'Scalable Computing,' *Computer Science Today: Recent Trends and Developments*, J. van Leeuwen (Ed.), LNCS 1000, pp. 46–61, Springer-Verlag, 1995.

[9] McColl, W.F., 'Universal Computing,' *Proc. 2nd Euro-Par Conference*, LNCS 1123, pp. 25–36, Springer-Verlag, 1996.

[10] Patt, Y.N., 'The I/O Subsystem – A Candidate for Improvement,' *IEEE Computer*, 27(3), pp. 15–16, 1994.

[11] Sibeyn, J.F., 'Deterministic Routing and Sorting on Rings,' *Proc. 8th International Parallel Processing Symposium*, pp. 406–410, IEEE, 1994.

[12] Sibeyn, J.F., 'Better Trade-offs for Parallel List Ranking,' submitted to *Algorithmica*, 1996.

[13] Sibeyn, J.F., T. Harris, 'Exploiting Locality in LT-RAM Computation,' *Proc. 4th Scandinavian Workshop on Algorithm Theory*, LNCS 824, pp. 338–349, Springer-Verlag, 1994.

[14] Tishkin, A., 'The Bulk-Synchronous Parallel Random Access Machine,' *Proc. 2nd Euro-Par Conference*, LNCS 1124, pp. 327–338, Springer-Verlag, 1996.

[15] Valiant, L.G., 'A Bridging Model for Parallel Computation,' *Communications of the ACM*, 33(8), pp. 103–111, 1990.

Topological Chaos for Elementary Cellular Automata*

Gianpiero Cattaneo[1], Michele Finelli[2] and Luciano Margara[2,3]

[1] Dipartimento di Scienze dell'Informazione, Università di Milano, Italy.
[2] Dipartimento di Scienze dell'Informazione, Università di Bologna, Italy.
[3] Istituto di Matematica Computazionale del CNR, Pisa, Italy.

Abstract. We apply the definition of chaos given by Devaney for discrete time dynamical systems to the case of elementary cellular automata, i.e., 1-dimensional binary cellular automata with radius 1. A discrete time dynamical system is chaotic according to the Devaney's definition of chaos if it is topologically transitive, is sensitive to initial conditions, and has dense periodic orbits. We enucleate an easy-to-check property of the local rule on which a cellular automaton is based which is a necessary condition for chaotic behavior. We prove that this property is also sufficient for a large class of elementary cellular automata. The main contribution of this paper is the formal proof of chaoticity for many non additive elementary cellular automata. Finally, we prove that the above mentioned property does not remain a necessary condition for chaoticity in the case of non elementary cellular automata.

1 Introduction

The notion of chaos is very appealing, and it has intrigued many scientists (see [1, 2, 6, 10, 13] for some works on the properties that characterize a chaotic process). There are simple deterministic dynamical systems that exhibit unpredictable behavior. Though counterintuitive, this fact has a very clear explanation. The lack of *infinite precision* in describing the state of the system causes a loss of *information* which is dramatic for some processes which quickly loose their deterministic nature to assume a non deterministic (unpredictable) one.

A chaotic phenomenon can indeed be viewed as a deterministic one, in the presence of infinite precision, and as a nondeterministic one, in the presence of finite precision constraints.

Thus one should look at chaotic processes as at processes merged into time, space, and precision bounds, which are the key resources in the science of computing. A nice way in which one can analyze this finite/infinite dichotomy is by using cellular automata (CA) models. Consider the 1-dimensional CA (X, σ), where $X = \{0,1\}^{\mathbf{Z}}$ and σ is the shift map on X. In order to completely describe the elements of X, we need to operate on sequences of binary digits of infinite length. Assume for a moment that this is possible. Then the shift map is

* Partially supported by MURST 40% and 60% funds.

completely predictable, i.e., one can completely describe $\sigma^n(x)$, for any $x \in X$ and for any integer n. In practice, only finite objects can be computationally manipulated. Let $x \in X$. Assume we know a portion of x of length n (the portion between the two vertical lines in Figure 1). One can easily verify that $\sigma^n(x)$ completely depends on the unknown portion of x. In other words, if we have finite precision, the shift map becomes unpredictable, as a consequence of the combination of the finite precision representation of x and the *sensitivity* of σ.

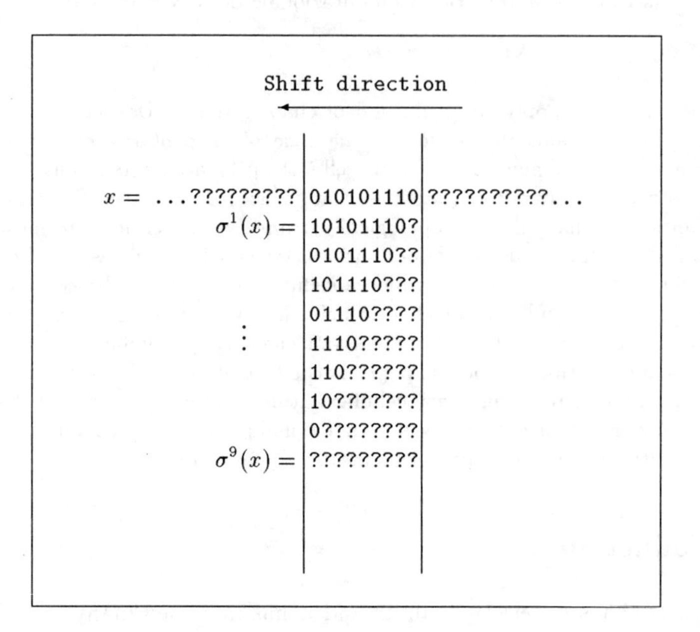

Fig. 1. Finite precision combined with sensitivity to initial conditions causes unpredictability after a few iterations (x represents the state of the CA at time step 0, and $\sigma^i(x)$ the state at time step i).

In the case of discrete time dynamical systems (DTDS) defined on a metric space, many definitions of chaos are based on the notion of sensitivity (see for example [6, 9]).

We now recall the definition of sensitivity to initial conditions for a DTDS (X, F). Here, we assume that X is equipped with a distance d and that the map F is continuous on X according to the metric topology induced by d.

Definition 1. A DTDS (X, F) is sensitive to initial conditions if and only if there exists $\delta > 0$ such that for any $x \in X$ and for any neighborhood $N(x)$ of x, there is a point $y \in N(x)$ and a natural number n, such that $d(F^n(x), F^n(y)) > \delta$. δ is called the sensitivity constant.

Intuitively, a map is sensitive to initial conditions, or simply sensitive, if there exist points arbitrarily close to x which eventually separate from x by at least δ under iteration of F. We emphasize that not all points near x need eventually separate from x, but there must be at least one such point in every neighborhood of x. If a map possesses sensitive dependence on initial conditions, then for all practical purposes, the dynamics of the map defies numerical approximation. Small errors in computation which are introduced by round-off may become magnified upon iteration. The results of numerical computation of an orbit, no matter how accurate, may be completely different from the real orbit.

In the case of continuous dynamical systems defined on a metric space, there are many possible definitions of chaos, ranging from measure theoretic notions of randomness in ergodic theory to the topological approach we will adopt here.

We now recall two other properties which are central to topological chaos theory: *topological transitivity* and *denseness of periodic orbits.*

Definition 2. A dynamical system (X, F) is topologically transitive if and only if for all non empty open subsets U and V of X there exists a natural number n such that $F^n(U) \cap V \neq \emptyset$.

Intuitively, a topologically transitive map has points which eventually move under iteration from one arbitrarily small neighborhood to any other. As a consequence, the dynamical system cannot be decomposed into two disjoint open sets which are invariant under the map.

Definition 3. Let $P(F) = \{x \in X |\ \exists n \in \mathbf{N} : F^n(x) = x\}$ be the set of the periodic points of F. A dynamical system (X, F) has dense periodic orbits if and only if $P(F)$ is a dense subset of X, i.e., for any $x \in X$ and $\epsilon > 0$, there exists $y \in P(F)$ such that $d(x, y) < \epsilon$.

Denseness of periodic orbits is often referred to as the *element of regularity* a chaotic dynamical system must exhibit.

The popular book by Devaney [6] isolates the above defined three components as being the essential features of chaos. They are formulated for a continuous map F, $F : X \to X$, on some metric space (X, d).

Definition 4. Let F, $F : X \to X$, be a continuous map on a metric space (X, d). Then the dynamical system (X, F) is chaotic according to the Devaney's definition of chaos if and only if

(1) F is topologically transitive,
(2) F has dense periodic orbits, and
(3) F is sensitive to initial conditions.

In [5] one of the authors proved that, in the case of CA, topological transitivity implies sensitivity to initial conditions. As a consequence of the above mentioned result, to prove that a CA is chaotic in the sense of Devaney, one has to prove that it is topologically transitive and has dense periodic orbits.

In this paper we apply the Devaney's definition of chaos to the class of elementary cellular automata (ECA), i.e., binary 1-dimensional CA with radius 1. To this extent, we introduce the notion of *permutivity* of a map in a certain variable. A boolean map f is permutive in the variable x_i if $f(\ldots, x_i, \ldots) = 1 - f(\ldots, 1 - x_i, \ldots)$. In other words, f is permutive in the variable x_i if any change of the value of x_i causes a change of the output produced by f, independently of the values assumed by the other variables.

The main results of this paper can be summarized as follows.

- An ECA based on a local rule f is topologically transitive if and only if f is permutive either in the first (leftmost) or in the last (rightmost) variable.
- All the ECA based on a local rule f which is of the form $(1+g)$ mod 2 where g is an additive local rule are chaotic in the sense of Devaney.
- All the ECA based on a local rule f which is permutive either in the leftmost or in the rightmost variable and such that $f(0,0,0) \neq f(1,1,1)$ are chaotic in the sense of Devaney.
- There exists a chaotic (in the sense of Devaney) CA based on a local rule f with radius 1 which is not permutive in any variable.
- There exists a chaotic (in the sense of Devaney) CA defined on a binary set of states based on a local rule f which is not permutive in any variable.

In the case of 1-dimensional CA, there have been many attempts of classification according to their asymptotic behavior (see for example [3, 7, 12, 14]), but none of them completely captures the notion of chaos. As an example, Wolfram divides 1-dimensional CA in four classes according to the outcome of a large number of experiments. Wolfram's classification scheme, which does not rely on a precise mathematical definition, has been formalized by Culik and Yu [4] who split CA in three classes of increasing complexity. Unfortunately membership in each of these classes is shown to be undecidable. We wish to emphasize that in this paper we propose the first classification of the ECA rule space based on a widely accepted rigorous mathematical definition of chaos.

The rest of this paper is organized as follows. In Section 2 we give basic notation and definitions. In Section 3 we classify ECA rule space according to the Devaney's definition of chaos. In Section 4 we prove that leftmost and/or rightmost permutivity is not a necessary condition for chaotic behavior of non elementary CA.

2 Notations and Definitions

Let $\mathcal{A} = \{0, 1, \ldots, m-1\}$ be a finite alphabet and f, $f : \mathcal{A}^{2k+1} \to \mathcal{A}$, be any map. A 1-dimensional CA based on the *local rule* f is a pair $(\mathcal{A}^{\mathbf{Z}}, F)$, where

$$\mathcal{A}^{\mathbf{Z}} = \{c \mid c : \mathbf{Z} \to \mathcal{A}\}$$

is the space of configurations and F, $F : \mathcal{A}^{\mathbf{Z}} \to \mathcal{A}^{\mathbf{Z}}$, is defined as follows. For every $i \in \mathbf{Z}$

$$F[c](i) = f(c(i-k), \ldots, c(i+k)), \; c \in \mathcal{A}^{\mathbf{Z}}.$$

We assume that f depends on the $2k+1$ variables $x_{-k}, \ldots, x_k$ and we say that k is the *radius* of f. Throughout the paper, $F[c]$ will denote the result of the application of the map F to the configuration c and $c(i)$ will denote the ith element of the configuration c. We recursively define $F^n[c]$ by $F^n[c] = F[F^{n-1}[c]]$, where $F^0[c] = c$.

We now give the definition of permutive local rule and that of leftmost [rightmost] permutive local rule, respectively.

Definition 5. [8] f is permutive in x_i, $-k \le i \le k$, if and only if, for any given sequence

$$\overline{x}_{-k}, \ldots, \overline{x}_{i-1}, \overline{x}_{i+1}, \ldots, \overline{x}_k \in \mathcal{A}^{2k},$$

we have

$$\{f(\overline{x}_{-k}, \ldots, \overline{x}_{i-1}, x_i, \overline{x}_{i+1}, \ldots, \overline{x}_k) : \; x_i \in \mathcal{A}\} = \mathcal{A}.$$

Definition 6. f is leftmost [rightmost] permutive if and only if there exists an integer i, $-k \le i \le k$, such that

- $i \ne 0$,
- f is permutive in the ith variable, and
- f does not depend on x_j, $j < i$, [$j > i$].

Let g, $g : \mathcal{A} \to \mathcal{A}$, be any map. We say that a local rule f, $f : \mathcal{A}^{2k+1} \to \mathcal{A}$, is *trivial* if it satisfies $f(x_{-k}, \ldots, x_k) = g(x_0)$. Trivial CA (CA based on a trivial local rule) exhibit a very simple behavior. We define the following notion of distance (known as *Tychonoff distance*) over the space of configurations $\mathcal{A}^{\mathbf{Z}}$. For every $a, b \in \mathcal{A}^{\mathbf{Z}}$

$$d(a,b) = \sum_{i=-\infty}^{+\infty} \frac{1}{m^{|i|}} |a(i) - b(i)|,$$

where m denotes the cardinality of $\mathcal{A}$. It is easy to verify that d is a metric on $\mathcal{A}^{\mathbf{Z}}$ and that the metric topology induced by d coincides with the product topology induced by the discrete topology of $\mathcal{A}$. With this topology, $\mathcal{A}^{\mathbf{Z}}$ is a compact and totally disconnected space and F is a (uniformly) continuous map.

Definition 7. A 1-dimensional CA based on a local rule f, $f : \mathcal{A}^{2k+1} \to \mathcal{A}$, is an elementary CA (ECA) if and only if $k = 1$ and $\mathcal{A} = \{0, 1\}$.

We enumerate the $2^{2^3} = 256$ different ECA as follows. The ECA based on the local rule f is associated to the natural number n_f, where

$$n_f = f(0,0,0) \cdot 2^0 + f(0,0,1) \cdot 2^1 + \cdots + f(1,1,0) \cdot 2^6 + f(1,1,1) \cdot 2^7.$$

3 Chaotic Elementary Cellular Automata

In this section we classify ECA rule space according to the Devaney's Definition of Chaos.

3.1 Topological transitivity

We recall the following result due to one of the authors.

Theorem 8. *[11] Let $(\mathcal{A}^{\mathbf{Z}}, F)$ be any leftmost [rightmost] permutive 1-dimensional CA defined on a finite alphabet $\mathcal{A}$ of prime cardinality. Then $(\mathcal{A}^{\mathbf{Z}}, F)$ is topologically transitive.*

As a consequence of Theorem 8, we have that all the leftmost [rightmost] permutive ECA are topologically transitive.

We now prove that if an ECA is neither leftmost nor rightmost permutive, then it is not surjective and then not topologically transitive. Let $\mathcal{A} = \{0, \ldots, m-1\}$ be any finite alphabet. Let f be any local rule of radius $k \geq 0$ defined on $\mathcal{A}$. We define f_n, $f_n : \mathcal{A}^{n+2k} \to \mathcal{A}^n$, as follows. For every $c \in \mathcal{A}^{n+2k}$

$$f_n[c](i) = f(c(i), \ldots, c(i+2k)),\ 1 \leq i \leq n.$$

We denote by $f_n^{-1}[a]$ the set of the predecessors of $a \in \mathcal{A}^n$ according to the map f_n and by $\#(f_n^{-1}[a])$ its cardinality. We say that a finite configuration c_n of length n is *circular* if and only if $c_n(1) = c_n(n-1)$ and $c_n(2) = c_n(n)$.

We recall the following result due to Hedlund.

Theorem 9. *[8] Let $(\mathcal{A}^{Z}, F)$ be the CA based on the local rule f, $f : \mathcal{A}^{2k+1} \to \mathcal{A}$, with radius $k \geq 0$. Then the two following statements are equivalent.*

(i) F is surjective.
(ii) For every $n \geq 1$, and for every $a \in \mathcal{A}^n$, $\#(f_n^{-1}[a]) = m^{2k}$.

Let f, $f : \mathcal{A}^{2k+1} \to \mathcal{A}$, be any local rule with radius $k \geq 0$. We say that f is *balanced* if for any $t \in \mathcal{A}$ $\#(f_1^{-1}[t]) = m^{2k}$. We have the following result.

Theorem 10. *Let $(\{0,1\}^{\mathbf{Z}}, F)$ be a non trivial CA based on the local rule f, $f : \{0,1\}^3 \to \{0,1\}$. If F is surjective, then f is either leftmost or rightmost permutive.*

Proof. Assume that f is a balanced local rule which is neither leftmost nor rightmost permutive. Then there exist two finite configurations c_1 and c_2 of length 2 such that $f_1[0c_1] = f_1[1c_1] = a$ and $f_1[c_20] = f_1[c_21] = b$, where $a, b \in \{0,1\}$. Consider the ECA for which $c_1 = c_2$. We have that $\#(f_2^{-1}[ab]) \geq 4$. Since f is balanced, one can readily verify that $\#(f_2^{-1}[ab]) > 4$, and then, by Theorem 9, we have that F is not surjective. Consider now the case $c_1 \neq c_2$. There are only

4 balanced non trivial ECA which are neither leftmost nor rightmost permutive for which $c_1 \neq c_2$. They are ECA 43,113,142, and 212. For these ECA, it is easy to check that $\#(f_3^{-1}[010]) = 3$. Again, from Theorem 9, we have that F is not surjective.

The next Corollary is a direct consequence of Theorems 8 and 10.

Corollary 11. *Let $(\{0,1\}^{\mathbf{Z}}, F)$ be an ECA based on the local rule f. Then, the two following statements are equivalent.*

(i) F is topologically transitive.
(ii) f is either leftmost or rightmost permutive (or both).

3.2 Denseness of Periodic Orbits

In this section we investigate the property of having dense periodic orbits. It has been shown in [11] that additive 1-dimensional cellular automata defined on any finite alphabet of prime cardinality have dense periodic orbits. Here we prove that a large class of non additive ECA have dense periodic orbits.

We first consider the subclass of ECA based on a local rule f of the form $(1 + g) \bmod 2$, where g is an additive local rule.

Let $\cdots a \cdots \in \{0,1\}^{\mathbf{Z}}$ denote the all a's constant configuration, where $a \in \{0,1\}$. We have the following result

Theorem 12. *Let f_i, $i \in \{0,1\}$, be an additive local rule such that $f_i(1,1,1) = i$. Let $g_i = (1 + f_i) \bmod 2$, $i \in \{0,1\}$. Let F_i and G_i be the ECA based on f_i and g_i, respectively. Then*

(i) $x \in \{0,1\}^{\mathbf{Z}}$ is periodic for F_0 if and only if $(\cdots 1 \cdots + x) \bmod 2$ is periodic for G_0,
(ii) $x \in \{0,1\}^{\mathbf{Z}}$ is periodic for F_1 if and only if x is periodic for G_1.

Proof. Let x be a periodic point for F_i, $i \in \{0,1\}$, i.e., there exists $n \in \mathbf{N}$ that $F_i^n[x] = x$.

Since F_i is additive and by the definition of G_i we have

$$
\begin{aligned}
G_i^n[x] &= G_i^{n-1}[G[x]] \\
&= G_i^{n-1}[(\cdots 1 \cdots + F_i[x]) \bmod 2] \\
&= G_i^{n-2}[G_i[(\cdots 1 \cdots + F_i[x]) \bmod 2]] \\
&= G_i^{n-2}[(\cdots 1 \cdots + F_i[(\cdots 1 \cdots + F_i[x]) \bmod 2]) \bmod 2] \\
&= G_i^{n-2}[(\cdots 1 \cdots + F_i[\cdots 1 \cdots] + F_i^2[x]) \bmod 2] \\
&\;\;\vdots \\
&= (\cdots 1 \cdots + F_i[\cdots 1 \cdots] + \cdots + F_i^{n-1}[\cdots 1 \cdots] + F_i^n[x]) \bmod 2. \qquad (1)
\end{aligned}
$$

Case (i). From Equation 1 we have

$$G_0^n[x] = (\cdots 1 \cdots + x) \bmod 2,$$
$$G_0^{2n}[x] = (\cdots 1 \cdots + x) \bmod 2.$$

Case (ii). Since $2n$ is even, from Equation 1 we have $G_1^{2n}[x] = x$.

The following corollary is a direct consequence of Theorems 12.

Corollary 13. *Let* $(\{0,1\}^{\mathbf{Z}}, F)$ *be an ECA based on a local rule* f *of the form* $(1+g) \bmod 2$, *where* f *is a non trivial additive local rule. Then* $P(F)$ *is a dense subset of* $\{0,1\}^{\mathbf{Z}}$.

We now study another subclass of non additive ECA,

those based on a local rule f (which is not of the form $(1+g) \bmod 2$ with g additive) which satisfies the following two properties.

(a) f is either rightmost or leftmost permutive, and
(b) $f(0,0,0) \neq f(1,1,1)$.

We proceed as follows. We first prove that ECA 75 has dense periodic orbits and then we derive the same result for the other ECA of the above described class.

Theorem 14. *Let* f *be the local rule 75. Then every finite configuration of odd length has exactly 1 circular predecessor.*

Proof. Let $(\{0,1\}^{\mathbf{Z}}, F)$ be the ECA based on the local rule f. Consider a finite configuration c_n of odd length n. Since F is surjective we have that $\#(f_n^{-1}[c_n]) = 4$. We say that a set of predecessors of a finite configuration is *admissible* if and only if at least one of them is circular. Consider the sets $S_3^1, \ldots, S_3^8$ of predecessors of the configurations $c_3^1, \ldots, c_3^8$ of length 3. One can verify that all of them are admissible. As an example, the set of predecessors of $c_3^2 = 010$ given by

$$S_3^2 = \{11100, 11101, 10111, 10010\}$$

is admissible since 10010 is a circular predecessor of 010, i.e., $f_3[10010] = 010$.

Let $T_3^1, \ldots, T_3^8$ be the sets of finite configurations of length 4 obtained from $S_3^1, \ldots, S_3^8$ by taking the first and the last 2 bits of each configuration of length 5. As an example, we have

$$T_3^8 = \{11-00, 11-01, 10-11, 10-10\}.$$

We say that T_i^j is *admissible* if and only if at least one of its element is of the form $ab - ab$. Consider now the sets of predecessors of all the configurations of length 5 (they can be easily obtained from the sets of predecessors of the configurations of length 3). All of them are admissible. The sets $T_5^1, \ldots, T_5^{32}$ can be obtained as above by taking the first and the last 2 bits of each configuration of length 7 or they can be directly obtained from the sets $T_3^1, \ldots, T_3^8$.

Consider now the sequence

$$\mathcal{T} = T_1^1, \ldots, T_1^4, T_3^1, \ldots, T_3^8, T_5^1, \ldots, T_5^{32}, \ldots$$

Since each T_i^j is a finite size object (it consists of 4 configurations of length 4) and since each T_i^j can be deterministically constructed starting from another T_h^k with $h = i - 2$, we conclude that after a finite number of steps no new set T_i^j can be generated.

With the aid of a personal computer we have verified that there exists a finite number $k \geq 1$ such that (i) all the sets $T_{2h+1}^1, \ldots, T_{2h+1}^{2^{2h+1}}$ of finite configurations of length $2h+1$ with $0 < h < k$ are admissible and (ii)

$$\bigcup_{i=1}^{2^{2k+1}} T_{2k+1}^i \subseteq \bigcup_{i=0}^{k-1} \bigcup_{j=1}^{2^{2i+1}} T_{2i+1}^j.$$

Properties (i) and (ii) ensure that all the sets of predecessors of finite configurations of odd length are admissible. As a consequence, every configuration of odd length has at least (and then exactly) one circular predecessor.

The following corollary is a direct consequence of Theorem 14.

Corollary 15. *Let f be the elementary local rule 75 and $(\{0,1\}^{\mathbf{Z}}, F)$ be the ECA based on f. Then $P(F)$ is a dense subset of $\{0,1\}^{\mathbf{Z}}$.*

Proof. Let $c_n \in \{0,1\}^n$ be any finite configuration of odd length n. Let $c \in \{0,1\}^{\mathbf{Z}}$ be defined by $c = \cdots c_n c_n c_n \cdots$. From Theorem 14 we have that c is a periodic configuration for F. Since c_n is an arbitrary configuration of odd length we have that the periodic points of F are dense in $\{0,1\}^{\mathbf{Z}}$.

We now prove that ECA 180 has dense periodic orbits.

Corollary 16. *ECA 180 has dense periodic orbits.*

Proof. Let f and g denote the local rules 75 and 180, respectively. We prove that each finite configuration $c_n \in \{0,1\}^n$, with n odd, has a circular predecessor according to the map g_n. Let $\overline{c_n}$ be defined by

$$\overline{c_n}(i) = \begin{cases} 1 \text{ if } c_n(i) = 0, \\ 0 \text{ if } c_n(i) = 1. \end{cases}$$

By Theorem 14 there exists a circular configuration $b \in \{0,1\}^{n+2}$ such that $F[b] = \overline{c_n}$. Since $f(x,y,z) = (1 + g(x,y,z)) \bmod 2$, we have that $g_n[b] = c_n$. This completes the proof.

We now prove that all the remaining leftmost or rightmost ECA have dense periodic orbits.

Corollary 17. *ECA 45, 89, and 101 have dense periodic orbits.*

Proof. (Sketch) Let f_1, f_2, f_3, and f_4 denote local rules 75, 45, 89, and 101. It is easy to verify that

$$f_1(x_{-1}, x_0, x_1) = 1 - f_2(1 - x_{-1}, 1 - x_0, 1 - x_1), \tag{2}$$
$$f_1(x_{-1}, x_0, x_1) = f_3(x_1, x_0, x_{-1}), \tag{3}$$
$$f_1(x_{-1}, x_0, x_1) = 1 - f_4(1 - x_1, 1 - x_0, 1 - x_{-1}). \tag{4}$$

As a consequence of Equations 2,3, and 4 ECA based on the local rules f_1, f_2, f_3, and f_4 enjoy the same set of topological and metric properties.

Corollary 18. *ECA 210, 166, and 154 have dense periodic orbits.*

Proof. (Sketch) Let f_1, f_2, f_3, and f_4 be the local rules 180, 210, 166, and 154. It is easy to verify that

$$f_1(x_{-1}, x_0, x_1) = 1 - f_2(1 - x_{-1}, 1 - x_0, 1 - x_1), \tag{5}$$
$$f_1(x_{-1}, x_0, x_1) = f_3(x_1, x_0, x_{-1}), \tag{6}$$
$$f_1(x_{-1}, x_0, x_1) = 1 - f_4(1 - x_1, 1 - x_0, 1 - x_{-1}). \tag{7}$$

As a consequence of Equations 5,6, and 7 ECA based on the local rules f_1, f_2, f_3, and f_4 enjoy the same set of topological and metric properties.

Since ECA 75, 45, 210, 180, 166, 89, 154, 101 are either leftmost or rightmost permutive, by Theorem 8 we have that they are topologically transitive and then chaotic in the sense of Devaney.

3.3 ECA rule space classification

In order to give a complete classification of the ECA rule space according to the Devaney's definition of chaos it remains to answer the following question.

Open Problem 19 *Leftmost or rightmost permutive ECA based on a local rule f, (not of the form $(1+g) \bmod 2$ with g additive) such that $f(0,0,0) = f(1,1,1)$ (ECA 30, 86, 106, 120, 135, 149, 169, 225) have dense periodic orbits ?*

We conjecture that the answer to Open Problem 19 is Yes, i.e., ECA 30, 86, 106, 120, 135, 149, 169, 225 have dense periodic orbits and then they are chaotic in the sense of Devaney. In this case, chaos for ECA would reduce to the property of being either leftmost or rightmost permutive.

4 Permutivity Vs Chaos for General Cellular Automata

In this section we discuss the relation between leftmost and/or rightmost permutivity and the Devaney's definition of chaos in the case of non elementary CA (NECA). We prove that

(i) There exist a chaotic CA based on a local rule f with radius 1 which is not permutive in any input variable of f.
(ii) There exist a chaotic CA defined on a binary set of states based on a local rule f which is not permutive in any input variable of f.

Proof of (i) (Sketch). We introduce a particular equivalence relation between dynamical systems: the *topological conjugacy*. Let (X, F) and (Y, G) be two dynamical systems defined over the spaces X and Y, respectively. We say that (X, F) is topologically conjugate to (Y, G) if there exists a homeomorphism H, $H : X \to Y$, such that $G \circ H = H \circ F$, where $F \circ G$ denotes the composition of functions F and G. It can be proved that if (X, F) and (Y, G) are topologically conjugate, then they satisfy the same topological and set theoretic properties.

We now construct a NECA which is topologically conjugate with the chaotic ECA 90. The conjugacy is given by the injective binary CA based on the local rule h defined by

$$h(x_{-1}, x_0, x_1, x_2) = x_0 + x_{-1}x_2(1 + x_1).$$

It is easy to verify that $h \circ h = Id$, where Id is the identity local rule. Let g be the local rule defined by $g = h \circ f_{90} \circ h$

It is easy to verify that g is neither leftmost nor rightmost permutive. Since the binary CA based on g is topologically conjugate to the ECA 90, it satisfies the same topological properties satisfied by ECA 90 and then it is chaotic in the sense of Devaney.

Proof of (ii) (Sketch). We now construct a NECA with radius 1 over an alphabet of cardinality 512 which is chaotic in the sense of Devaney but it is neither leftmost nor rightmost permutive. We proceed as follows. Let $\mathcal{A} = \{0, 1, \ldots, 511\}$. For each $a \in \mathcal{A}$ consider the sequence $a_0, \ldots, a_8$, $a_j \in \{0, 1\}$, such that $\sum_{i=0}^{8} a_i 2^i = a$. Let f, $f : \mathcal{A}^3 \to \mathcal{A}$, be the local rule defined by

$$f(a, b, c) = d \Leftrightarrow g_9[a_6, a_7, a_8, b_0, \ldots, b_8, c_0, \ldots, c_5] = d_0, \ldots, d_8.$$

It takes a little effort to verify that the NECA based on the local rule f is topologically conjugated to the ECA 90 and then it is chaotic in the sense of Devaney. Since g is neither leftmost nor rightmost permutive, we conclude that f is neither leftmost nor rightmost permutive.

5 Conclusions

We have classified elementary cellular automata rule space according to one of the most popular definition of chaos given for general discrete time dynamical systems: the Devaney's definition of chaos. We wish to emphasize that this is, to our knowledge, the first classification of elementary cellular automata rule space according to a rigorous mathematical definition of chaos. We are currently applying the Devaney's definition of chaos to the case of general cellular automata.

References

1. D. Assaf, IV and W. A. Coppel, Definition of Chaos. *The American Mathematical Monthly, 865*, 1992.
2. J. Banks, J. Brooks, G. Cairns, G. Davis, and P. Stacey, On the Devaney's Definition of Chaos. *The American Mathematical Monthly, 332-334*, 1992.
3. K. Culik, L. P. Hurd, and S. Yu, Computation theoretic aspects of CA. *Physica D 45, 357-378*,1990.
4. K.Culik and S. Yu, Undecidability of CA Classification Schemes. *Complex Systems 2(2), 177-190*, 1988.
5. B. Codenotti and L. Margara, Transitive Cellular Automata are Sensitive. *The American Mathematical Monthly, Vol. 103, 58-62*, 1996.
6. R. L. Devaney, An Introduction to Chaotic Dynamical Systems. *Addison Wesley*, 1989.
7. H. A. Gutowitz, A Hierarchycal Classification of Cellular Automata. *Physica D 45, 136-156*, 1990.
8. G. A. Hedlund, Endomorphism and Automorphism of the Shift Dynamical System. *Mathematical System Theory 3(4), 320-375*, 1970.
9. C. Knudsen, Aspects of Noninvertible Dynamics and Chaos, *Ph.D. Thesis*, 1994.
10. C. Knudsen, Chaos Without Nonperiodicity, *The American Mathematical Monthly, 563-565*, 1994.
11. P. Favati, G. Lotti and L. Margara, Additive cellular Automata are chaotic According to Devaney's Definition of Chaos. *To appear on Theoretical Computer Science.*
12. K. Sutner, Classifying Circular Cellular Automata. *Physica D 45, 386-395*, 1990.
13. M. Vellekoop and R. Berglund, On Intervals, Transitivity = Chaos. *The American Mathematical Monthly, 353-355*, 1994.
14. S. Wolfram, Theory and Application of Cellular Automata. *Word Scientific Publishing Co., Singapore*, 1986.

On the Complexity of Balanced Boolean Functions

A. Bernasconi

Dipartimento di Informatica, Università di Pisa
and Istituto di Matematica Computazionale del C.N.R. (Pisa).
e-mail: bernasco@di.unipi.it .

Abstract. This paper introduces the notions of *balanced* and *strongly balanced Boolean functions* and examines the complexity of these functions using harmonic analysis on the hypercube. The results are applied to derive a lower bound related to AC^0 functions.

1 Introduction

One of the best-known results in Circuit Complexity is that constant depth circuits require exponential size to compute the parity function (see [3] and [4]). Here we generalize this result to a class of functions, which we call *strongly balanced functions*.

We say that a Boolean function $f : \{0,1\}^n \to \{0,1\}$ is *balanced* if it takes the value 1 on exactly half of the input strings; *balanced up to the level* k (shortly k*-balanced*) if any subfunction depending on k variables is balanced; and *strongly balanced* if f is balanced up to a level k such that $n - k = \Omega\left((\log n)^d\right)$ for a constant d. These definitions are made precise below.

Our main result is that AC^0-circuits cannot compute strongly balanced functions. More precisely, we prove that a circuit of constant depth d require size $\Omega\left(2^{\frac{(n-k)^{1/d}}{20}}\right)$ to compute a k-balanced function. Note that this class of functions contains the parity and its complement as very special cases, since they are the only functions balanced up to $k = 1$ (see section 4). Moreover, it should be pointed out that our result allows us to gain insights to the structural properties of AC^0 functions.

The main tool of the proof is the harmonic analysis on the hypercube, that yields an interesting spectral characterization of k-balanced functions, together with a result proved in [7], stating that AC^0 functions have almost all of their *power spectrum* on the low-order coefficients.

The paper is organized as follows. In section 2 we provide some of the notation we use, and recall some basic definitions. In section 3 we give the necessary background on Fourier transform on the hypercube, and review the results by Linial et al. (see [7]) about the spectral characterization of AC^0 functions. Section 4 is devoted to the definition of balanced and k-balanced Boolean functions. In section 5 we prove our main result: we first give a spectral characterization of k-balanced functions, and then prove that strongly balanced functions cannot be computed by AC^0 circuits. In section 6 we construct, as an example, a special family of k balanced and non-degenerated Boolean functions. Finally, in section 7 we provide a framework for future research.

2 Basic definitions

First of all, we provide some of the notation we use.

Given a Boolean function f on n binary variables, we will use different kinds of notation: the *classical notation*, where the input string is given by n binary variables; the *set notation*, based on the correspondence between the set $\{0,1\}^n$ and and the power set of $\{1,2,\ldots,n\}$; the 2^n-tuple *vector representation* $f = (f_0\, f_1\, \ldots f_{2^n-1})$, where $f_i = f(x(i))$ and $x(i)$ is the binary expansion of i. Unless otherwise specified, the indexing of vectors and matrices starts from 0 rather than 1.

We will use the notation $|f|$ to denote the number of strings accepted by the function f, i.e. $|f| = |\{w \in \{0,1\}^n \mid f(w) = 1\}|$.

Given a binary string $w \in \{0,1\}^n$, we denote with $w^{(i)}$ the string obtained from w by flipping its i-th bit ($1 \le i \le n$), i.e. w and $w^{(i)}$ differ only on the i-th bit, and by $|w|$ the number of ones in w, which is sometimes called *cardinality* of the string because of the correspondence between sets of positive integers and strings over the alphabet $\{0,1\}$. Finally, all the logarithms are to the base 2.

We now review some basic definitions.

AC^0 circuits

An AC^0 circuit consists of AND, OR and NOT gates, with inputs $x_1,\ldots,x_n$. Fan-in to the gates is unbounded. The size of the circuit (i.e. the number of the gates) is bounded by a polynomial in n, and its depth is bounded by a constant (for a more detailed description, see [4]).
The set of functions computable by an AC^0 circuit of depth d is denoted by $AC^0[d]$.

Sensitivity

The sensitivity of a Boolean function is a measure of how the value of the function is sensitive to changes in the input.
More precisely, let f be a Boolean function depending on n binary variables. The *sensitivity* of f on the string $w \in \{0,1\}^n$, $s_w(f)$, is the number of locations i for which $f(w) \neq f(w^{(i)})$, i.e.

$$s_w(f) = \sum_{i=1}^{n} (f(w) - f(w^{(i)}))^2 .$$

The *average sensitivity* of f, $s(f)$, is the average of $s_w(f)$ over all $w \in \{0,1\}^n$, taken with respect to the uniform distribution:

$$s(f) = \frac{1}{2^n} \sum_{w} s_w(f) = \mathbf{E}[s_w(f)] .$$

This measure has been discussed in the literature under various names, e.g. it is sometimes called 'critical complexity' (see [8] and [9]).

Restriction

A *restriction* ρ is a mapping of the input variables to the set $\{0, 1, \star\}$, where

- $\rho(x_i) = 0$ means that we substitute the value 0 for x_i;
- $\rho(x_i) = 1$ means that we substitute the value 1 for x_i;
- $\rho(x_i) = \star$ means that x_i remains a variable.

Given a function f on n binary variables, we will denote by f_ρ the function obtained from f by applying the restriction ρ; f_ρ will be a function of the variables x_i for which $\rho(x_i) = \star$.

The *domain* of a restriction ρ, $dom(\rho)$, is the set of variables mapped to 0 or 1 by ρ. The *size* of a restriction ρ, $size(\rho)$, is defined as the number of variables which were given the value $\star$, i.e. $size(\rho) = n - |dom(\rho)|$.

3 Abstract harmonic analysis and AC^0 functions

We give some background on abstract harmonic analysis on the hypercube. We refer to [1] and [6] for a more detailed exposition.

We consider Boolean functions as 0-1 valued real functions defined on the domain $\{0\,,\,1\}^n$. They are a vector space of dimension 2^n, and the set of functions $\{f_x(y) = (1 \text{ iff } x = y)\}$, where x ranges over $\{0\,,\,1\}^n$ is a basis. Another basis is given by the functions $\{g_S(x) = \sum_{i \in S} x_i\}$, where $S \subseteq \{1, 2, \ldots, n\}$ and the sum is modulo 2. The Fourier coefficients of f are the coefficients of f in this basis.

More precisely, consider the space $\mathcal{F}$ of all the two-valued functions on $\{0\,,\,1\}^n$. The domain of $\mathcal{F}$ is a locally compact Abelian group and the elements of its range, i.e. 0 and 1, can be added and multiplied as complex numbers. The above properties allow one to analyze $\mathcal{F}$ by using tools from harmonic analysis. The functions $Q_w(x) = (-1)^{w_1 x_1}(-1)^{w_2 x_2} \ldots (-1)^{w_n x_n} = (-1)^{w^T x}$ are known as *Fourier transform kernel functions*, and the set $\{Q_w | w \in \{0\,,\,1\}^n\}$ is an orthogonal basis for $\mathcal{F}$.

We can now define the *Abstract Fourier Transform* of a Boolean function f as the rational valued function f^* which defines the coefficients of f with respect to the basis $\{Q_w(x), w \in \{0\,,\,1\}^n\}$, i.e.,

$$f^*(w) = 2^{-n} \sum_x Q_w(x) f(x)\,.$$

Then

$$f(x) = \sum_w Q_w(x) f^*(w)$$

is the Fourier expansion of f. Moreover, given a Fourier coefficient $f^*(w)$, we define its *order* as the cardinality of the string w.

It is interesting to note that the zero-order Fourier coefficient is equal to the probability that the function takes the value 1, while the other Fourier coefficients measure the correlation between the function and the parity of subsets of its input bits.

Using the binary 2^n-tuple representation for the functions f and f^*, and considering the natural ordering of the n-tuples x and w, one can derive a convenient matrix formulation for the transform pair. Let us consider a $2^n \times 2^n$ matrix H_n whose (i,j)-th entry h_{ij} satisfies $h_{ij} = (-1)^{\underline{i}^T \underline{j}}$, where $\underline{i}^T \underline{j}$ denotes the inner product of the binary expansions of i and j. If $f = [f_0\, f_1 \ldots f_{2^n-1}]^T$ and $f^* = [f_0^*\, f_1^* \ldots f_{2^n-1}^*]^T$, then, from the fact that $H_n^{-1} = 2^{-n} H_n$, we get $f = H_n f^*$ and $f^* = 2^{-n} H_n f$.

Note that the matrix H_n is the Hadamard symmetric transform matrix [6] and can be recursively defined as

$$H_1 = \begin{pmatrix} 1 & 1 \\ 1 & -1 \end{pmatrix}, \qquad H_n = \begin{pmatrix} H_{n-1} & H_{n-1} \\ H_{n-1} & -H_{n-1} \end{pmatrix}.$$

There exists an interesting link between harmonic analysis and the notions of sensitivity (see [2] for more details): the average sensitivity of a function f can be defined as a weighed sum of its Fourier coefficients.

Lemma 1 $s(f) = 4 \sum_w |w| (f^*(w))^2$.

■

We now present an interesting application of harmonic analysis to circuit complexity, due to Linial et al. (see [7]).

As we have already mentioned, one of the best known results in circuit complexity is that AC^0 circuits require exponential size to compute the parity function. More precisely, AC^0-circuits cannot even approximate the parity function. This fact has a direct consequence on the Fourier transform, because, as we have already mentioned, the Fourier coefficients measure the correlation between the function and the parity of subsets of its input bits. Consequently, each high order Fourier coefficient of an AC^0 function must be very small (where "high order" means coefficients corresponding to strings of large cardinality). By exploiting this fact, Linial et al. were able to prove that not only is each individual high order coefficient small, but in fact the sum of squares (i.e. the *power spectrum*) associated with all high Fourier coefficients is very small.

Lemma 2 ([7]) *Let f be a Boolean function on n variables computable by a Boolean circuit of depth d and size M, and let t be any integer. Then*

$$\sum_{|w|>t} (f^*(w))^2 \leq \frac{1}{2} M\, 2^{-\frac{t^{1/d}}{20}}.$$

■

An application of this lemma, together with lemma 1 implies:

Lemma 3 ([7]) *For any function $f \in AC^0[d]$, we have $s(f) = O((\log n)^d)$.*

■

Thus, every function in AC^0 has low average sensitivity to its input. Changing one bit of the input is very unlikely to change the value of the function, when the original input and the bit are chosen at random.

4 Balanced and k-balanced Boolean functions

Let f be a Boolean function depending on n variables. We say that the f is *balanced* if the probability that f takes the value 1 is equal to $\frac{1}{2}$.

Definition 1 *A Boolean function $f : \{0,1\}^n \rightarrow \{0,1\}$ is* balanced *if $|f| = 2^{n-1}$, that is if it takes the value 1 on exactly half of the input strings.*

We would like to enforce this definition introducing the notion of functions *balanced up to the level k* (shortly *k-balanced* functions). In order to do this, we will make use of the notion of *restriction* (see section 2). Let $1 \leq k \leq n$.

Definition 2 *We say that a Boolean function $f : \{0,1\}^n \rightarrow \{0,1\}$ is* balanced up to the level k, *if for any restriction ρ such that $size(\rho) = k$, the resulting subfunction f_ρ is balanced.*

Note that if a function is k-balanced, then it is also ℓ-balanced, for any $\ell \geq k$. We finally define the class of *strongly balanced* Boolean functions.

Definition 3 *A Boolean function f depending on n variables is* strongly balanced *if there exists a constant d such that f is balanced up to a level k s.t. $n - k = \Omega\left((\log n)^d\right)$.*

5 Spectral characterization and complexity

In this section we prove our main result, stating that strongly balanced functions cannot be computed by AC^0 circuits. We first derive a spectral characterization of k-balanced functions, and then use it, together with lemma 2, to determine a lower bound on the size required by a depth d circuit to compute them. Finally, an easy application of this bound will provide our thesis.

Theorem 4 *A Boolean function $f : \{0,1\}^n \rightarrow \{0,1\}$ is balanced up to the level k if and only if the following two properties hold:*

(1) f is balanced;
(2) for any string w such that $0 < |w| \leq n - k$, $f^*(w) = 0$.

Proof.

- If f is k-balanced, then property (1) holds by definition. Thus, we only need to prove property (2).
  Let $\mu \equiv (\mu_1, \mu_2, \ldots, \mu_n)$ be a Boolean string such that $0 < |\mu| = n-\ell \leq n-k$. Moreover, let $U = \{i \mid \mu_i = 1\}$. For any string $u \in \{0,1\}^{n-\ell}$, let f_u denote

the subfunction defined by the restriction that assigns to the variables x_i such that $i \in U$, the $(n-\ell)$ values taken from the string u, and leaves undetermined the other ℓ variables.
Then, we have

$$\begin{aligned} f^*(\mu) &= \frac{1}{2^n} \sum_w (-1)^{\mu^T w} f(w) \\ &= \frac{1}{2^n} \sum_w (-1)^{\sum_{i \in U} w_i} f(w) \\ &= \frac{1}{2^n} \sum_{u \in \{0,1\}^{n-\ell}} \left[(-1)^{|u|} \sum_{v \in \{0,1\}^\ell} f_u(v) \right] \\ &= \frac{1}{2^n} \sum_{u \in \{0,1\}^{n-\ell}} \left[(-1)^{|u|} \, |f_u| \right] . \end{aligned}$$

For any $u \in \{0,1\}^{n-\ell}$, the subfunction f_u depends on $\ell \geq k$ variables and, since f is balanced up to k, we have $|f_u| = 2^{\ell-1}$. Thus, we get

$$f^*(\mu) = \frac{2^{\ell-1}}{2^n} \sum_{u \in \{0,1\}^{n-\ell}} (-1)^{|u|} = 0 .$$

– We now prove that if properties (1) and (2) hold, then f is k-balanced. Let us choose $(n-k)$ variables out of n, and let U be the set of the indices of these $(n-k)$ variables. For any $u \in \{0,1\}^{n-k}$, let f_u denote the subfunction obtained from f by assigning to the variables in the set U, the $(n-k)$ values taken from the string u, and leaving undetermined the other k variables. For any u, f_u depends on k variables. We show that any such subfunction accepts exactly 2^{k-1} inputs, i.e. for any string u, $|f_u| = 2^{k-1}$.
Let $f_\#$ denote the vector of the cardinality of the 2^{n-k} subfunctions f_u, and let f_U^* denote the vector of the Fourier coefficients related to the 2^{n-k} strings $w \equiv (w_1, w_2, \ldots, w_n)$ such that $w_i = 0$ for any $i \notin U$. Note that all the 2^{n-k} coefficients in the vector f_U^* are of order less or equal to $n-k$. Because of the recursive definition of Hadamard matrices, it turns out that

$$f_U^* = \frac{1}{2^n} H_{n-k} \, f_\# .$$

From properties (1) and (2), from the fact that the zero order Fourier coefficient is equal to the probability that the function takes the value 1, it then follows

$$f_U^* = \frac{1}{2} \begin{pmatrix} 1 \\ 0 \\ \vdots \\ 0 \end{pmatrix} ,$$

from which

$$f_{\#} = 2^n \; H_{n-k}^{-1} \; f_U^* = \frac{2^n}{2^{n-k}} \; H_{n-k} \; f_U^*$$

$$= 2^{k-1} \; H_{n-k} \begin{pmatrix} 1 \\ 0 \\ \vdots \\ 0 \end{pmatrix} = 2^{k-1} \begin{pmatrix} 1 \\ 1 \\ \vdots \\ 1 \end{pmatrix} .$$

Thus, the theorem follows by repeating the same argument for all the $\binom{n}{k}$ choices of the set U.

∎

This spectral characterization allows us to conclude that the parity function and its complement represent special cases among the class of k-balanced functions, since they are the only two functions balanced up to $k = 1$.

Corollary 5 *The only two functions balanced up to $k = 1$ are the parity function and its complement.*

Proof. Let f be a function balanced up to the level 1, and let f_0^* denote the zero-order Fourier coefficient. Then from the *Parseval's identity* $\sum_v (f^*(v))^2 = 2^{-n} \sum_v f(v) = f_0^*$, and from theorem 4, it follows that the Fourier coefficients of f must have the following absolute values:

$$|f^*(w)| = \begin{cases} \frac{1}{2} & \text{if} \quad |w| = 0 \\ 0 & \text{if} \quad 0 < |w| < n \\ \frac{1}{2} & \text{if} \quad |w| = n \, . \end{cases}$$

Thus, the corollary follows, since the only functions with such a kind of spectrum are the parity and its complement (see [1] for more details).

∎

At this point we are able to state and prove a lower bound on the size needed by a depth d circuit to compute a k balanced function, establishing in this way a connection between complexity and structural properties of Boolean functions.

Theorem 6 *Let f be a k-balanced Boolean function depending on n variables computable by a circuit of constant depth d and size M. Then*

$$M \geq \frac{1}{2} 2^{\frac{(n-k)^{1/d}}{20}} .$$

Proof. An application of lemma 2 yields the following inequality:

$$M \geq 2^{\frac{t^{1/d}}{20}} \, 2 \sum_{|w|>t} (f^*(w))^2 .$$

Let us choose $t = n - k$. From the fact that $f^*(w) = 0$ for any $0 < |w| \leq n - k$ (see theorem 4) it follows

$$\sum_{|w|>n-k} (f^*(w))^2 = \sum_{w:|w|\neq 0} (f^*(w))^2 = \sum_{w} (f^*(w))^2 - (f_0^*)^2 ,$$

where f_0^* denotes the zero-order Fourier coefficient. Then, by using the Parseval's identity $\sum_v (f^*(v))^2 = f_0^* = \frac{1}{2}$, we get $\sum_{|w|>n-k} (f^*(w))^2 = \frac{1}{4}$, and the thesis immediately follows:

$$M \geq 2^{\frac{(n-k)^{1/d}}{20}} 2 \sum_{|w|>t} (f^*(w))^2 = \frac{1}{2} 2^{\frac{(n-k)^{1/d}}{20}} .$$

■

Our main result, stating that strongly balanced functions do not belong to the set $AC^0[d]$, follows immediately as a corollary of theorem 6.

Corollary 7 *A strongly balanced function f requires superpolynomial size to be computed by a depth d circuit.*

Proof. Easily follows from definition 3 and theorem 6.

■

Note how the lower bound to the size can become exponential:

Corollary 8 *Depth d circuits require exponential size to compute k-balanced functions whenever k is s.t. $n - k = \Omega(n^\varepsilon)$, for any positive constant $\varepsilon < 1$.*

Proof. Immediate from theorem 6.

■

We conclude this section by evaluating the average sensitivity of k-balanced functions.

Lemma 9

1. *If f is balanced up to $k < n$, then $s(f) > n - k$.*
2. *If f is balanced up to n, then $s(f) \geq 1$.*

Proof.

1. Let p be the probability that f takes the value 1 and let f_0^* denote the zero-order Fourier coefficient of f. The thesis follows by applying theorem 4 and lemma 1:

$$\begin{aligned} s(f) &= 4 \sum_{w} |w| (f^*(w))^2 > 4\,(n-k) \sum_{w:\ |w|\neq 0} (f^*(w))^2 \\ &= 4\,(n-k) \left(\sum_{w} (f^*(w))^2 - (f_0^*)^2 \right) = 4\,(n-k)\ (p - p^2) \\ &= n - k . \end{aligned}$$

2. As before, the thesis follows from theorem 4 and lemma 1:

$$s(f) = 4\sum_{w} |w|(f^*(w))^2 \geq 4 \sum_{w:\ |w|\neq 0} (f^*(w))^2$$

$$= 4\left(\sum_{w}(f^*(w))^2 - (f_0^*)^2\right) = 4\ (p - p^2) = 1\,.$$

■

For a strongly balanced function f, we immediately get $s(f) = \Omega((\log n)^d)$, and this is again an indication that $f \notin AC^0[d]$ (see lemma 3).

6 A special family of balanced functions

In this section we construct, as an example, a special family of k-balanced and non-degenerated Boolean functions.

Let $1 < k \leq n$, and let $h_{(0)} : \{0,1\}^k \to \{0,1\}$ be defined as follows

$$h_{(0)}(w_1, w_2, \ldots, w_k) = \begin{cases} \bigwedge_{i=2}^{k} w_i & \text{if } \ w_1 = 0 \\ \bigvee_{i=2}^{k} w_i & \text{if } \ w_1 = 1\,. \end{cases}$$

Then, we recursively compute a function f_k, depending on n variables, as follows. Given $h_{(i)}$, depending on $k+i$ variables, we define $h_{(i+1)}$ as follows:

$$h_{(i+1)}(w_1, w_2, \ldots, w_{k+i+1}) = \begin{cases} h_{(i)}(w_2, w_3, \ldots, w_{k+i+1}) & \text{if } \ w_1 = 0 \\ \neg h_{(i)}(w_2, w_3, \ldots, w_{k+i+1}) & \text{if } \ w_1 = 1\,. \end{cases}$$

Finally, we take $f_k = h_{(n-k)}$.

Proposition 10 *Let $\alpha \in \{0,1\}^{n-k}$, $\beta \in \{0,1\}^k$, with $\beta \equiv (\beta_1, \beta_2, \ldots, \beta_k)$, and $w = \alpha\beta \in \{0,1\}^n$. The Fourier spectrum of f_k has the following structure:*

$$f_k^*(\alpha\beta) = \begin{cases} \frac{1}{2} & |\alpha| = |\beta| = 0 \\ 0 & |\alpha| < n-k \\ 0 & |\alpha| = n-k,\ |\beta| \text{ even} \\ -\frac{1}{2^{k-1}} & |\alpha| = n-k,\ \beta_1 = 0,\ |\beta| \text{ odd} \\ \frac{1}{2^{k-1}} - \frac{1}{2} & |\alpha| = n-k,\ |\beta| = \beta_1 = 1 \\ \frac{1}{2^{k-1}} & |\alpha| = n-k,\ \beta_1 = 1,\ |\beta| \neq 1 \text{ odd}\,. \end{cases}$$

Proof. Let us consider the construction of f_k, and, for any $i \geq 1$, let $\alpha \in \{0,1\}^i$ and $\beta \in \{0,1\}^k$. It is possible to prove, by an easy induction on i, that

$$h_{(i)}^*(\alpha\beta) = \begin{cases} \frac{1}{2} & |\alpha| = |\beta| = 0 \\ 0 & |\alpha| < i,\ |\beta| > 0 \\ h_{(0)}^*(\beta) - \frac{1}{2} & |\alpha| = i,\ |\beta| = 0 \\ h_{(0)}^*(\beta) & |\alpha| = i,\ |\beta| > 0\,. \end{cases}$$

Then, the theorem follows since $f_k \equiv h_{(n-k)}$ and since for $\beta \equiv (\beta_1, \beta_2, \ldots, \beta_k) \in \{0,1\}^n$ we have

$$h^*_{(0)}(\beta) = \begin{cases} \frac{1}{2} & |\beta| = 0 \\ 0 & |\beta| \neq 0 \text{ even} \\ -\frac{1}{2^{k-1}} & \beta_1 = 0,\ |\beta| \text{ odd} \\ \frac{1}{2^{k-1}} - \frac{1}{2} & \beta_1 = |\beta| = 1 \\ \frac{1}{2^{k-1}} & \beta_1 = 1,\ |\beta| \neq 1 \text{ odd}. \end{cases}$$

■

Proposition 11 *The function f_k is balanced up to k.*

Proof. From proposition 10, it follows that $f^*(w) = 0$ for any string w such that $0 < |w| \leq n - k$, while $f^*_0 = \frac{1}{2}$. Thus, the thesis follows from theorem 4.

■

7 Conclusion

Any attempt to find connections between mathematical properties and complexity has a strong relevance to the field of Complexity Theory. This is due to the lack of mathematical techniques to prove lower bounds for general models of computation. This work represents a step in this direction: we find a property that makes Boolean functions "hard" to compute and show how the Fourier transform could be used as a mathematical tool for the analysis of Boolean functions complexity. Further work to be done includes a deeper analysis of the connections between structural properties of Boolean functions and complexity and a characterization of the Fourier spectrum of functions computable by NC^1 circuits.

Acknowledgements

I would like to thank Bruno Codenotti and Janos Simon for very helpful discussions and comments.

References

1. A. Bernasconi, B. Codenotti. *Sensitivity of Boolean Functions, Abstract Harmonic Analysis and Circuit Complexity.* ICSI Technical Report TR-93-030 (1993).
2. A. Bernasconi, B. Codenotti, J. Simon. *On the Fourier Analysis of Boolean Functions.* Submitted for publication (1996).
3. M. Furst, J. Saxe, M. Sipser. *Parity, circuits, and the polynomial-time hierarchy.* Math. Syst. Theory, Vol. 17 (1984), pp. 13-27.
4. J. Håstad. *Computational limitations for small depth circuits.* Ph.D. Dissertation, MIT Press, Cambridge, Mass. (1986).
5. S.L. Hurst, D.M. Miller, J.C. Muzio. *Spectral Method of Boolean Function Complexity.* Electronics Letters, Vol. 18 (33) (1982), pp. 572-574.

6. R. J. LECHNER. *Harmonic Analysis of Switching Functions.* In *Recent Development in Switching Theory*, Academic Press (1971), pp. 122-229.
7. N. LINIAL, Y. MANSOUR, N. NISAN. *Constant Depth Circuits, Fourier Transform, and Learnability.* Journal of the ACM, Vol. 40 (3) (1993), pp. 607-620.
8. H .U. SIMON. *A tight $\Omega(\log\log n)$ bound on the time for parallel RAM's to compute nondegenerate Boolean functions.* FCT'83, Lecture Notes in Computer Science **158** (1983).
9. I. WEGENER. *The complexity of Boolean functions.* Wiley-Teubner Series in Comp. Sci., New York – Stuttgart (1987).

On Sets with Easy Certificates and the Existence of One-Way Permutations*

Lane A. Hemaspaandra[1]** and *Jörg Rothe*[2]*** and *Gerd Wechsung*[2]***

[1] Department of Computer Science, University of Rochester, Rochester, NY 14627, USA
[2] Institut für Informatik, Friedrich-Schiller-Universität Jena, 07743 Jena, Germany

Abstract

Can easy sets only have easy certificate schemes? In this paper, we study the class of sets that, for all NP certificate schemes (i.e., NP machines), always have easy acceptance certificates (i.e., accepting paths) that can be computed in polynomial time. We also study the class of sets that, for all NP certificate schemes, infinitely often have easy acceptance certificates. We give structural conditions that control the size of these classes. We also provide negative results showing that some of our positive claims are optimal. Our negative results are proven using a novel observation: The classic "wide spacing" oracle construction technique instantly yields non-bi-immunity results.

Easy certificate classes are also a useful notion in the study of whether one-way functions exist. This is one of the most important open questions in cryptology. We extend the results of Grollmann and Selman [GS88] by obtaining a complete characterization regarding the existence of a certain type of one-way function—(partial) one-way permutations—in terms of easy certificate classes. By Grädel's recent results about one-way functions [Grä94], this also links statements about easy certificates of NP sets with statements in finite model theory. In addition, we give a condition necessary and sufficient for the existence of (total) one-way permutations.

1 Introduction

Borodin and Demers [BD76] proved the following result.

Theorem 1 [BD76] *If* NP $\cap$ coNP $\neq$ P, *then there exists a set* $L \subseteq$ SAT *such that* $L \in$ P, *yet for no polynomial-time computable function* f *does it hold that: for each* $F \in L$, $f(F)$ *outputs a satisfying assignment of* F.

* Supported in part by grants NSF-INT-9513368/DAAD-315-PRO-fo-ab, NSF-CCR-9322513, NSF-CCR-8957604, and NSF-INT-9116781/JSPS-ENG-207.

** Email: lane@cs.rochester.edu. Work done in part while visiting Friedrich-Schiller-Universität Jena.

*** Email: rothe@informatik.uni-jena.de and wechsung@minet.uni-jena.de. Work done in part while visiting the University of Rochester and Le Moyne College, Syracuse, NY.

That is, under a hypothesis most complexity theoreticians would guess to be true, it follows that there is a set of satisfiable formulas for which it is easy to determine they are satisfiable, yet it is hard to determine why (i.e., via what satisfying assignment) they are satisfiable. Motivated by their work, this paper seeks to study, complexity-theoretically, the classes of sets that do or do not have easy certificates. In particular, we are interested in the following four classes. $\mathrm{EASY}_{\forall}^{\forall}$ is the class of sets L such that for each NP machine M accepting them, there is a polynomial-time computable function f_M such that for each $x \in L$, $f_M(x)$ outputs an accepting path of $M(x)$. That is, $\mathrm{EASY}_{\forall}^{\forall}$ is the class of sets that for all certificate schemes, have easy certificates for all elements of the set. We can analogously define $\mathrm{EASY}_{\mathrm{io}}^{\forall}$, $\mathrm{EASY}_{\forall}^{\exists}$, and $\mathrm{EASY}_{\mathrm{io}}^{\exists}$. However, we note that $\mathrm{EASY}_{\forall}^{\exists} = \mathrm{P}$ and $\mathrm{EASY}_{\mathrm{io}}^{\exists}$ equals the class of non-P-immune NP sets. Regarding the two $\mathrm{EASY}_{\ldots}^{\forall}$ classes, we provide equivalent characterizations of the classes in terms of relative generalized Kolmogorov complexity, showing that they are robust. We also provide structural conditions—regarding immunity and class collapses—that put upper and lower bounds on the sizes of these two classes. Finally, we provide negative results showing that some of our positive claims are optimal with regard to relativizations. Our negative results are proven using a novel observation: We show that the classic "wide spacing" oracle construction technique yields *instant* non-bi-immunity results. Furthermore, we establish a result that improves upon Baker, Gill, and Solovay's classic result that $\mathrm{NP} \neq \mathrm{P} = \mathrm{NP} \cap \mathrm{coNP}$ holds in some relativized world [BGS75].

Another strong motivation for the study of easy certificate classes is presented in Section 3. An important question that has always attracted researchers is: What makes NP-complete problems intractable? One possible source of their potential intractability is the fact that there are many possible sets of solutions: The search space is exponential so the cardinality of the set of sets of solutions is double-exponential in the input size. Another possible source of NP's complexity is that all solutions (even if there are just a few of them) may be random in the sense of Kolmogorov complexity and thus hard to find. For both reasons one may try to "remove" the difficulty from NP by considering subclasses of NP that, by definition, contain only easy sets with respect to either type of difficulty. NP's subclasses UP (unambiguous polynomial time) [Val76] and FewP (ambiguity-bounded polynomial time) [All86, AR88] both implicitly reduce the richness of the class of potential solutions to $2^{n^{\mathcal{O}(1)}}$. To single out those NP sets that, for all NP machines accepting them, have easy solutions—i.e., solutions of small Kolmogorov complexity—for all instances in the set, $\mathrm{EASY}_{\forall}^{\forall}$ is the appropriate concept as noted above. Interestingly, both these concepts of easy NP sets (to wit, UP and $\mathrm{EASY}_{\forall}^{\forall}$) have their own connection to the invertibility of certain types of one-way functions, as will be stated below. Intuitively, a one-way function is a function that is easy to compute but hard to invert. One-way functions play a central role in complexity-theoretic cryptography [GS88], where the open question of whether such functions do or do not exist is of central importance.

Many types of one-way functions have been studied in the literature. Most notable among such results is Grollmann and Selman's characterization of the existence of certain types of *injective* one-way functions by conditions such as $\mathrm{P} \neq \mathrm{UP}$ or $\mathrm{P} \neq \mathrm{UP} \cap \mathrm{coUP}$ [GS88] (see also [Ko85]). Allender extended their results by proving that *poly-one* one-way functions exist if and only if $\mathrm{P} \neq \mathrm{FewP}$ [All86]. Fenner et al. [FFNR96] proved the existence of *surjective many-one* one-way functions equivalent to $\mathrm{P} \not\subseteq \mathrm{EASY}_{\forall}^{\forall}$. In Section 3, a characterization of the existence of *injective and surjective* one-way functions is given by separating P from a class, denoted $\mathrm{EASY}_{\forall}^{\forall}(\mathrm{UP})$, which combines the restriction of unambiguous computation with the constraint required by $\mathrm{EASY}_{\forall}^{\forall}$. Thus, $\mathrm{EASY}_{\forall}^{\forall}(\mathrm{UP})$ *simultaneously* reduces the solution space of NP problems to at most one solution and requires that this one solution can be found and printed out in polynomial time, if it exists. Furthermore, the existence of *surjective poly-one* one-way functions is shown to be equivalent to the separation of P and $\mathrm{EASY}_{\forall}^{\forall}(\mathrm{FewP})$ (which is the polynomially ambiguity-bounded analog of $\mathrm{EASY}_{\forall}^{\forall}(\mathrm{UP})$). Our work is connected to the seemingly unrelated field of (finite model) logic; from Grädel's [Grä94] recent results about one-way functions, we obtain as a corollary equivalences between statements about easy certificates of NP sets and statements in finite model theory such as that the weak definability principle in a logic on finite structures fails to hold. Finally, we show that the existence of total injective one-way functions with a P-rankable range is a condition necessary and sufficient for the existence of one-way permutations.

2 Easy Certificate Classes

2.1 Definitions and Robustness

For the standard notations and the complexity-theoretical concepts used in this paper we refer to some standard text book on computational complexity. Fix the alphabet $\Sigma = \{0,1\}$. Let $\langle\cdot,\cdot\rangle : \Sigma^* \times \Sigma^* \to \Sigma^*$ denote a standard pairing function. We will abbreviate "polynomial-time deterministic (nondeterministic) Turing machine" by DPM (NPM). For any Turing machine M, $L(M)$ denotes the set of strings accepted by M, and the notation $M(x)$ means "M on input x." For any NPM N and any input x, we assume that all paths of $N(x)$ are suitably encoded by strings over Σ and we denote the set of accepting paths of $N(x)$ by $\mathrm{acc}_N(x)$. As a notational convention, for any NPM N, we will say "N has always (respectively, N has infinitely often) easy certificates" to mean that (the encoding of) an accepting path of $N(x)$ can be printed in polynomial time for each string $x \in L(N)$ (respectively, for infinitely many $x \in L(N)$). Similarly, we will say "N has only (respectively, N has infinitely often) hard certificates" to mean that no FP function is able to output (the encoding of) an accepting path of $N(x)$ for each string $x \in L(N)$ (respectively, for infinitely many $x \in L(N)$). P (respectively, NP) is the class of all sets that are accepted by some DPM (NPM). Let FINITE be the class of all finite sets. FP denotes the class of all polynomial-time computable functions. UP and FewP are the classes of those NP sets that have at most one (at most polynomially many) solutions for all instances in the set [Val76, All86]. For any complexity class $\mathcal{C}$, a set L is said to be $\mathcal{C}$-*immune* if L is infinite and no infinite subset of L is in $\mathcal{C}$. Let $\mathcal{C}$-immune denote the class of all $\mathcal{C}$-immune sets. A set L is said to be $\mathcal{C}$-*bi-immune* if both L and its complement, $\overline{L}$, are $\mathcal{C}$-immune. Let $\mathcal{C}$-bi-immune denote the class of all $\mathcal{C}$-bi-immune sets. For classes $\mathcal{C}$ and $\mathcal{D}$ of sets, $\mathcal{D}$ is said to be $\mathcal{C}$-immune (respectively, $\mathcal{C}$-bi-immune) if $\mathcal{D} \cap (\mathcal{C}\text{-immune}) \neq \emptyset$ (respectively, if $\mathcal{D} \cap (\mathcal{C}\text{-bi-immune}) \neq \emptyset$).

Definition 2 $\mathrm{EASY}_{\forall}^{\forall}$ *is the class of all sets L that either are finite, or (a) $L \in \mathrm{NP}$, and (b) for every NPM N such that $L(N) = L$, there exists an FP function f_N such* *that, for all $x \in L$, $f_N(x) \in acc_N(x)$. In a similar vein, define the complexity classes* $\mathrm{EASY}_{\mathrm{io}}^{\forall}$, $\mathrm{EASY}_{\forall}^{\exists}$, *and* $\mathrm{EASY}_{\mathrm{io}}^{\exists}$, *where an upper index "$\exists$" (respectively, a lower index "io") means that the above condition (b) is only required to hold for* some *NPM N (respectively, for* infinitely many *inputs).*

The inclusion relations between the four classes defined above are stated below. Moreover, it can be shown that $\mathrm{EASY}_{\forall}^{\exists} = \mathrm{P}$ and $\mathrm{EASY}_{\mathrm{io}}^{\exists}$ equals the class of all non-P-immune NP sets (for the proof, see the full version of this paper [HRW95]), and we therefore will not further discuss these two classes in this paper.

Theorem 3 *1.* $\mathrm{EASY}_{\forall}^{\exists} = \mathrm{P}$ *and* $\mathrm{EASY}_{\mathrm{io}}^{\exists} = \overline{\text{P-immune}} \cap \mathrm{NP}$.
2. $\mathrm{FINITE} \subseteq \mathrm{EASY}_{\forall}^{\forall} \subseteq \mathrm{EASY}_{\mathrm{io}}^{\forall} \subseteq \mathrm{EASY}_{\mathrm{io}}^{\exists} \subseteq \mathrm{NP}$.
3. $\mathrm{EASY}_{\forall}^{\forall} \subseteq \mathrm{EASY}_{\forall}^{\exists} \subseteq \mathrm{EASY}_{\mathrm{io}}^{\exists}$.

Next, we show that these notions are robust by providing characterizations of the classes in terms of Kolmogorov complexity. The Kolmogorov complexity of finite strings was introduced independently by Kolmogorov [Kol65] and Chaitin [Cha66]. Roughly speaking, the Kolmogorov complexity of a finite binary string x is the length of a shortest program that generates x. Intuitively, if a string x can be generated by a program shorter than x itself, then x can be "compressed." The notion of *generalized* Kolmogorov complexity ([Adl79, Har83, Sip83], see the survey [LV90]) is a version of Kolmogorov complexity that provides information about not only whether and how far a string can be compressed, but also how fast it can be "restored."

We now give the definition of (unconditional and conditional) generalized Kolmogorov complexity. Let U be a fixed universal Turing machine. For functions s and t mapping $\mathbb{N}$ to $\mathbb{N}$ and for any string z, define the *conditional generalized Kolmogorov complexity* (under condition z), denoted $\mathrm{K}[s(n), t(n) \mid z]$, in which the information of z is given for free and does not count for the complexity, as follows. A string x of length n is in $\mathrm{K}[s(n), t(n) \mid z]$ if and only if there exists a string y, $|y| \leq s(n)$, such that $U(\langle y, z \rangle)$ outputs x in at most $t(n)$ steps. In particular, the *unconditional generalized Kolmogorov complexity* is defined by $\mathrm{K}[s(n), t(n)] \stackrel{\mathrm{df}}{=} \mathrm{K}[s(n), t(n) \mid \epsilon]$, where ϵ denotes the empty string. Of particular interest in this paper are certificates (more precisely, strings encoding accepting paths of NPMs) that have *small* Kolmogorov complexity, i.e., strings whose Kolmogorov resource bounds are $s(n) = k \log n$ and $t(n) = n^k$ for some constant k. The P-printable[3] sets are closely related to sets of strings having small unconditional generalized Kolmogorov complexity: A set S is P-printable if and only if $S \in \mathrm{P}$ and $S \subseteq \mathrm{K}[k \log n, n^k]$ for some constant k [AR88]. Below we note a similar connection between the sets in $\mathrm{EASY}_{\forall}^{\forall}$ and the sets of certificates having small *conditional* generalized Kolmogorov complexity, thus showing the robustness of these notions. An analogous characterization holds true for $\mathrm{EASY}_{\mathrm{io}}^{\forall}$.

Observation 4 *$L \in \mathrm{EASY}_{\forall}^{\forall}$ if and only if for each normalized NPM N accepting L there is a constant k (which may depend on N) such that for each string $x \in L$ it holds that* $\mathrm{acc}_N(x) \cap \mathrm{K}[k \log n, n^k \mid x] \neq \emptyset$.

[3] A set S is P-*printable* if there exists a DPM M such that for each length n, M on input 1^n prints all elements of S having length at most n [HY84]. This notion is important in the field of data compression.

2.2 Positive Results

In this section, we prove a number of implications and equivalences among certain statements about sets in NP that, in terms of immunity and class collapses, put upper and lower bounds on the sizes of the easy certificate classes. Figure 1 in the appendix summarizes all known such implications, some of which follow easily from the definitions. Theorem 5 below gives the non-trivial implications.

Theorem 5 *1. If* NP ∩ coNP *is* P-*bi-immune, then* $\mathrm{EASY}^{\forall}_{\mathrm{io}} = \mathrm{FINITE}$.
2. *If* NP *is* P-*immune, then* $\mathrm{EASY}^{\forall}_{\mathrm{io}} \neq \mathrm{NP}$.
3. *If* $\mathrm{EASY}^{\forall}_{\mathrm{io}} \neq \mathrm{NP}$, *then there exists an infinite* P *set having no infinite* P-*printable subset.*
4. [All92] *If there exists an infinite* P *set having no infinite* P-*printable subset, then* $\mathrm{P} \neq \mathrm{NP}$.
5. [BD76] *If* $\mathrm{NP} \cap \mathrm{coNP} \neq \mathrm{P}$, *then* $\mathrm{P} \not\subseteq \mathrm{EASY}^{\forall}_{\forall}$.

Proof. (1) Let Q be any P-bi-immune set such that $Q \in \mathrm{NP} \cap \mathrm{coNP}$ via NPMs N_Q and $N_{\overline{Q}}$, that is, $L(N_Q) = Q$ and $L(N_{\overline{Q}}) = \overline{Q}$. By way of contradiction, assume there exists an infinite set L in $\mathrm{EASY}^{\forall}_{\mathrm{io}}$. Let N be any NPM accepting L. Consider the following NPM $\widehat{N}$ for L. Given x, $\widehat{N}$ runs $N(x)$ and rejects on all rejecting paths of $N(x)$. On all accepting paths of $N(x)$, $\widehat{N}$ nondeterministically guesses whether $x \in Q$ or $x \in \overline{Q}$, simultaneously guessing certificates (i.e., accepting paths of $N_Q(x)$ or $N_{\overline{Q}}(x)$) for whichever guess was made, and accepts on each accepting path of $N_Q(x)$ or $N_{\overline{Q}}(x)$. Clearly, $L(\widehat{N}) = L$. By our assumption that L is an infinite set in $\mathrm{EASY}^{\forall}_{\mathrm{io}}$, $\widehat{N}$ has easy certificates for infinitely many inputs. Let $f_{\widehat{N}}$ be an FP function that infinitely often outputs an easy certificate of $\widehat{N}$. Let $\widehat{L} \stackrel{\mathrm{df}}{=} \{x \mid f_{\widehat{N}}(x) \text{ outputs an easy certificate of } \widehat{N}(x)\}$. Note that $\widehat{L}$ is an infinite subset of L, and that for any input x, it can be checked in polynomial time whether x belongs to $Q \cap \widehat{L}$ or $\overline{Q} \cap \widehat{L}$, respectively, by simply checking whether the string printed by $f_{\widehat{N}}$ indeed certifies either $x \in Q \cap \widehat{L}$ or $x \in \overline{Q} \cap \widehat{L}$. Thus, either $Q \cap \widehat{L}$ or $\overline{Q} \cap \widehat{L}$ must be an infinite set in P, which contradicts that Q is P-bi-immune. Hence, every set in $\mathrm{EASY}^{\forall}_{\mathrm{io}}$ is finite.

The proof of the remaining parts is omitted due to space limitations. □

For the proof of the following theorem we refer to the full version of this paper [HRW95]. We note that the first part of the theorem is easily obtained by combining a result of Adleman [Adl79] (see also [Tra84] for a discussion of Levin's related work) with Observation 4.

Theorem 6 *1.* $\mathrm{P} \neq \mathrm{NP}$ *if and only if* $\mathrm{EASY}^{\forall}_{\forall} \neq \mathrm{NP}$.
2. $\mathrm{EASY}^{\forall}_{\mathrm{io}} \subseteq \mathrm{FINITE} \cup (\mathrm{NP} - \mathrm{P})$ *if and only if* $\Sigma^* \notin \mathrm{EASY}^{\forall}_{\mathrm{io}}$.

Finally, we note that the authors of [FFNR96] have shown a number of very interesting conditions, including "$\Sigma^* \notin \mathrm{EASY}^{\forall}_{\forall}$" and "there exists an honest polynomial-time computable *onto* function that is not polynomial-time invertible," to be all equivalent to the statement "$\mathrm{P} \not\subseteq \mathrm{EASY}^{\forall}_{\forall}$." Section 3 will extend the results of [FFNR96] and further elaborate the close connection between certain properties of easy certificate classes and the complexity of inverting surjective functions. In addition to the equivalences given above, Selman has found another characterization of $\mathrm{P} = \mathrm{EASY}_{\forall}^{\forall}$ in terms of the question of whether $\mathrm{EASY}_{\forall}^{\forall}$ is closed under complementation (for the proof and discussion, see [HRW95]).

Claim 7 [Sel95] *The following are equivalent.*

1. $\mathrm{P} = \mathrm{EASY}_{\forall}^{\forall}$.
2. $\mathrm{EASY}_{\forall}^{\forall}$ *is closed under complementation.*
3. *There exists a set* L *in* P *such that* $L \in \mathrm{EASY}_{\forall}^{\forall}$ *and* $\overline{L} \in \mathrm{EASY}_{\forall}^{\forall}$.

2.3 Negative Results

In this section, we show that some of the results from the previous section are optimal with respect to relativizable techniques. That is, for some of the implications displayed in Figure 1, we construct an oracle relative to which the reverse of that implication fails.

One main technical contribution in the proof of Theorem 8 (omitted here) is that we give a novel application of the classic "wide spacing" oracle construction technique: We show that this technique *instantly* yields the non-P-bi-immunity of NP relative to some oracle (see the full version of this paper [HRW95]).

Theorem 8 *There exists a recursive oracle set* A *such that*

$$\mathrm{NP}^A = \mathrm{PSPACE}^A \text{ and } \mathrm{NP}^A \text{ is } \mathrm{P}^A\text{-immune and } \mathrm{NP}^A \text{ is not } \mathrm{P}^A\text{-bi-immune.}$$

Note that, relative to the oracle A in the above theorem, *simultaneously* the reverse of arrows (1a) and (1c) in Figure 1 fails and $\mathrm{FINITE} \neq \mathrm{EASY}_{\forall}^{\forall} \neq \mathrm{P}$, i.e., the reverse of arrow (10) in Figure 1 fails to hold. In fact, via using a Kolmogorov complexity based oracle construction, we can even show that there exists an oracle D such that $\mathrm{P}^D \neq (\mathrm{EASY}_{\mathrm{io}}^{\forall})^D$. The proof of Corollary 9 can be found in the full version of this paper [HRW95].

Corollary 9 (to the proof of Theorem 8) *There exists a recursive oracle* A *such that the following simultaneously holds:* $\mathrm{FINITE} \neq (\mathrm{EASY}_{\forall}^{\forall})^A \neq \mathrm{P}^A \neq \mathrm{NP}^A \cap \mathrm{coNP}^A$ *and* NP^A *is* P^A*-immune, yet* $\mathrm{NP}^A \cap \mathrm{coNP}^A$ *is not* P^A*-bi-immune.*

Baker, Gill, and Solovay proved that there exists an oracle E relative to which $\mathrm{P}^E \neq \mathrm{NP}^E$, yet $\mathrm{P}^E = \mathrm{NP}^E \cap \mathrm{coNP}^E$ [BGS75]. Due to the priority argument they apply, this proof is the most complicated of all proofs presented in their paper. Theorem 10 below improves upon their result by showing that the implication $(\mathrm{P} \neq \mathrm{NP} \Longrightarrow \mathrm{P} \not\subseteq \mathrm{EASY}_{\forall}^{\forall})$ (which is the reverse of arrow (6) in Figure 1) fails to hold in some relativized world. It is worth noting that thus already the trivial part of the implication chain $(\mathrm{P} \neq \mathrm{NP} \cap \mathrm{coNP} \Longrightarrow \mathrm{P} \not\subseteq \mathrm{EASY}_{\forall}^{\forall})$ and $(\mathrm{P} \not\subseteq \mathrm{EASY}_{\forall}^{\forall} \Longrightarrow \mathrm{P} \neq \mathrm{NP})$ (arrows (5) and (6) in Figure 1) is shown to be irreversible up to the limits of relativizing techniques. Corollary 11 then follows from Theorem 10 and the fact that the proof of Theorem 5.5 relativizes, that is, stating the contrapositive of Theorem 5.5: For every oracle B, if $\mathrm{P}^B \subseteq (\mathrm{EASY}_{\forall}^{\forall})^B$, then $\mathrm{P}^B = \mathrm{NP}^B \cap \mathrm{coNP}^B$.

Theorem 10 $(\exists A)\,[\mathrm{NP}^A \neq \mathrm{P}^A = (\mathrm{EASY}_\forall^\forall)^A]$.

Corollary 11 [BGS75] $(\exists A)\,[\mathrm{NP}^A \neq \mathrm{P}^A = \mathrm{NP}^A \cap \mathrm{coNP}^A]$.

3 One-Way Functions and Easy Certificate Classes

First, let us extend Definition 2 to classes other than NP.

Definition 12 *For $\mathcal{C} \in \{NP, UP, FewP\}$, define* $\mathrm{EASY}_\forall^\forall(\mathcal{C})$ *to be the class of all sets L that either are finite, or (a) $L \in \mathcal{C}$, and (b) for every $\mathcal{C}$-machine N such that $L(N) = L$, there exists an FP function f_N such that, for all $x \in L$, $f_N(x) \in acc_N(x)$. As in the previous section, let* $\mathrm{EASY}_\forall^\forall$ *denote* $\mathrm{EASY}_\forall^\forall(\mathrm{NP})$.

Claim 13 $\mathrm{EASY}_\forall^\forall \subseteq \mathrm{EASY}_\forall^\forall(\mathrm{FewP}) \subseteq \mathrm{EASY}_\forall^\forall(\mathrm{UP}) \subseteq \mathrm{P} \subseteq \mathrm{UP} \subseteq \mathrm{FewP} \subseteq \mathrm{NP}$.

Definition 14 *For any function $f : \Sigma^* \to \Sigma^*$, let $dom(f)$ and $range(f)$ denote the domain and range of f, respectively. f is* honest *if there is a polynomial p such that for any $y \in range(f)$ and for any $x \in dom(f)$, if $y = f(x)$, then $|x| \leq p(|y|)$. f is* poly-one *if there is a polynomial p such that $\|f^{-1}(y)\| \leq p(|y|)$ for each $y \in range(f)$. A (many-one) function f is said to be* FP-invertible *if there is a function $g \in$* FP *such that for any $y \in range(f)$, $g(y)$ prints* some *value of $f^{-1}(y)$. An honest function f is said to be a* one-one *(*poly-one *or* many-one*)* one-way function *if f is one-one (poly-one or many-one), $f \in$ FP, and f is not FP-invertible. If $f : \Sigma^* \to \Sigma^*$ is a total, surjective, and one-one one-way function, f is called a* one-way permutation.

3.1 Characterizing the Existence of Partial One-Way Permutations

The authors of [FFNR96] have characterized the existence of surjective *many-one* one-way functions by the condition $\mathrm{P} \not\subseteq \mathrm{EASY}_\forall^\forall$. Furthermore, they have given relativized evidence that $\mathrm{P} \not\subseteq \mathrm{EASY}_\forall^\forall$ does not imply $\mathrm{P} \neq \mathrm{NP} \cap \mathrm{coNP}$ [FR94, FFNR96]. In this section, we give analogous characterizations of the existence of surjective one-one one-way functions and surjective poly-one one-way functions by separating P from $\mathrm{EASY}_\forall^\forall(\mathrm{UP})$ and $\mathrm{EASY}_\forall^\forall(\mathrm{FewP})$. Note that the type of function discussed in item (2) of Theorem 15 below is the partial-function analog of a (total) one-way permutation.

Theorem 15 *The following are equivalent.*

1. $\mathrm{EASY}_\forall^\forall(\mathrm{UP}) \neq \mathrm{P}$.
2. *There exists a partial one-one one-way function f with $range(f) = \Sigma^*$.*
3. $\Sigma^* \notin \mathrm{EASY}_\forall^\forall(\mathrm{UP})$.
4. $\mathrm{EASY}_\forall^\forall(\mathrm{UP})$ *is not closed under complementation.*

By Grollmann and Selman's characterization of the existence of partial one-one one-way functions with $\mathrm{range}(f) = \Sigma^*$ [GS88], we immediately have Corollary 16, which has previously been proven directly by Hartmanis and Hemaspaandra (then Hemachandra) [HH88], using different notation. As a point of interest, we note that Corollary 16 proves that separating P from a certain class containing P is equivalent to separating P from a certain class contained in P.

Corollary 16 [HH88] $\mathrm{P} \neq \mathrm{UP} \cap \mathrm{coUP}$ *if and only if* $\mathrm{EASY}_{\forall}^{\forall}(\mathrm{UP}) \neq \mathrm{P}$.

A seemingly unrelated connection comes from finite model theory. Grädel [Grä94] has recently shown that $\mathrm{P} = \mathrm{UP} \cap \mathrm{coUP}$ if and only if the weak definability principle holds for every first order logic $\mathcal{L}$ on finite structures that captures P. The weak definability principle says: Every totally defined query (on the set of finite structures of the relations of a first order logic $\mathcal{L}$) that is implicitly definable in $\mathcal{L}$ is also explicitly definable in $\mathcal{L}$ (see [Grä94] for those notions not defined here).

Corollary 17 $\mathrm{EASY}_{\forall}^{\forall}(\mathrm{UP}) \neq \mathrm{P}$ *if and only if the weak definability principle fails for some first order logic* $\mathcal{L}$ *on finite structures that captures P.*

Now we characterize the existence of surjective poly-one one-way functions by separating P and $\mathrm{EASY}_{\forall}^{\forall}(\mathrm{FewP})$. The proof is omitted.

Theorem 18 *The following are equivalent.*

1. *There exists a partial surjective poly-one one-way function.*
2. *There exists a total surjective poly-one one-way function.*
3. *There exists a total poly-one one-way function* f *with range*$(f) \in \mathrm{P}$.
4. *There exists a partial poly-one one-way function* f *with range*$(f) \in \mathrm{P}$.
5. $\mathrm{EASY}_{\forall}^{\forall}(\mathrm{FewP}) \neq \mathrm{P}$.
6. $\Sigma^* \notin \mathrm{EASY}_{\forall}^{\forall}(\mathrm{FewP})$.
7. $\mathrm{EASY}_{\forall}^{\forall}(\mathrm{FewP})$ *is not closed under complementation.*

Note that $\mathrm{P} \neq \mathrm{FewP}$ is clearly implied by each of the conditions of Theorem 18. Note also that $\mathrm{P} \neq \mathrm{FewP} \cap \mathrm{coFewP}$ clearly implies each of the conditions of Theorem 18, though it is not known whether the converse holds. We conjecture that it does not (equivalently, we conjecture that the converse of the FewP analog of the Borodin-Demers theorem does not hold). Thus, the conditions of Theorem 18 are intermediate (potentially strictly) between the conditions $\mathrm{P} \neq \mathrm{FewP} \cap \mathrm{coFewP}$ and $\mathrm{P} \neq \mathrm{FewP}$.[4]

[4] Regarding the condition $\mathrm{P} \neq \mathrm{FewP}$, Allender [All86] showed that the following conditions are all equivalent: (a) $\mathrm{P} \neq \mathrm{FewP}$, (b) there exists a total poly-one one-way function, and (c) there exists a total poly-one weak one-way function. Weak one-way functions mean the following. A poly-one function f is *strongly* FP-*invertible* if there is a function $g \in \mathrm{FP}$ such that for every $y \in \mathrm{range}(f)$, $g(y)$ prints *all* elements of $f^{-1}(y)$. A function f is called a *weak one-way function* if $f \in \mathrm{FP}$, f is poly-one, f is honest, and f is not strongly FP-invertible. Similarly, it is not hard to see, e.g., from Allender's proof, that also equivalent to (a), (b), and (c) are each of these conditions: (d) there exists a total poly-one weak one-way function f with range$(f) \in \mathrm{P}$, and (e) there exists a partial poly-one one-way function. We note that the following condition is also equivalent to each of (a)–(e): (f) there exists a total surjective poly-one weak one-way function. This is true for the following reasons. Clearly (f) implies (d). Also, (e) implies (f) as if h is a function satisfying (e), then h' satisfies (f), where

$$h'(x) \stackrel{\mathrm{df}}{=} \begin{cases} \epsilon & \text{if } x \in \{\epsilon, 0, 1\} \\ h(z)0 & \text{if } x = z00 \text{ and } h(z) \neq \bot \\ z1 & \text{if } (x = z00 \wedge h(z) = \bot) \text{ or } x = z11 \\ z0 & \text{if } x = z01 \text{ or } x = z10. \end{cases}$$

3.2 Characterizing the Existence of Total One-Way Permutations

For many types of one-way functions, the existence question has been characterized in the literature as equivalent to the separation of suitable complexity classes. Such a characterization for the existence of one-way permutations, however, is still missing. To date, the result closest to this goal is the above-mentioned characterization of the existence of a *partial* one-way permutation f by the condition $\mathrm{P} \neq \mathrm{UP} \cap \mathrm{coUP}$ [GS88] (see also Theorem 15). Since f is not total, f is not a permutation of Σ^* (even though f is a bijection mapping a subset of Σ^* onto Σ^*). Thus, $\mathrm{P} \neq \mathrm{UP} \cap \mathrm{coUP}$ potentially is a strictly weaker condition than the existence of a (total) one-way permutation. Of course, such a function f can be made total [GS88], but only at the cost of loss of surjectivity (even though such a total one-way function created from f still has a range in P). However, we will show below that the existence of one-way permutations is equivalent to the existence of total injective one-way functions whose range is P-rankable. A set A is P-*rankable* if there exists a function *rank* $\in$ FP so that for each $x \in \Sigma^*$, $\mathit{rank}(x) = \|\{w \in A \,|\, w \leq_{\text{lex}} x\}\|$ [GS91].

Theorem 19 *One-way permutations exist if and only if there exist total one-one one-way functions whose range is P-rankable.*

Proof. The "only if" direction is immediate, since Σ^* is P-rankable.

For the converse, suppose there exists a total one-one one-way function f whose range is P-rankable. We will define a one-way permutation h. Intuitively, the idea is to fill in the holes in the range of f, using its P-rankability. Let $T = \mathrm{range}(f)$ be P-rankable. For each n, let $\mathit{holes}(n) \stackrel{\text{df}}{=} 2^n - \|T^{=n}\|$. Note that since T is P-rankable, the function *holes* is in FP. Let us introduce some useful notation. For each string x, let $k(x)$ be the lexicographical position of x among the length $|x|$ strings; e.g., $k(000) = 1$ and $k(111) = 8$. For each string x and each $j \in \mathbb{N}$, let $x - j$ denote the string that in lexicographical order comes j places before x. For each set A and each $k \in \mathbb{N}$, let $A_{[k]}$ be the kth string of A in lexicographical order. Now define

$$h(x) \stackrel{\text{df}}{=} \begin{cases} f(x - \sum_{i=0}^{|x|} \mathit{holes}(i)) & \text{if } k(x) > \mathit{holes}(|x|) \\ (\overline{T} \cap \Sigma^{|x|})_{[k(x)]} & \text{if } k(x) \leq \mathit{holes}(|x|). \end{cases}$$

Since T is P-rankable and $f \in$ FP, we have $h \in$ FP. Clearly, h is honest and injective, h is total, and $\mathrm{range}(h) = \Sigma^*$. If one could invert h in polynomial time, then f would also be FP-invertible, as the P-rankability of T allows one to find the string in the range of f that should be inverted with respect to h, and after inverting we shift the inverse with respect to h, say z, by $\sum_{i=0}^{|z|} \mathit{holes}(i)$ positions to obtain the true inverse with respect to f. Hence, h is a one-way permutation. □

Note that P-rankability of the range of f suffices to give us Theorem 19, and Theorem 19 is stated in this way. However, even weaker notions would work. Without going into precise details, we remark that one just needs a function that, from some easily found and countable set of places, is an honest *address function* (see [GHK92]) for the complement of the range of f. Of course, the ultimate goal is to find a characterization of the existence of one-way permutations in terms of a separation of suitable complexity classes.

The tables below summarize the results that are known from the literature and from this section. Note that for one-one functions, it does not make sense to ask whether they are *weak* one-way functions, since for one-one functions, FP-invertibility and strong FP-invertibility are clearly identical notions.

Partial functions	one-one	poly-one
no restriction	$P \neq UP$ [GS88]	$P \neq FewP$ (Footnote 4)
surjective	$P \neq EASY_{\forall}^{\forall}(UP)$ (Thm. 15)	$P \neq EASY_{\forall}^{\forall}(FewP)$ (Thm. 18)
range in P	$P \neq EASY_{\forall}^{\forall}(UP)$ (Thm. 15)	$P \neq EASY_{\forall}^{\forall}(FewP)$ (Thm. 18)

Total functions	one-one	poly-one
no restriction	$P \neq UP$ [GS88]	$P \neq FewP$ [All86]
surjective	open question (but note Thm. 19)	$P \neq EASY_{\forall}^{\forall}(FewP)$ (Thm. 18)
range in P	$P \neq EASY_{\forall}^{\forall}(UP)$ (Thm. 15)	$P \neq EASY_{\forall}^{\forall}(FewP)$ (Thm. 18)
weak	—	$P \neq FewP$ [All86]
surj. & weak	—	$P \neq FewP$ (Footnote 4)
P-range & weak	—	$P \neq FewP$ (Footnote 4)

Acknowledgments

We thank Eric Allender for pointing out that if $NP \cap coNP$ is P-bi-immune, then $EASY_{\forall}^{\forall} = FINITE$, for pointing out an alternative proof of Theorem 5.4, and for generally inspiring this line of research. We acknowledge interesting discussions with Alan Selman, Lance Fortnow, and Erich Grädel on this subject. In particular, we are indebted to Alan Selman for generously permitting us to include Claim 7 and to Lance Fortnow for providing us with an advance copy of [FFNR96].

References

[Adl79] L. Adleman. Time, space, and randomness. Technical Report MIT/LCS/TM-131, MIT, Cambridge, MA, April 1979.

[All86] E. Allender. The complexity of sparse sets in P. In *Proceedings of the 1st Structure in Complexity Theory Conference*, pages 1–11. Springer-Verlag *Lecture Notes in Computer Science #223*, June 1986.

[All92] E. Allender. Applications of time-bounded Kolmogorov complexity in complexity theory. In O. Watanabe, editor, *Kolmogorov Complexity and Computational Complexity*, EATCS Monographs on Theoretical Computer Science, pages 4–22. Springer-Verlag, 1992.

[AR88] E. Allender and R. Rubinstein. P-printable sets. *SIAM Journal on Computing*, 17(6):1193–1202, 1988.

[BD76] A. Borodin and A. Demers. Some comments on functional self-reducibility and the NP hierarchy. Technical Report TR 76-284, Cornell Department of Computer Science, Ithaca, NY, July 1976.

[BGS75] T. Baker, J. Gill, and R. Solovay. Relativizations of the P=?NP question. *SIAM Journal on Computing*, 4(4):431–442, 1975.

[Cha66] G. Chaitin. On the length of programs for computing finite binary sequences. *Journal of the ACM*, 13:547–569, 1966.

[FFNR96] S. Fenner, L. Fortnow, A. Naik, and J. Rogers. On inverting onto functions. In *Proceedings of the 11th Annual IEEE Conference on Computational Complexity*, pages 213–222. IEEE Computer Society Press, May 1996.

[FR94] L. Fortnow and J. Rogers. Separability and one-way functions. In *Proceedings of the 5th International Symposium on Algorithms and Computation*, pages 396–404. Springer-Verlag *Lecture Notes in Computer Science #834*, August 1994.

[GHK92] J. Goldsmith, L. Hemachandra, and K. Kunen. Polynomial-time compression. *Computational Complexity*, 2(1):18–39, 1992.

[Grä94] E. Grädel. Definability on finite structures and the existence of one-way functions. *Methods of Logic in Computer Science*, 1:299–314, 1994.

[GS88] J. Grollmann and A. Selman. Complexity measures for public-key cryptosystems. *SIAM Journal on Computing*, 17(2):309–335, 1988.

[GS91] A. Goldberg and M. Sipser. Compression and ranking. *SIAM Journal on Computing*, 20(3):524–536, 1991.

[Har83] J. Hartmanis. Generalized Kolmogorov complexity and the structure of feasible computations. In *Proceedings of the 24th IEEE Symposium on Foundations of Computer Science*, pages 439–445. IEEE Computer Society Press, 1983.

[HH88] J. Hartmanis and L. Hemachandra. Complexity classes without machines: On complete languages for UP. *Theoretical Computer Science*, 58:129–142, 1988.

[HRW95] L. Hemaspaandra, J. Rothe, and G. Wechsung. Easy sets and hard certificate schemes. Technical Report Math/95/5, Institut für Informatik, Friedrich-Schiller-Universität Jena, Jena, Germany, 1995.

[HY84] J. Hartmanis and Y. Yesha. Computation times of NP sets of different densities. *Theoretical Computer Science*, 34:17–32, 1984.

[Ko85] K. Ko. On some natural complete operators. *Theoretical Computer Science*, 37:1–30, 1985.

[Kol65] A. Kolmogorov. Three approaches for defining the concept of information quantity. *Prob. Inform. Trans.*, 1:1–7, 1965.

[LV90] M. Li and P. Vitányi. Applications of Kolmogorov complexity in the theory of computation. In A. Selman, editor, *Complexity Theory Retrospective*, pages 147–203. Springer-Verlag, 1990.

[Sel95] A. Selman. Personal Communication, May, 1995.

[Sip83] M. Sipser. A complexity theoretic approach to randomness. In *Proceedings of the 15th ACM Symposium on Theory of Computing*, pages 330–335, 1983.

[Tra84] B. Trakhtenbrot. A survey of Russian approaches to *perebor* (brute-force search) algorithms. *Annals of the History of Computing*, 6(4):384–400, 1984.

[Val76] L. Valiant. The relative complexity of checking and evaluating. *Information Processing Letters*, 5:20–23, 1976.

A Implications Between Statements About NP Sets

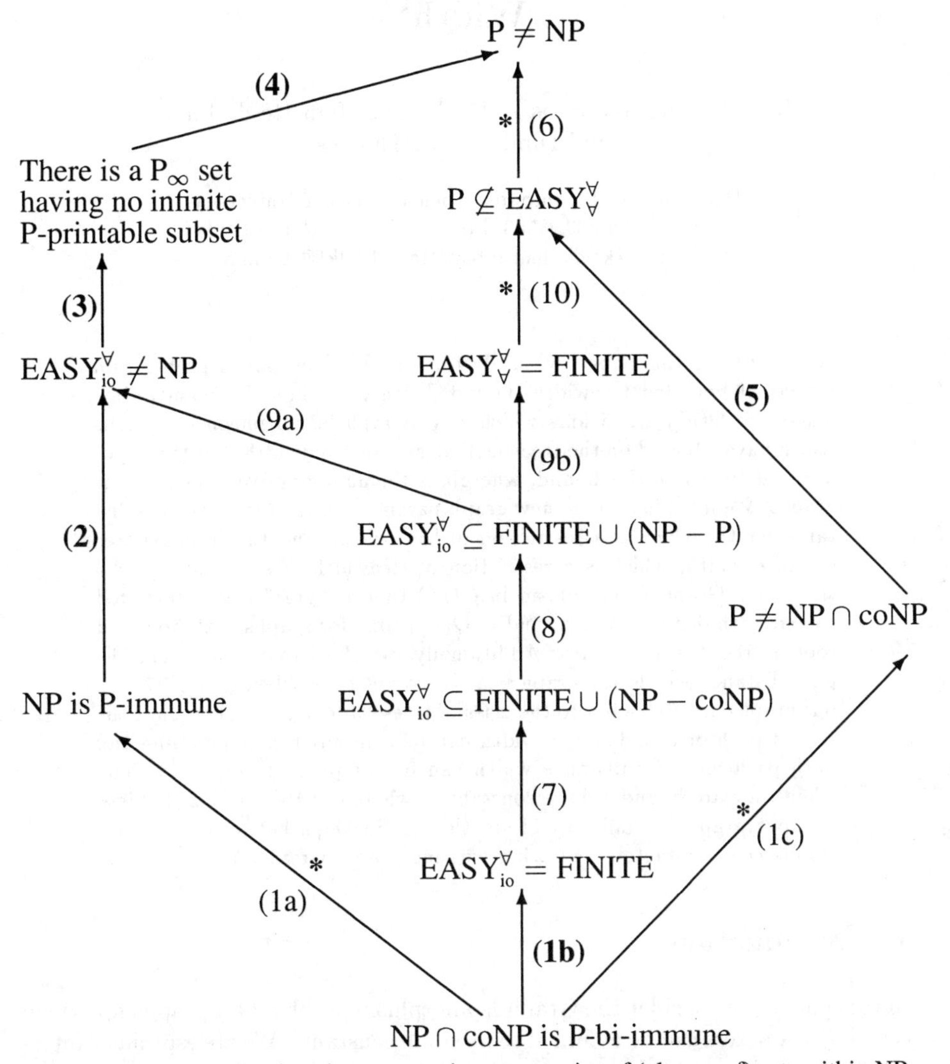

Figure 1: Some implications between various properties of (classes of) sets within NP.

Key: Arrows labeled by boldface numbers indicate the non-trivial implications that are shown in Theorem 5. Arrows marked by a "*" represent those implications that are not invertible up to the limits of relativizations (as stated in Section 2.3). P_∞ denotes the class P − FINITE of all infinite P sets.

Isomorphism for Graphs of Bounded Distance Width*

Koichi Yamazaki, Hans L. Bodlaender, Babette de Fluiter, and Dimitrios M. Thilikos

Department of Computer Science, Utrecht University,
P.O. Box 80.089, 3508 TB Utrecht, the Netherlands
E-mail: {koichi,hansb,babette,sedthilk}@cs.ruu.nl

Abstract. In this paper, we study the Graph Isomorphism problem on graphs of bounded treewidth, bounded degree or bounded bandwidth. There are $O(n^c)$ algorithms which solve Graph Isomorphism on graphs which have a bound on the treewidth, degree or bandwidth, but the exponent c depends on this bound, where n is the number of vertices in input graphs. We introduce some new graph parameters: the (rooted) path distance width, which is a restriction of bandwidth, and the (rooted) tree distance width, which is a restriction of treewidth. We give algorithms that solve Graph Isomorphism in $O(n^2)$ time for graphs with bounded rooted path distance width, and in $O(n^3)$ time for graphs with bounded rooted tree distance width. Additionally, we show that computing the path distance width of a graph is a NP-complete problem even if the input graphs are restricted to the class of trees. Moreover we show that the rooted path or tree distance width can be computed in $O(ne)$ time and both path and tree distance width can be computed in $O(n^{k+1})$ time, when they are bounded by a constant k, where e is the number of edges in input graphs. Finally, we study the relationships between the newly introduced parameters and other existing graph parameters.

1 Introduction

In this paper, we consider the Graph Isomorphism problem, for graphs for which certain parameters can be assumed to be small constants. We are especially interested in Graph Isomorphism on graphs where one of the parameters treewidth, bandwidth, genus, or degree of the graph is bounded, as there are many interesting graph classes which have a bound on one of these parameters (see e.g. [1]). For these parameters, it has been shown that when the parameter is a constant, then Graph Isomorphism is polynomial time solvable (see [8] for bounded degree, [7, 9] for bounded genus, and [2] for bounded treewidth). However, in each

* The first author was supported by Ministry of Education, Science, Sports and Culture of Japan as Oversea's Research Scholar. The third author was supported by the Foundation for Computer Science (S.I.O.N) of the Netherlands Organization for Scientific Research (N.W.O.). The last author was supported by the Training and Mobility of Researchers (TMR) Program, (EU contract no ERBFMBICT950198).

of these three cases, the exponent of the algorithm grows with the parameter. Thus, a question is, whether algorithms exist for Graph Isomorphism with a running time $O(f(k)n^c)$, c a small constant, k the maximum degree / treewidth / bandwidth / ... / of the graph; in other words, whether Graph Isomorphism is *fixed parameter tractable* (in the sense of the fixed parameter complexity theory of Downey and Fellows, see e.g. [5, 6]), with the maximum degree / treewidth / bandwidth / ... / as parameter.

Thus, we are looking for answers to the questions whether Graph Isomorphism is fixed parameter tractable for the case that the parameter is the maximum degree, treewidth or bandwidth of the graph. These questions are apparently hard. In this paper, we are able to solve some interesting special cases of these problems.

For this, several graph parameters are introduced: the *(rooted) path distance width*, and the *(rooted) tree distance width*. Intuitively, a graph has small rooted path distance width if, for some root vertex v, all the vertices with the same distance from v have small cardinality. If, instead of a single vertex v we consider a set of vertices, then we get the notion of path distance width. The notion of path distance width can be seen to have a close relation to bandwidth; the notion of tree distance width is a tree-like generalisation of this notion, with a close relationship to treewidth.

This paper is organised as follows. In Section 2 we prove that the rooted path (tree) distance width of a graph can be computed in $O(ne)$ time, that computing the path distance width of a graph is NP-hard, but if the path or tree distance width is at most some fixed constant k, then the minimum path (tree) distance width can be computed in $O(n^{k+1})$ time. The main results of the paper can be found in Section 3: it is shown that Graph Isomorphism is solvable in $O(n^2)$ time for graphs with bounded rooted path distance width, thus solving a significant special case of Graph Isomorphism for graphs of bounded bandwidth. Furthermore, it is shown that Graph Isomorphism is solvable in $O(n^3)$ time for graphs with bounded rooted tree distance width, which solves a special case for Graph Isomorphism for graphs of bounded treewidth. In Section 4, the relations between the different considered parameters are investigated.

2 Definitions and complexity results for distance width

The graphs we consider are simple, undirected and connected, and contain no self loops. For a graph G, we denote the set of vertices of G by $V(G)$ and the set of edges by $E(G)$. For a subset S of $V(G)$, we denote the subgraph induced by S by $G[S]$. For two graphs G and H, a function $f : V(G) \to V(H)$ is called an *isomorphism* (from G to H) if f is a bijection and for each $v, w \in V(G)$, $\{v, w\} \in E(G)$ iff $\{f(v), f(w)\} \in E(H)$. Two graphs G and H are *isomorphic* if there is an isomorphism from G to H. The Graph Isomorphism problem is the problem of checking for two given graphs whether they are isomorphic.

A *graph parameter* is a function which maps each graph to a positive integer. We first review a number of graph parameters.

A *tree decomposition* of a graph G is a pair $(\{X_i \mid i \in I\}, T = (I, F))$, where $\{X_i \mid i \in I\}$ is a collection of subsets of $V(G)$ and T is a tree, such that

- $\bigcup_{i \in I} X_i = V(G)$,
- for each edge $\{v, w\} \in E(G)$, there is an $i \in I$ such that $v, w \in X_i$, and
- for each $v \in V(G)$ the set of nodes $\{i \mid v \in X_i\}$ forms a subtree of T.

The width of a tree decomposition $(\{X_i \mid i \in I\}, T = (I, F))$ equals $\max_{i \in I}(|X_i| - 1)$. The *treewidth* of a graph G is the minimum width over all tree decompositions of G. The corresponding graph parameter (which is the function that maps each graph to its treewidth) is denoted by $\mathcal{TW}$.
A *(linear) layout* of a graph G is a bijection $f : V(G) \rightarrow \{1, \cdots, |V(G)|\}$.
The *bandwidth* of a layout f of a graph G is defined as $\max_{\{u,v\} \in E(G)} |f(u) - f(v)|$.
The bandwidth of a graph G is the minimum bandwidth over all layouts of G. The corresponding graph parameter is denoted by $\mathcal{BW}$.
For a given graph G and two vertices $u, v \in V(G)$, $d_G(u, v)$ denotes the distance between u and v, which is the number of edges on a shortest path between u and v. For a set $S \subseteq V(G)$ and a vertex $w \in V(G)$, $d_G(S, w)$ denotes $\min_{v \in S} d_G(v, w)$.
A *tree distance decomposition* of a graph G is a triple $(\{X_i \mid i \in I\}, T = (I, F), r)$, where

- $\bigcup_{i \in I} X_i = V(G)$ and for all $i \neq j$, $X_i \cap X_j = \emptyset$,
- for each $v \in V(G)$, if $v \in X_i$, then $d_G(X_r, v) = d_T(r, i)$, and
- for each edge $\{v, w\} \in E(G)$, there are $i, j \in I$ such that $v \in X_i$, $w \in X_j$ and either $i = j$ or $\{i, j\} \in F$.

Node r is called the root of the tree T, and X_r is called the root set of the tree distance decomposition. The width of a tree distance decomposition $(\{X_i \mid i \in I\}, T, r)$ is equal to $\max_{i \in I}|X_i|$. The *tree distance width* of a graph G is the minimum width over all possible tree distance decompositions of G. The corresponding graph parameter is denoted by $\mathcal{TDW}$.
A *rooted tree distance decomposition* of a graph G is a tree distance decomposition $(\{X_i \mid i \in I\}, T = (I, F), r)$ of G in which $|X_r| = 1$. The *rooted tree distance width* of a graph G is the minimum width over all rooted tree distance decompositions. The corresponding graph parameter is denoted by $\mathcal{RTDW}$.
The *(rooted) path distance decomposition* and the parameter of *(rooted) path distance width* of a graph $G = (V, E)$ are defined as the (rooted) tree distance decomposition and (rooted) tree distance width above with the difference that the tree T is a path (i.e. two nodes in T have degree one, and all others have degree two). For reasons of simplicity we will denote a (rooted) path distance decomposition as $(X_0, X_1, \ldots, X_t)$, where X_0 is the root set of the decomposition. We denote the corresponding graph parameters by $\mathcal{PDW}$, and $\mathcal{RPDW}$, respectively.

It is easy to check that for any graph G, $\mathcal{TW}(G) \leq 2\mathcal{TDW}(G) - 1$, $\mathcal{TDW}(G) \leq \mathcal{RTDW}(G)$, $\mathcal{BW}(G) \leq 2\mathcal{PDW}(G) - 1$ and $\mathcal{PDW}(G) \leq \mathcal{RPDW}(G)$. Hence fixed parameter tractability of Graph Isomorphism for $\mathcal{TW}$ implies the same for $\mathcal{TDW}$ and $\mathcal{RTDW}$, and fixed parameter tractability of Graph Isomorphism for $\mathcal{BW}$ implies the same for $\mathcal{PDW}$ and $\mathcal{RPDW}$. Also, showing that for Graph Isomorphism is fixed parameter tractable for e.g. $\mathcal{RPDW}$ might give more insight in whether it is fixed parameter tractable for $\mathcal{BW}$. Therefore, we study the complexity of Graph Isomorphism on graphs for which $\mathcal{PDW}$, $\mathcal{RPDW}$, $\mathcal{TDW}$ or $\mathcal{RTDW}$ is bounded. (See also Section 4.)

For a given graph G and $S \subseteq V(G)$, there is a unique path distance decomposition of G with root set S, and this decomposition can be found in $O(|E(G)|)$ time: for each vertex $v \in V(G)$, compute $d_G(S, v)$ using breadth first search. Then for each possible distance d, make a node X_d containing all vertices with distance d to S.

For a given graph G and set $S \subseteq V(G)$, there may be more than one tree distance decomposition with root set S. However there is a unique *minimal* tree distance decomposition: a tree distance decomposition $(\{X_i \mid i \in I\}, T = (I, F), r)$ is minimal if for each $i \in I$, the subgraph $G[Y_i]$ of G is connected, where Y_i is the union of all nodes X_j for which j is in the subtree of T rooted at i. It is fairly easy to see that for a given root set S, such a minimal tree distance decomposition is unique, and that it can be found with procedure GET-TDD presented in Figure 1, which can be made to run in $O(|E(G)|)$ time. Thus we have the following result (presented without proof).

Theorem 2.1 *Given a graph G and a set $S \subseteq V(G)$, we can compute in $O(|E(G)|)$ time the unique path distance decomposition with root set S, or the unique minimal tree distance decomposition with root set S.*

From Theorem 2.1 and the fact that, if a graph G has (rooted) path or tree distance width at most k, then $|E(G)| = O(k|V(G)|)$, we get the following result.

Corollary 2.2 *There is an algorithm that computes a rooted path (tree) distance decomposition of minimum width of a graph G in $O(k|V(G)|^2)$ time, where k is the rooted path (tree) distance width of the graph.*

Theorem 2.3 *The following problem is NP-complete even if the input graphs are trees. Given a graph G and an integer k, does G have path distance width at most k?*

Due to space restrictions we omit the proof of the above theorem in this extended abstract. We also conjecture that the problem of computing tree distance width is NP-complete.

Using Theorem 2.1, observing that there are $O(n^k)$ choices for the root set S, we get the following result about the complexity of path (tree) distance width when it is bounded by a fixed constant k.

Corollary 2.4 *There is an algorithm that computes a path (tree) distance decomposition of minimum width of a graph G in $O(k|V(G)|^{k+1})$ time, where k denotes the path (tree) distance width of the graph.*

```
Procedure GET-TDD
input:  a graph G = (V, E) and a root set S
output: the minimal tree distance decomposition ({X_i | i ∈ I}, T = (I, F), r),
        where (X_r = S)
 1:  for all v ∈ V set distance(v) = d_G(S, v);
 2:  m := max_{v∈V} distance(v);
 3:  I := ∅, F := ∅, and h := 0;
 4:  for all i, 0 ≤ i ≤ m + 1 set V_i = {v ∈ V | distance(v) = i};
 5:  for i := m down to 0 do
 6:     Compute the connected components of G[{v ∈ V | i ≤ distance(v) ≤ i + 1}]
        /* We call the connected components S_1, ..., S_t */;
 7:     for j := 1 to t do
 8:        X_{h+j} := S_j − V_{i+1};
 9:        Add edges {v, u}, v, u ∈ X_{h+j} to E(G) so that G[X_{h+j}] becomes
           connected;
10:        I := I ∪ {h + j};
11:        F := F ∪ {{h + j, k} | X_k ⊂ S_j ∧ k ≤ h};
12:     od
13:     h := h + t;
14:  od
15:  end.
```

Fig. 1. Procedure GET-TDD(G, S).

Finally, we conjecture that the problem of deciding if the path (tree) distance width of a graph is no more than some fixed parameter k, is not fixed parameter tractable i.e. it is hard for some class of the W hierarchy defined by Downey and Fellows in [5, 6].

3 Graph isomorphism for graphs of bounded distance width

In this section, we show that the isomorphism problem is fixed parameter tractable for graphs of bounded rooted path distance width or bounded rooted tree distance width. We present two algorithms testing isomorphism of two input graphs of which the rooted path distance width and the rooted tree distance width, respectively, are bounded by a constant k. The running time of the first algorithm is $O(n^2)$, and of the latter one $O(n^3)$, where n is the number of vertices in input graphs.

Theorem 3.1 *There is an algorithm which, for fixed k, checks whether two input graphs G and H of rooted path distance width at most k are isomorphic, and runs in $O(|V(G)|^2)$ time.*

Proof: For input graphs G and H, the algorithm works as follows. There are two phases. In the first phase, a rooted path distance decomposition $D^G = (X_0, X_1, \ldots, X_t)$ of minimum width is computed for G. By Corollary 2.2, this takes $O(|V(G)|^2)$ time. Let k denote the width of D^G.

In the second phase of the algorithm, we compute for each $w \in V(H)$ the unique rooted path distance decomposition $D^H = (Y_0, Y_1, \ldots, Y_s)$ of H with root set $\{w\}$. If the width of D^H equals k and $s = t$, then we test whether decompositions D^G and D^H are isomorphic: D^G and D^H are said to be isomorphic if there is an isomorphism $f : V(G) \to V(H)$ from G to H, such that for each i, $1 \leq i \leq t$, and each $x \in X_i$, $f(x) \in Y_i$. If D^G and D^H are isomorphic, then we may conclude that G and H are isomorphic. On the other hand, if there is no rooted path distance decomposition of H which is isomorphic to D^G, then G and H can not be isomorphic. So the given algorithm correctly computes whether G and H are isomorphic.

We now describe how to test whether two rooted path decompositions are isomorphic. Suppose $D^G = (X_0, X_1, \ldots, X_t)$ is a rooted path distance decomposition of G with root set X_0, and $D^H = (Y_0, Y_1, \ldots, Y_s)$ is a decomposition of H with root set Y_0, such that $s = t$ and D^G and D^H have the same width. For each i, $1 \leq i \leq t$, we compute the set R_i, which contains all isomorphisms from $G[X_i]$ to $H[Y_i]$, which are *extendible*, i.e. R_i contains all isomorphisms f from $G[X_i]$ to $H[Y_i]$ for which there is an isomorphism g from $G[X_i \cup \ldots \cup X_t]$ to $H[Y_i \cup \ldots, \cup Y_t]$, such that $\forall x \in X_i,\ f(x) = g(x) \wedge \forall j,\ i \leq j \leq t, \forall x \in X_j\ g(x) \in Y_j$. Now, R_0 is not empty iff D^G and D^H are isomorphic. The set R_0 can be computed in a bottom-up way, by first computing R_t, and then, for each i, $0 \leq i < t$, computing R_i from R_{i+1}.

The set R_t can be easily computed: if $|X_t| \neq |Y_t|$, then $R_t = \emptyset$. Otherwise, for each bijection $f : X_t \to Y_t$, check if it is an isomorphism from $G[X_t]$ to $H[Y_t]$. If so, put f in R_t. This takes constant time, since there are at most $k!$ such bijections.

For each i, $0 \leq i < t$, we can compute R_i as follows. First check if $|X_i| = |Y_i|$. If not, then $R_i = \emptyset$. Otherwise, for each bijection $f : X_i \to Y_i$, check if there is a $g \in R_{i+1}$, such that $f \cup g : X_i \cup X_{i+1} \to Y_i \cup Y_{i+1}$ is an isomorphism from $G[X_i \cup X_{i+1}]$ to $H[Y_i \cup Y_{i+1}]$. If so, then put f in R_i. This can again be done in constant time. It is quite easy to see that R_i is computed correctly.

The running time of phase two of the algorithm is $O(|V(G)|^2)$: computing the rooted path distance decomposition of H for each possible root set takes $O(|V(G)|^2)$ time since $|E(G)| = O(|V(G)|)$. Furthermore, checking if two decompositions are isomorphic can be done in $O(|V(G)|)$ time: for each i, computing R_i takes constant time, and there are at most $O(|V(G)|)$ nodes. □

Theorem 3.2 *There is an algorithm which, for fixed k, checks whether two input graphs G and H of rooted tree distance width at most k are isomorphic, and runs in $O(|V(G)|^3)$ time.*

Proof: The basic structure of the algorithm is the same as for graphs of bounded rooted path distance width: there are again two phases. In the first phase, a minimal rooted tree distance decomposition $D^G = (\{X_i \mid i \in I^G\}, T^G = (I^G, F^G), r^G)$ of minimum width is computed for G. By Corollary 2.2, this can be done in $O(|V(G)|^2)$ time. Let k denote the width of D^G.

In the second phase of the algorithm, we compute for each $w \in V(H)$ the unique minimal rooted tree distance decomposition D^H of H with root set $\{w\}$. Let $D^H = (\{Y_i \mid i \in I^H\}, T^H = (I^H, F^H), r^H)$ denote this decomposition. If the width of D^H equals k, then we test whether decompositions D^G and D^H are isomorphic: D^G and D^H are said to be isomorphic if there is an isomorphism $f : V(G) \to V(H)$ from G to H and an isomorphism $g : I^G \to I^H$ from T^G to T^H, such that $g(r^G) = r^H$ and for each i, $i \in I^G$, and each $x \in X_i$, $f(x) \in Y_{g(i)}$. By the same argument as for the rooted path distance decompositions, we may conclude that G and H are isomorphic if D^G and D^H are isomorphic. On the other hand, if there is no minimal rooted tree distance decomposition of H which is isomorphic to D^G, then G and H can not be isomorphic. So the given algorithm correctly computes whether G and H are isomorphic.

We now give the algorithm that tests whether two rooted tree distance decompositions are isomorphic.

Suppose $D^G = (\{X_i \mid i \in I^G\}, T^G = (I^G, F^G), r^G)$ and $D^H = (\{Y_i \mid i \in I^H\}, T^H = (I^H, F^H), r^H)$ are minimal rooted tree distance decompositions of graphs G and H, respectively. Note that D^G and D^H can not be isomorphic if T^G and T^H are not isomorphic. Therefore, we first test whether T^G and T^H are isomorphic, using the algorithm from [4]. This takes $O(|V(G)|)$ time. Now suppose T^G and T^H are isomorphic. The *depth* of a node in a rooted tree is its distance to the root (so the root has depth zero). Let m denote the maximum depth of a node in T^G (and hence in T^H). The nodes with depth d are called the nodes on level d. Now, for each level d, $0 \leq d \leq m$, and each pair of nodes i, j, with $i \in I^G$ and $j \in I^H$, both on level d, we compute the set $R_d^{i,j}$ of isomorphisms $f : X_i \to Y_j$ from $G[X_i]$ to $H[Y_i]$ which are *extendible* in the following sense. Let T_i^G denote the rooted tree distance decomposition formed by the subtree of T^G rooted at node i, and T_j^H the rooted tree distance decomposition formed by the rooted subtree of T^H with root node j. Furthermore, let G_i denote the subgraph of G induced by all vertices in the nodes of T_i^G, and similarly, let H_j denote the subgraph of H induced by all vertices in the nodes of T_j^H. We say an isomorphism f from $G[X_i]$ to $H[Y_j]$ is extendible if there is an isomorphism $g' : V(T_i^G) \to V(T_j^H)$ from T_i^G to T_j^H and an isomorphism $f' : V(G_i) \to V(H_j)$ from G_i to H_j such that

- $g'(i) = j$,
- for each $a \in V(T_i^G)$ and each $v \in X_a$, $f'(v) \in Y_{g'(a)}$, and furthermore,
- for each $v \in X_i$, $f(v) = f'(v)$ (see Figure 2).

Now, $R_0^{r^G, r^H}$ is not empty iff D^G and D^H are isomorphic. The set $R_0^{r^G, r^H}$ can be computed in a bottom-up way, by first computing $R_m^{i,j}$ for all nodes $i \in I^G$ and $j \in I^H$ on level m, and then, for each d, $0 \leq d < m$, computing $R_d^{i,j}$ for all nodes $i \in I^G$ and $j \in I^H$ on level d, by using the values of $R_{d+1}^{i',j'}$ for all children i' of i and j' of j.

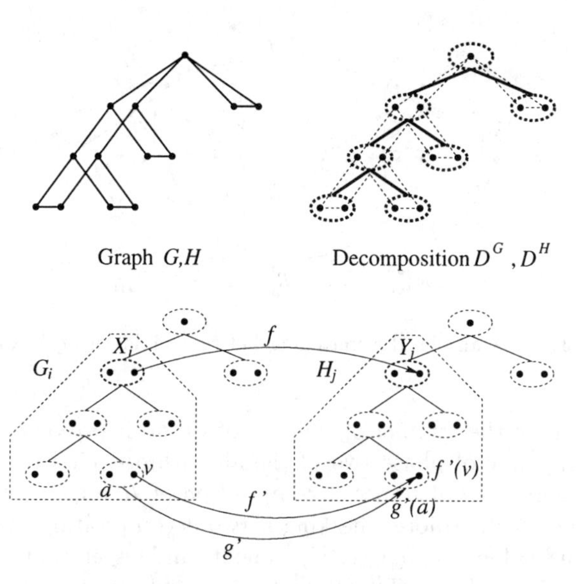

Fig. 2. An example of extendibility of isomorphism from $G[X_i]$ to $H[Y_j]$.

The sets $R_m^{i,j}$ can be easily computed: if $|X_i| \neq |Y_j|$, then $R_m^{i,j} = \emptyset$. Otherwise, for each bijection $f : X_i \to Y_j$, check if it is an isomorphism. If so, put f in $R_m^{i,j}$. This takes constant time for each i and j.

For each d, $0 \leq d < m$, and each $i \in I^G$ and $j \in I^H$ on level d, we can compute $R_d^{i,j}$ as follows. First check if $|X_i| = |Y_j|$ and the number of children of i equals the number of children of j. If not, then $R_d^{i,j} = \emptyset$. Otherwise, for each isomorphism $f : X_i \to Y_j$ from $G[X_i]$ to $H[Y_j]$, we do the following. We try to make a matching between the children of i and the children of j, i.e. we try to match each child i' of i to a child j' of j, in such a way that there is a $g \in R_{d+1}^{i',j'}$ for which $f \cup g : X_i \cup X_{i'} \to Y_j \cup Y_{j'}$ is an isomorphism from $G[X_i \cup X_{i'}]$ to $H[Y_j \cup Y_{j'}]$. It can be seen that, if there is such a matching, then f is extendible (see Figure 3).

We now show how to compute a matching as described above. We try to match the children of i one by one to a child of j as follows. Take a child i' of i which has not yet been matched. For each unmatched child j' of j, try to match i' to j'. As soon as a j' is found which can be matched to i', then match i' to j', and go on with the next child of i. If there is no j' which can be matched to i', then there can be no matching between the children of i and of j, and hence f is not extendible. (We can actually use this 'greedy matching algorithm', and need not use the standard 'maximum flow' matching algorithm, because of transitivity of isomorphism.)

Now, if each child of i is matched to a child of j, then we add f to $R_d^{i,j}$. It can be seen that $R_d^{i,j}$ is computed correctly this way. The time it takes to compute

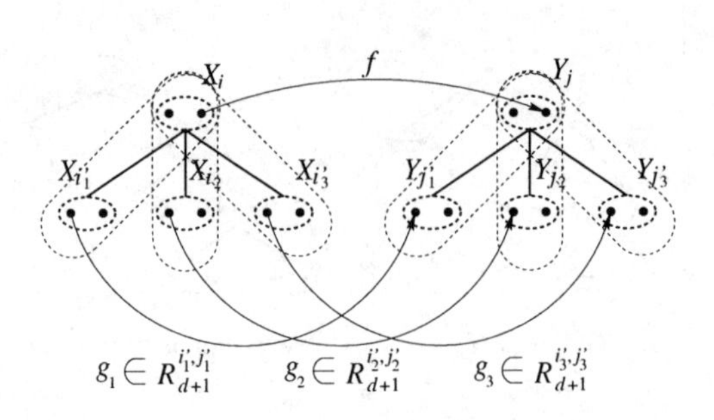

Fig. 3. An example of a matching between children of X_i and Y_j.

$R_d^{i,j}$ is quadratic in the number of children of i (and hence of j).

The running time of phase two of the algorithm is $O(|V(G)|^3)$: computing the minimal rooted tree distance decomposition of H for a root set $\{w\}$ takes $O(|V(G)|)$ time. Furthermore, checking if two decompositions D^G and D^H are isomorphic can be done in $O(|V(G)|^2)$ time: it can be seen that each edge in T^G is compared to each edge in T^H at most a constant number of times. Since the number of different root sets for H is $|V(H)|$, the running time of $O(|V(G)|^3)$ now follows. □

Note that a simple modification of our algorithms additionally gives the following result.

Corollary 3.3 *There is an algorithm which, for fixed k, checks whether two input graphs G, H of path or tree distance width at most k are isomorphic, and runs in $O(|V(G)|^{k+1})$ or $O(|V(G)|^{k+2})$ time, respectively.*

4 Relationships among classes of bounded width

In this section, first, we give the definitions of some known graph parameters in order to investigate their relations with (rooted) path distance width and (rooted) tree distance width.

A *strong tree decomposition* (see also [10]) of a graph G is a pair $(\{X_i \mid i \in I\}, T = (I, F))$, where $\{X_i \mid i \in I\}$ is a collection of subsets of $V(G)$ and T is a tree, such that

- $\bigcup_{i \in I} X_i = V(G)$ and for all $i \neq j$, $X_i \cap X_j = \emptyset$,
- for each edge $\{v, w\} \in E(G)$, there are $i, j \in I$ with $v \in X_i$ and $w \in X_j$, such that either $i = j$ or $\{i, j\} \in F$.

The width of a strong tree decomposition $(\{X_i \mid i \in I\}, T = (I, F))$ equals $\max_{i \in I}(|X_i|)$. The *strong treewidth* of a graph G is the minimum width over all possible strong tree decompositions of G. The corresponding graph parameter is denoted by $\mathcal{STW}$.

A *connected strong tree decomposition* of a graph G is a strong tree decomposition $(\{X_i \mid i \in I\}, T = (I, F))$ of G such that for each $i \in I$, $G[X_i]$ is connected.

The *connected strong treewidth* of G is the minimum width over all connected strong tree decompositions of G. The corresponding graph parameter is denoted by $\mathcal{CSTW}$.
A *path decomposition* of a graph G is a tree decomposition $(\{X_i \mid i \in I\}, T = (I, F))$ in which T is a path. The pathwidth of a graph G is the minimum width over all path decompositions of G. The corresponding graph parameter is denoted by $\mathcal{PW}$.
In the same way, we can define the notions of *strong pathwidth*, and *connected strong pathwidth*. We denote the corresponding graph parameters by $\mathcal{SPW}$ and $\mathcal{CSPW}$, respectively.
The *cutwidth* of a layout f of a graph G is defined as

$$\max_{1 \leq i < |V(G)|} |\{\{u, v\} \in E(G) : f(u) \leq i < f(v)\}|.$$

The cutwidth of a graph G is the minimum cutwidth over all layouts of G. The corresponding graph parameter is denoted by $\mathcal{CW}$.
For a given graph G, a *subdivision* is the operation which adds a new vertex u to G and replaces an edge $e = \{v, w\} \in E(G)$ by two edges $\{v, u\}$ and $\{u, w\}$ (i.e. it splits an edge of G into two edges). A refinement of a graph G is a graph G' which is obtained from G by a number of subsequent subdivisions.
The *topological bandwidth* of a graph G is the minimum bandwidth over all refinements of G. The corresponding graph parameter is denoted by $\mathcal{TBW}$.
By $\mathcal{D}$ we denote the graph parameter which maps each graph to the maximum degree of any vertex in the graph.

Let f and f' be two graph parameters. We say that f' *covers* f, denoted by $f \preceq f'$, if there is a function $g : \mathbf{N} \to \mathbf{N}$, such that for each graph G and each integer k, if $f(G) \leq k$ then $f'(G) \leq g(k)$ (we also say that f is covered by f'). For instance, if we take $f = \mathcal{BW}$ and $f' = \mathcal{CW}$, then $f \preceq f'$: for each graph G, $\mathcal{CW}(G) \leq \mathcal{BW}(G)(\mathcal{BW}(G) - 1)/2$ [1]. Hence if we take $g(k) = k(k-1)/2$, then for each graph G and each integer k, if $\mathcal{BW}(G) \leq k$ then $\mathcal{CW}(G) \leq g(k)$.

If a graph parameter f is not covered by a parameter f', we denote this by $f \not\preceq f'$. If $f \preceq f'$ but $f' \not\preceq f$, then we say that f' *strictly covers* f, denoted by $f \prec f'$. If $f \preceq f'$ and $f' \preceq f$, then we say $f \approx f'$. If $f \not\preceq f'$ and $f' \not\preceq f$, then we say that f and f' are not related, and we denote this by $f \not\approx f'$ (note that saying that $f \not\approx f'$ is not equivalent to saying that $f \approx f'$ does not hold). It is easy to see that $\prec$, $\preceq$ and $\approx$ are transitive relations.

The notion of covering is interesting in the following sense. Suppose we have a graph problem P (for example the isomorphism problem), and we have two graph parameters f and f', such that $f \preceq f'$. If problem P is fixed parameter tractable for parameter f', then we can conclude immediately that P is fixed parameter tractable for f. On the other hand, if we can show that problem P is fixed parameter tractable for parameter f, then this might help to get more insight in whether P is fixed parameter tractable for parameter f'.

We now give a number of relations for the graph parameters that are defined in Section 2.

Theorem 4.1 *The following relations hold (see also Figure 4).*

(1) $\mathcal{TW} \not\approx \mathcal{D}$	(2) $\mathcal{CW} \prec \mathcal{TW}$	(3) $\mathcal{CW} \prec \mathcal{D}$
(4) $\mathcal{CW} \approx \mathcal{TBW}$	(5) $\mathcal{BW} \prec \mathcal{CW}$	(6) $\mathcal{SPW} \approx \mathcal{BW}$
(7) $\mathcal{RTDW} \not\preceq \mathcal{BW}$	(8) $\mathcal{RPDW} \prec \mathcal{PDW}$	(9) $\mathcal{PDW} \prec \mathcal{BW}$
(10) $\mathcal{CSPW} \prec \mathcal{RPDW}$	(11) $\mathcal{CSTW} \not\approx \mathcal{TDW}$	(12) $\mathcal{RTDW} \not\approx \mathcal{CSTW}$

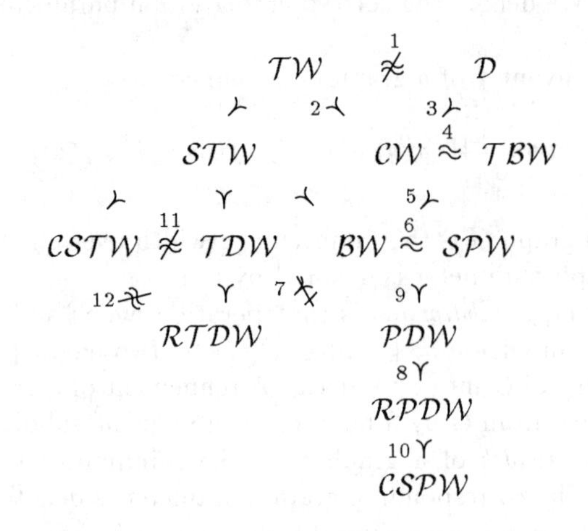

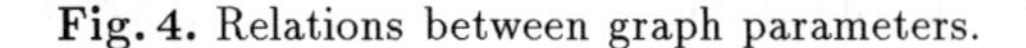

Fig. 4. Relations between graph parameters.

Proof. We omit the proofs. It is not difficult to prove (1) — (8). □

An immediate consequence of the above relations and Theorem 3.1 is that graph isomorphism is fixed parameter tractable (can be solved in $O(n^2)$) when the input graphs with n vertices have bounded connected strong pathwidth.

An interesting open problem is to find, in the hierarchy depicted in Figure 4, the boundary between the parameters that give fixed parameter tractability for Graph Isomorphism, and the parameters that (probably) do not, i.e. for which Graph Isomorphism is $W[t]$-hard for some t (as defined by [5, 6]): at present, we only know that Graph Isomorphism is fixed parameter tractable for parameters $\mathcal{RTDW}$, $\mathcal{RPDW}$ and $\mathcal{CSPW}$, but for all other parameters in the figure, the problem is still open.

5 Conclusions

The following results were shown in this paper:

- NP-completeness of deciding the path distance width of a graph,

- A quadratic algorithm for Graph Isomorphism on graphs with bounded rooted path distance width,
- A cubic algorithm for Graph Isomorphism on graphs with bounded rooted tree distance width,
- Relationships between several graph parameters.

We were motivated by the following question: Is graph isomorphism fixed parameter tractable, when the bandwidth of a graph is the parameter in consideration (i.e., is assumed to be a small constant)? Note that, if not, then also graph isomorphism is not fixed parameter tractable when the degree is taken as parameter. However, if Graph Isomorphism is fixed parameter tractable for bandwidth, then it is still an open question whether Graph Isomorphism is fixed parameter tractable for bounded degree. Also, an interesting question is the fixed parameter tractability of graph isomorphism taking the strong pathwidth in consideration — such fixed parameter tractability would imply the same result with bandwidth as parameter. These three problems are left as interesting open problems. However, from the results of this paper, we have that graph isomorphism is fixed parameter tractable with connected strong pathwidth as parameter.

References

1. H. L. Bodlaender, A partial k-arboretum of graphs with bounded treewidth, Technical Report UU-CS-1996-02, Dept. of Computer Science, Utrecht University, The Netherlands.
2. H. L. Bodlaender, Polynomial algorithms for graph isomorphism and chromatic index on partial k-trees, *J. Algorithms*, 11 (1990), pp. 631–643.
3. F. R. K. Chung and P. D. Seymour, Graphs with small bandwidth and cutwidth, *Discrete Math.*, 75 (1989), pp. 113–119.
4. C. J. Colbourn and K. S. Booth. Linear time automorphism algorithms for trees, interval graphs, and planar graphs. *SIAM J. Comput.*, 10 (1981), pp. 203–225.
5. R. G. Downey and M. R. Fellows, Fixed-parameter tractability and completeness I: Basic Results, *SIAM J. Comput.*, 24 (1995), pp. 873–921.
6. R. G. Downey and M. R. Fellows, Fixed-parameter tractability and completeness II: On completeness for $W[1]$, *Theor. Comput. Sci.* 141 (1995), pp. 109–131.
7. L. S. Filotti and J. N. Mayer, A polynomial time algorithm for determining the isomorphism of graphs of fixed genus, *Proc. 12th Ann. ACM Symp. Theory of Computing,* Los Angeles (1980), pp. 236–243.
8. E. M. Luks, Isomorphism of graphs of bounded valence can be tested in polynomial time, *JCSS*, 25 (1982), pp. 42–65.
9. G. L. Miller, Isomorphism testing for graphs of bounded genus, *Proc. 12th Ann. ACM Symp. Theory of Computing,* Los Angeles (1980), pp. 225–235.
10. D. Seese, Tree-partite graphs and the complexity of algorithms, Proc. Int. Conf. on Fundamentals of Computation Theory, LNCS 199, (1985), pp. 412–421.

Hardness of Approximating Problems on Cubic Graphs*

Paola Alimonti[1] and Viggo Kann[2]

[1] Dip. Informatics Sistemistica
University of Rome "la Sapienza", Via Salaria 113, 00198 Rome, Italy
E-mail: alimon@dis.uniroma1.it
[2] Numerical Analysis and Computing Science
Royal Institute of Technology
S-100 44 Stockholm, Sweden
E-mail: viggo@nada.kth.se

Abstract. Four fundamental graph problems, Minimum vertex cover, Maximum independent set, Minimum dominating set and Maximum cut, are shown to be APX-complete even for cubic graphs. This means that unless P=NP these problems do not admit any polynomial time approximation scheme on input graphs of degree bounded by three.

1 Introduction

Among combinatorial optimization problems that are computationally hard to solve, NP-hard optimization problems on graphs have a great relevance both from the theoretical and practical point of view.

Despite the apparent simplicity of cubic and at-most cubic graphs, several NP-hard graph problems remain NP-hard even if restricted to these classes of graphs, but become polynomial time solvable for graphs of degree 2 [10, 12].

Since one can be almost certain that NP-hard problems cannot be efficiently solved, one has to restrict oneself to compute approximate solutions. Therefore it would be desirable to identify if and how much boundedness of the graph degree is helpful in approximation.

It is well known that the variation of NP-hard graph problems in which the degree of the graph is bounded by a constant often allows to achieve different results with respect to the approximation properties. Namely, problems that for general graphs cannot be approximated within constant approximation ratio (e.g. Maximum independent set, Minimum dominating set and Minimum independent dominating set) have been shown to be in APX (i.e. approximable within *some* constant) for bounded degree graphs. For some NP-hard optimization problems that are approximable for general graphs (e.g. Minimum vertex

* This work was supported by the CEE project ALCOM-IT ESPRIT LTR, project no. 20244, "Algorithms and Complexity in Information Technology"; the Italian Project "Algoritmi, Modelli di Calcolo e Strutture Informative", Ministero dell'Università e della Ricerca Scientifica e Tecnologica, and by TFR. Parts of this work were done when the first author was visiting the Royal Institute of Technology.

cover) better approximation ratios have been achieved for graphs of low degree [4, 5, 6, 7, 13, 15, 16, 17, 18].

Nevertheless, many graph problems are APX-hard even if the degree of the graph is bounded by some constant, and therefore they can be approximated within some constant factor of the optimum, but cannot be approximated within *any* constant (PTAS) [14, 15, 16, 18].

Some problems are known to be APX-hard even for cubic or at-most-cubic graphs (e.g. Maximum 3-dimensional matching and Maximum independent dominating set [14, 15]). For several other graph problems it is just known that they are APX-hard for graphs of some bounded degree greater than 3 [18].

In this work we show APX-hardness results for several optimization problems on cubic or at-most-cubic graphs, namely for Minimum vertex cover, Maximum independent set (MAX IND SET), Minimum dominating set (MIN DOM SET), and Maximum cut (MAX CUT).

Surprisingly simple reductions are used for most of our results, but for showing the APX-completeness of MAX CUT on cubic graphs we need a quite complicated structure consisting of a chain of expander graphs. Expander graphs have been used in different ways in approximation preserving reductions [18, 3, 2, 9], and seem to be very useful. For a description of expander graphs and an algorithm constructing expander graphs we refer to Ajtai [1].

The remainder of the paper is organized as follows. In Sect. 2, we state basic definitions and notations. In Sect. 3, we show the APX-completeness of MIN VERTEX COVER, MAX IND SET and MIN DOM SET on cubic graphs. In Sect. 4, we prove the APX-completeness of MAX CUT on cubic graphs.

2 Definitions

Although various reductions preserving approximability within constants have been proposed (see [8]), the L-reduction defined in [18] is perhaps the easiest one to use. Given two NP optimization problems F and G and a polynomial time transformation f from instances of F to instances of G, we say that f is an *L-reduction* if there are positive constants α and β such that for every instance x of F

1. $opt_G(f(x)) \leq \alpha \cdot opt_F(x)$,
2. for every feasible solution y of $f(x)$ with objective value $m_G(f(x), y) = c_2$ we can in polynomial time find a solution y' of x with $m_F(x, y') = c_1$ such that $|opt_F(x) - c_1| \leq \beta \, |opt_G(f(x)) - c_2|$.

Using L-reductions (or reductions similar to them) one can show that a problem F is APX-complete, i.e., F is approximable within c for some c and every approximable problem can be L-reduced to F[3]. In this paper we will consider the following APX-complete problems.

[3] Previously the notion MAX SNP-completeness was used, but it is now more accurate to talk about APX-completeness. Every MAX SNP-complete problem is also APX-complete, even if a slightly different reduction has to be used, see for example [8].

MAX CUT$-B$

Instance: Graph $G = (V, E)$ of degree bounded by B.
Solution: A partition of V into two parts: a red part P_R and a green part P_G.
Measure: Cardinality of the set of edges that are cut, i.e. edges with one end point in P_R and one end point in P_G.

MAX IND SET$-B$

Instance: Graph $G = (V, E)$ of degree bounded by B.
Solution: An independent set for G, i.e., a subset $V' \subseteq V$ such that no two vertices in V' are joined by an edge in E.
Measure: Cardinality of the independent set, i.e., $|V'|$.

MIN DOM SET$-B$

Instance: Graph $G = (V, E)$ of degree bounded by B.
Solution: A dominating set for G, i.e., a subset $V' \subseteq V$ such that for all $u \in V - V'$ there is a $v \in V'$ for which $(u, v) \in E$.
Measure: Cardinality of the dominating set, i.e., $|V'|$.

MIN VERTEX COVER$-B$

Instance: Graph $G = (V, E)$ of degree bounded by B.
Solution: A vertex cover for G, i.e., a subset $V' \subseteq V$ such that for all $(u, v) \in E$ at least one of u and v is included in V'.
Measure: Cardinality of the vertex cover, i.e., $|V'|$.

MAX E3-SAT$-B$

Instance: Set of variables X, set of disjunctive clauses C over the variables X, where each clause consists of exactly three variables, and each variable occurs in at most B clauses.
Solution: Truth assignment of X.
Measure: Cardinality of the set of clauses from C that are satisfied by the truth assignment.

3 APX-completeness of some problems on cubic graphs

MIN VERTEX COVER$-B$, MAX IND SET$-B$ and MIN DOM SET$-B$ were known to be APX-complete for some bounded degree B, and explicit proofs have been obtained for MIN VERTEX COVER-4, MAX IND SET-4 and MIN DOM SET-8 [18, 15].

In the following we will shown that these problems remain APX-complete even if the degree of the graphs is bounded by 3.

Theorem 1. MIN VERTEX COVER-3 *is* APX-*complete.*

Proof. It is well known that MIN VERTEX COVER can be approximated within 2, and thus is included in APX.

Now we show that MIN VERTEX COVER−3 is APX-hard. Let f be the following L-reduction from MIN VERTEX COVER−4 to MIN VERTEX COVER−3.

Given a graph $G = (V, E)$ of bounded degree 4 construct an at-most-cubic graph $G' = (V', E')$ in the following way. Let v be a vertex of degree 4. Split $v \in V$ into two vertices v_1 and v_2 of degree 2, then add an extra vertex u and edges (v_1, u), and (v_2, u), see Fig. 1.

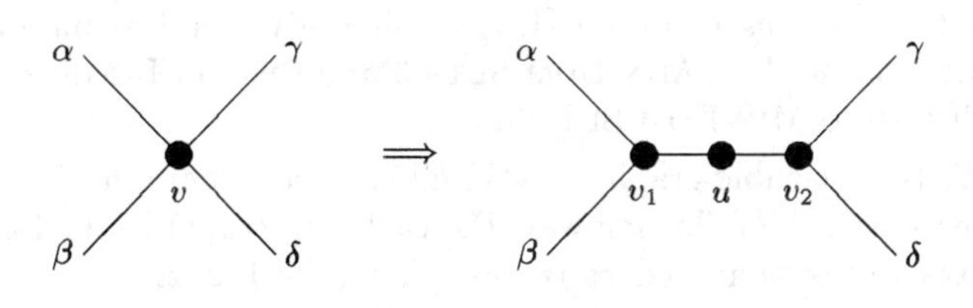

Fig. 1. The transformation of a degree 4 vertex used both in the reduction from MIN VERTEX COVER−4 to MIN VERTEX COVER−3 and the reduction from MAX IND SET−4 to MAX IND SET−3.

Given a vertex cover $C' \subseteq V'$ of $G' = f(G)$ we transform it back to a vertex cover $C \subseteq V$ of G in the following manner. Notice that for each triple of vertices $v_1, v_2, u \in V'$ coming from a vertex $v \in V$ of degree 4: either $u \in C'$, or at least two vertices (possibly one of them is u) belong to C'. If at least two vertices among v_1, v_2 and u are in C', we can substitute such vertices in C' by the pair v_1, v_2 and we still have a vertex cover of at most the same size. Then from every vertex cover C' we can construct a vertex cover C of size $|C'| - s$, where s is the number of vertices of degree 4 in G. Include in C any vertex of degree less than 4 that belongs to C' and any vertex $v \in V$ of degree 4 such that the vertices $v_1, v_2, \in V'$ belong to C'.

It is easy to see that from every vertex cover $C \subseteq V$ of G we can construct a vertex cover $C' \subseteq V'$ of $G' = f(G)$ of size exactly $|C| + s$. In C' we include every vertex in C that has degree smaller than 4, and for each vertex $v \in V$ of degree 4 we do as follows. If $v \in C$ then $v_1, v_2 \in C'$, if $v \notin C$ then $u \in C'$. Since G has bounded degree 4 we have $|C| \geq |V|/5 \geq s/5$. We see that $|C'| = |C| + s \leq 6 \cdot |C|$

Thus $opt(f(G)) \leq 6 \cdot opt(G)$ and we have shown that f is an L-reduction with $\alpha = 6$ and $\beta = 1$.

Theorem 2. MAX IND SET−3 *is* APX-*complete.*

This result was recently proved by Berman and Fujito [5] using a complex reduction from MAX E3-SAT−B. We can give a much simpler proof of this result using the same reduction as in the proof of Theorem 1. Analogously the reduction by Berman and Fujito could be used to show that MIN VERTEX COVER−3 is APX-complete.

Proof. (Outline) Since MAX IND SET−4 is APX-complete and since the complement of any vertex cover is an independent set the same transformation as above can be used to prove the theorem. We get an L-reduction from MAX IND SET−4 to MAX IND SET−3 with $\alpha = 6$ and $\beta = 1$.

Theorem 3. MIN DOM SET−3 *is* APX-*complete.*

Proof. It is well known that the variation of MIN DOM SET in which the degree of the graph is bounded by a constant belongs to APX. In the following we will prove that it remains NP-hard even if the degree of the graph is bounded by 3. Since L-reductions compose [18], we first give an L-reduction f_1 from MIN VERTEX COVER−3 to MIN DOM SET−6 and then an L-reduction f_2 from MIN DOM SET−6 to MIN DOM SET−3.

Given an at-most-cubic graph $G = (V, E)$ construct a graph $G' = (V', E')$ of bounded degree 6 in the following way. For each edge (u, v) in the former graph insert an extra vertex w and edges (u, w), (v, w), see Fig. 2.

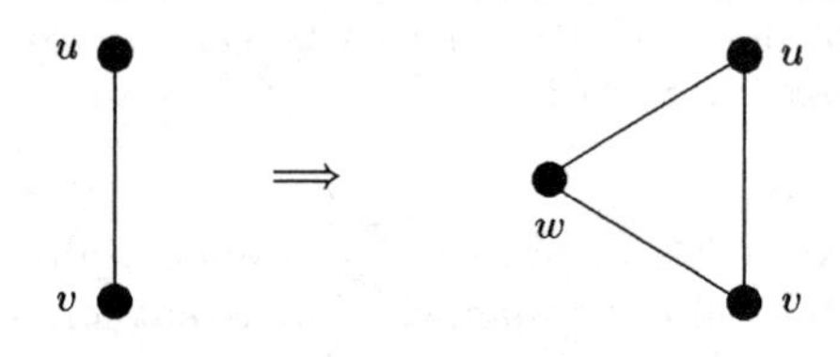

Fig. 2. The transformation of an edge in the reduction from MIN VERTEX COVER−3 to MIN DOM SET−6.

It is easy to see that every dominating set $D \subseteq V'$ of $G' = f_1(G)$ can be transformed into an equally good or better vertex cover $C \subseteq V$ of G by including in C the following vertices. For each vertex $v \in D$ such that $v \in V$, include v in C. For each vertex $w \in D$ such that $w \notin V$, choose a vertex v such that $(v, w) \in E'$ and include v in C.

Consider now a vertex cover $C \subseteq V$ of G. We can construct a dominating set $D \subseteq V'$ of G' of the same size by including in D exactly the same vertices. Thus f_1 is an L-reduction with $\alpha = \beta = 1$

Now we describe the L-reduction f_2 to MIN DOM SET−3. Given a graph $G = (V, E)$ of bounded degree 6 construct an at-most-cubic graph $G' = (V', E')$ in the following manner. Let $v \in V$ be a vertex of degree 4. We split v into two vertices v_1 and v_2 of degree 2. Then we add five extra vertices $u_{1,1}$, $u_{1,2}$, $u_{2,1}$, $u_{2,2}$, w, and edges $(v_1, u_{1,1})$, $(u_{1,1}, u_{1,2})$, $(u_{1,2}, w)$, $(v_2, u_{2,1})$, $(u_{2,1}, u_{2,2})$, $(u_{2,2}, w)$, see Fig. 3.

For each vertex $v \in V$ of degree 5 or 6 we do in a similar way except for the fact that we split v into three vertices, instead of two and extend the previous construction with a third leg from the center vertex w. More precisely we split

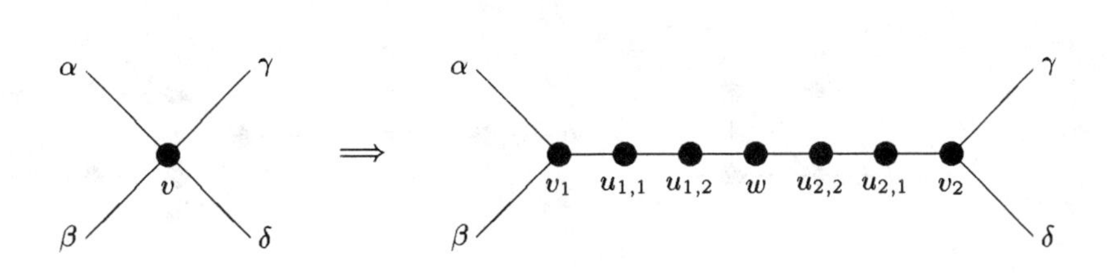

Fig. 3. The transformation of a degree 4 vertex in the reduction from MIN DOM SET–6 to MIN DOM SET–3.

v into v_1 and v_2 of degree 2 and v_3 of degree p, where p is 1 if the degree of v is 5 and 2 if the degree of v is 6. Then we add seven extra vertices $u_{1,1}$, $u_{1,2}$, $u_{2,1}$, $u_{2,2}$, $u_{3,1}$, $u_{3,2}$, w, and edges $(v_1, u_{1,1})$, $(u_{1,1}, u_{1,2})$, $(u_{1,2}, w)$, $(v_2, u_{2,1})$, $(u_{2,1}, u_{2,2})$, $(u_{2,2}, w)$, $(v_3, u_{3,1})$, $(u_{3,1}, u_{3,2})$, $(u_{3,2}, w)$.

It is easy to see that any dominating set $D' \subseteq V'$ of $G' = f_2(G)$ can be transformed back to a dominating set $D \subseteq V$ of G as follows. For each vertex $v \in V$ of degree 4: $v \in D$ if $|\{v_1, v_2, u_{1,1}, u_{1,2}, u_{2,1}, u_{2,2}, w\} \cap D'| \geq 3$ and $v \notin D$ if $|\{v_1, v_2, u_{1,1}, u_{1,2}, u_{2,1}, u_{2,2}, w\} \cap D'| = 2$. For each vertex $v \in V$ of degree greater than 4: $v \in D$ if $|\{v_1, v_2, v_3, u_{1,1}, u_{1,2}, u_{2,1}, u_{2,2}, u_{3,1}, u_{3,2}, w\} \cap D'| \geq 4$ and $v \notin D$ if $|\{v_1, v_2, v_3, u_{1,1}, u_{1,2}, u_{2,1}, u_{2,2}, u_{3,1}, u_{3,2}, w\} \cap D'| = 3$. It is clear that D is a dominating set of size $|D| \leq |D'| - 2 \cdot s_1 - 3 \cdot s_2$, where s_1 and s_2 are the number of vertices of degree 4 and greater than 4 in V, respectively.

Finally, given a dominating set $D \subseteq V$ of G we can construct a dominating set $D' \subseteq V'$ of $G' = f_2(G)$ such that $|D'| = |D| + 2 \cdot s_1 + 3 \cdot s_2$. Since G has bounded degree 6, we have $|D| \geq |V|/7$. Therefore $|D'| \leq |D| + 3 \cdot (s_1 + s_2) \leq 22 \cdot |D|$.

Thus, $opt(f_2(G)) \leq 22 \cdot opt(G)$ and we have shown that f_2 is an L-reduction with $\alpha = 22$ and $\beta = 1$.

The above results are still valid for cubic graphs, that is for graphs where every vertex has degree exactly three.

We simply show this for MIN VERTEX COVER–3.

Split each vertex v of degree two into two vertices, v_1 and v_2, of degree 1. Then add two extra vertices, u_1 and u_2, and edges (v_1, u_1), (v_1, u_2), (u_1, u_2), (v_2, u_1), and (v_2, u_2). For each vertex v of degree 1 add 6 extra vertices u_1, u_2, u_3, u_4, u_5, and u_6 and edges (v, u_1), (v, u_2), (u_1, u_2), (u_1, u_3), (u_2, u_5), (u_3, u_4), (u_3, u_6), (u_4, u_5), (u_5, u_6), and (u_4, u_6), see Fig. 4.

From every solution of size c in the at-most-cubic graph we can construct a solution in the cubic graph of size exactly $4s_1 + 2s_2 + c$, where s_1 and s_2 are the number of vertices of degree 1 and 2 in the at-most-cubic graph, respectively.

4 APX-completeness of MAX CUT on cubic graphs

MAX CUT can be approximated within 1.139 [11], and is therefore included in APX. We will first show that MAX CUT is APX-hard for multigraphs of degree 6, and then for simple graphs of degree 3.

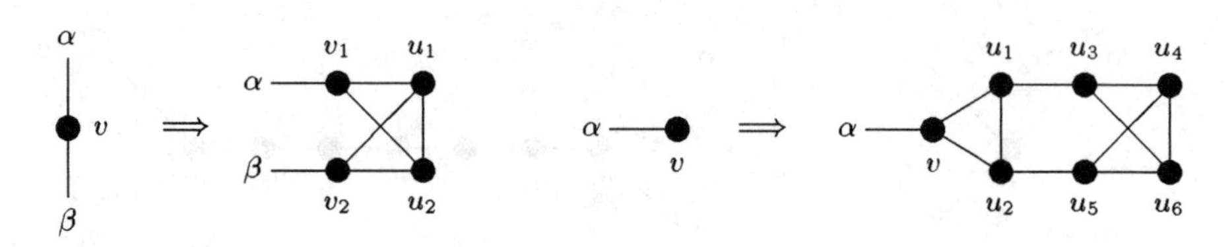

Fig. 4. The transformation of vertices of degree 2 and 1 in the reduction from at-most-cubic graphs to cubic graphs.

Theorem 4. MAX CUT−6 *for multigraphs is* APX-*complete.*

Proof. We will construct an L-reduction from MAX E3-SAT−3 to MAX CUT−6 for multigraphs.

Suppose we are given an instance of MAX E3-SAT−3 with n variables and m clauses. Without loss of generality we can assume that every variable occurs positively in at least one clause and negatively in at least one clause. This problem is known to be APX-complete [19].

Construct a multigraph with a vertex set consisting of two vertices named x_i and $\overline{x}_i$ (the *variable vertices*) for each variable, and four vertices named y_j, $\overline{y}_j$, b_{2j-1} and b_{2j} for each clause. For each clause $c_j = l_1 \vee l_2 \vee l_3$ we construct eight edges: (b_{2j-1}, l_1), (b_{2j}, y_j), (l_1, y_j), (l_2, l_3), $(l_2, \overline{y}_j)$, $(l_3, \overline{y}_j)$, and two parallel edges $(y_j, \overline{y}_j)$, see Fig. 5. We also, for every variable x_i, include two parallel edges between the vertices x_i and $\overline{x}_i$. The degree of the graph is 6.

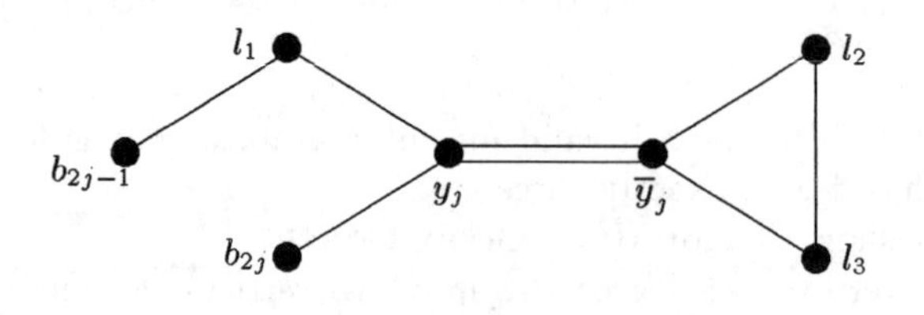

Fig. 5. The constructed edges from a clause $l_1 \vee l_2 \vee l_3$.

Consider a solution where all the b_i vertices in the graph are placed in the same part, say the red part. If we look at the subgraph in Fig. 5 we can see that if all the three l_k vertices are in the red part, then at most 4 edges can be cut, but if at least one of the l_k vertices is in the green part, then it is always possible to choose parts for y_j and $\overline{y}_j$ so that six edges are cut.

If x_i and $\overline{x}_i$ are placed in the same part we can move one of them to the other part without decreasing the number of cut edges. This is because at least one of x_i and $\overline{x}_i$ occurs just in one clause, and has therefore just two edges except the two edges connecting x_i and $\overline{x}_i$.

Now we have a correspondence between the values of the variables and the partition of the variable vertices—if the variable x_i is true then the vertex x_i is green and $\overline{x}_i$ is red, and if the variable x_i is false then the vertex x_i is red and $\overline{x}_i$ is green. Thus the number of cut edges will be $2n + 4m + 2s$ where s is the number of satisfied clauses in the corresponding MAX E3-SAT problem instance.

Recall that the above reasoning is valid only if all the b_j vertices are in the same part. In order to obtain this we construct a bipartite cubic expander between the b_j vertices and an equivalently large set of new vertices, called $c_{1,j}$. We then construct a chain of bipartite cubic expanders between $\{c_{i,j}\}$ and $\{c_{i+1,j}\}$ for $1 \leq i < k$, where k is a constant to be decided later. We thus have $k+1$ layers of vertices that are connected by expanders. The degree of each vertex is at most 6. Let N be the number of b_j vertices (which means that N is the number of vertices in any of the $k+1$ layers).

Ajtai has shown that cubic bipartite expander graphs of size N can be constructed in polynomial time in N [1]. Such an expander $G = (A \cup B, E)$ (where A and B are the two parts in the bipartition) has the property that for any subset $A' \subseteq A$ with $|A'| \leq |A|/2$, A' is connected to at least $(1+\alpha)|A'|$ vertices in B, and vice versa. α is some fixed positive constant.

Consider a solution and define the red part P_R as the part that contains most of the b_j vertices. We will show that we can move all the b_j vertices that are in the green part P_G to P_R without decreasing the size of the cut.

By moving the b_j vertices in P_G to P_R the number of cut edges not in the expander chain is decreased by at most $|\{b_j\} \cap P_G|$. On the other hand by putting all the vertices in the layers $\{b_j\}$ and $\{c_{i,j}\}$ for any even i in P_R, and all the vertices in the layers $\{c_{i,j}\}$ for any odd i in P_G, we will show that the cardinality of the cut in the expander chain is increased by at least the same number.

To show this, we calculate the gain that is guaranteed between any layers of vertices achieved by putting all the vertices as specified above. We will make use of the property of bipartite expanders described above.

We first consider the number g_0 of uncut edges between the b_j vertices and the $c_{1,j}$ vertices. By changing the partition as described above, the uncut edges will be cut, so the gain will be g_0. Let $m_0 = |\{b_j\} \cap P_G|$ and $m_1 = |\{c_{1,j}\} \cap P_R|$. By construction $m_0 \leq N/2$. We need to consider a few cases.

Case 1. $m_1 \leq N/2$

The m_0 vertices in $\{b_j\} \cap P_G$ are connected to at least $(1+\alpha) \cdot m_0$ vertices in $\{c_{1,j}\}$. Of these vertices at least $(1+\alpha) \cdot m_0 - m_1$ must be in $\{c_{1,j}\} \cap P_G$, which means that at least $(1+\alpha) \cdot m_0 - m_1$ of the edges from $\{b_j\} \cap P_G$ are uncut. Similarly $(1+\alpha) \cdot m_1 - m_0$ of the edges from $\{c_{1,j}\} \cap P_R$ to $\{b_j\}$ are uncut. Therefore $g_0 \geq (1+\alpha) \cdot m_0 - m_1 + (1+\alpha) \cdot m_1 - m_0 \geq \alpha(m_0 + m_1)$.

There are two subcases depending on if m_1, the number of red vertices in the $\{c_{1,j}\}$ layer, is greater or smaller than $m_0/(1+\alpha)$.

Case 1a. $m_1 \geq m_0/(1+\alpha)$

$$g_0 \geq \alpha \cdot (m_0 + (1-\alpha)m_0) \geq (2 \cdot \alpha - \alpha^2) \cdot m_0 \geq \alpha \cdot m_0 \ .$$

Case 1b. $m_1 < m_0/(1+\alpha)$

If we just look at the first type of noncut edges we get

$$g_0 \geq (1+\alpha) \cdot m_0 - m_1 \geq (1+\alpha-(1-\alpha+\alpha^2)) \cdot m_0 = (2 \cdot \alpha - \alpha^2) \cdot m_0 \geq \alpha \cdot m_0 \ .$$

Case 2. $m_1 > N/2$

Since the expander property is valid only for subsets of size at most $N/2$ we just look at the noncut edges from a subset of size $N/2$ of $\{c_{1,j}\} \cap P_R$. Then we get

$$g_0 \geq (1+\alpha) \cdot (N/2) - m_0 \geq (1+\alpha) \cdot m_0 - m_0 = \alpha \cdot m_0 \ .$$

Thus in all cases the number of uncut edges between the two layers of vertices is at least $\alpha \cdot m_0$.

By similarly counting noncut edges in the rest of the expander chain we will obtain that if at layer i the number m_i of vertices placed in the "wrong" part of the cut is no greater than $N/2$, the gain g_i between layer i and layer $i+1$ is at least $\alpha \cdot m_i$.

If in some layer i we have $m_i > N/2$, the gain g_i between layer i and layer $i+1$ can be calculated exactly as above, but with respect to the vertices placed in the right part instead of the wrong part. The gain then becomes $g_i \geq \alpha \cdot (N - m_i)$.

Therefore, if $(2/3) \cdot m_0 \leq m_i \leq N - (2/3) \cdot m_0$ for all i we will gain at least $(2/3)\alpha \cdot m_0$ in each layer. If we choose $k \geq 3/(2 \cdot \alpha)$ the total gain will become at least m_0.

In order to consider the cases where m_i is small or large for some i we compute the gain in another way. We observe that if $m_0 > m_1$ at least $3(m_0 - m_1)$ of the edges from $\{b_j\} \cap P_G$ to $\{c_{1,j}\}$ must be uncut. In the same way we get the gain $g_j \geq \max\{3(m_j - m_{j+1}), 0\}$ for the layer j. Suppose $m_i < (2/3) \cdot m_0$ for some i. Summing over all layers we get a total gain of at least $3(m_0 - m_i) > m_0$.

Finally, if we for some i have $m_i > N - (2/3) \cdot m_0$ we can do as for small m_i, but work in the other direction. We then get a total gain of at least $3(N - (2/3) \cdot m_0 - (N - m_0)) > m_0$, which completes the proof.

Theorem 5. MAX CUT–3 *is* APX-*complete.*

Proof. We give a reduction from MAX CUT–6 for multigraphs to MAX CUT–3 for simple graphs.

For a vertex v of degree $d, 2 \leq d \leq 4$ we do like follows: split the vertex into d *split vertices* of degree 1. Then add d extra vertices and construct a ring where every first vertex is one of the split vertices and every second vertex is a new vertex, see Fig. 6.

It is easy to see that if the split vertices are put in the same part of the partition and the new vertices are put in the other part, then every edge in the ring will be cut. Otherwise at least two of the edges in the ring will be uncut, and we can move one or two split vertices to the other part of the partition without decreasing the size of the cut.

For a vertex of degree 5 or 6 we do in a similar way, except that we need to construct two rings instead of just one. For every split vertex v_i we add two

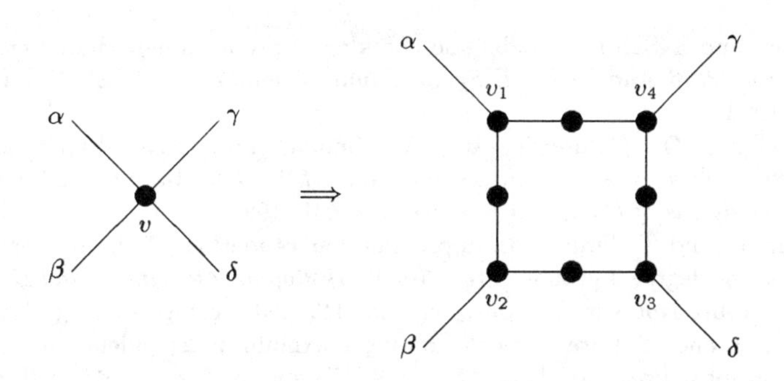

Fig. 6. The transformation of a degree 4 vertex in the reduction from Max Cut–6 to Max Cut–3.

new vertices (l_i and r_i) and edges (v_i, l_i) and (v_i, r_i). We then put the l_i vertices in a ring, interleaved with new vertices, and similarly we put the r_i vertices in another ring, interleaved with new vertices. It is possible to show that given a solution we can without decreasing the size of the cut modify it so that all the 5 or 6 split vertices, originating from the same vertex, are in the same part of the partition.

When putting all split vertices in the same part as the majority of them at most 3 of the edges from the split vertices to the rest of the graph may be uncut in the worst case. We will show how to gain at least four edges in the internal ring structure by putting all split vertices in the same part.

First we examine how the vertices in the two rings are partitioned, and observe that the size of the cut will not decrease if we partition the vertices in the two rings in the same way (by using the partition of the "best cut" ring for both rings), so we do that. Now, if the vertices in the rings (including the split vertices) are not partitioned alternately red and green we will gain at least two edges per ring in the cut by partition them so, and thereby the split vertices will become in the same part of the partition.

This is an L-reduction with $\alpha = 13$ and $\beta = 1$, and the constructed graph is clearly simple and of degree 3.

Max Cut is APX-hard even for cubic graphs. To show this we just have to extend the graph constructed in the above proof. For each node of degree 1 or 2 we simply add edges to gadgets of the following type: .

References

1. M. Ajtai. Recursive construction for 3-regular expanders. *Proc. 28th Ann. IEEE Symp. on Foundations of Comput. Sci.*, 295–304, 1987.
2. E. Amaldi and V. Kann, The complexity and approximability of finding maximum feasible subsystems of linear relations. *Theoret. Comput. Sci.* 147, 181–210, 1995.

3. S. Arora and S. Safra, Probabilistic checking of proofs; a new characterization of NP. *Proc. 33rd Annual IEEE Symp. Found. Comput. Sci.* IEEE Comput. Soc., 2–13, 1992.
4. M. Bellare, O. Goldreich, and M. Sudan, Free Bits, PCPs and non-approximability-towards tight results, *Proc. of the 36th Annual IEEE Conference on Foundations of Computer Science*, 422–431, 1995.
5. P. Berman and T. Fujito, On approximation properties of the independent set problem for degree 3 graphs, *Proc. 3rd Workshop on Algorithms and Data Structures*, Lecture Notes in Comput. Sci. 955, 449–460, Springer-Verlag, 1995.
6. P. Berman and M. Fürer, Approximating maximum independent set in bounded degree graphs, *Proc. 5th Ann. ACM-SIAM Symp. on Discrete Algorithms*, ACM-SIAM, 365–371, 1994.
7. P. Crescenzi and V. Kann, *A compendium of NP optimization problems*, Technical Report SI/RR-95/02, Dipartimento di Scienze dell'Informazione, Università di Roma "La Sapienza", 1995. The list is updated continuously. The latest version is available as `http://www.nada.kth.se/theory/problemlist.html`.
8. P. Crescenzi, V. Kann, R. Silvestri, and L. Trevisan, Structure in approximation classes, *Proc. 1st Ann. Int. Conf. Computing and Combinatorics*, Lecture Notes in Comput. Sci. 959, Springer-Verlag, 539–548, 1995.
9. P. Crescenzi, R. Silvestri, and L. Trevisan, To weight or not to weight: Where is the question? *Proc. 4th Israel Symp. Theory Comput. and Syst.*, 68–77, 1996.
10. M. R. Garey and D. S. Johnson, *Computers and Intractability: a guide to the theory of NP-completeness*, W. H. Freeman and Company, San Francisco, 1979.
11. M. X. Goemans and D. P. Williamson, Improved approximation algorithms for maximum cut and satisfiability problems using semidefinite programming, *J. ACM*, **42**, 1115–1145, 1995.
12. R. Greenlaw and R. Petreschi, Cubic graphs, *ACM Computing Surveys* **27**, 471–495, 1995.
13. M. M. Halldorsson, Approximating the minimum maximal independence number, *Inform. Process. Lett.* **46**, 169–172, 1993.
14. V. Kann, Maximum bounded 3-dimensional matching is MAX SNP-complete, *Inform. Process. Lett.* **37**, 27–35, 1991.
15. V. Kann, On the Approximability of NP-Complete Optimization Problems, *PhD Thesis*, Department of Numerical Analysis and Computing Science, Royal Institute of Technology, Stockholm, 1992.
16. V. Kann, Polynomially bounded minimization problems that are hard to approximate, *Nordic J. Computing* **1** 317–331, 1994.
17. C. Lund and M. Yannakakis, On the hardness of approximating minimization problems, *J. ACM* **41**, 960–981, 1994.
18. C. H. Papadimitriou and M. Yannakakis, Optimization, approximation, and complexity classes, *J. Comput. Syst. Sci.* **43**, 425–440, 1991.
19. C. H. Papadimitriou, *Computational Complexity*, Addison-Wesley, 1994.

Tree Contractions and Evolutionary Trees

Ming-Yang Kao*

Department of Computer Science
Duke University
Durham, NC 27708, U.S.A.
kao@cs.duke.edu.

Abstract. An *evolutionary tree* is a rooted tree where each internal vertex has at least two children and where the leaves are labeled with distinct symbols representing species. Evolutionary trees are useful for modeling the evolutionary history of species. An *agreement subtree* of two evolutionary trees is an evolutionary tree which is also a topological subtree of the two given trees. We give an algorithm to determine the largest possible number of leaves in any agreement subtree of two trees T_1 and T_2 with n leaves each. If the maximum degree d of these trees is bounded by a constant, the time complexity is $O(n \log^2 n)$ and is within a $\log n$ factor of optimal. For general d, this algorithm runs in $O(nd^2 \log d \log^2 n)$ time or alternatively in $O(nd\sqrt{d} \log^3 n)$ time.

1 Introduction

An *evolutionary tree* is a rooted tree where each internal vertex has at least two children and where the leaves are labeled with distinct symbols representing species. Evolutionary trees are useful for modeling the evolutionary history of species. Many mathematical biologists and computer scientists have been investigating how to construct and compare evolutionary trees [3, 5, 10, 13, 14, 15]. An *agreement subtree* of two evolutionary trees is an evolutionary tree which is also a topological subtree of the two given trees. A *maximum* agreement subtree is one with the largest possible number of leaves. Different theories about the evolutionary history of the same species often result in different evolutionary trees. A fundamental problem in computational biology is to determine how much two theories have in common. To a certain extent, this problem can be answered by computing a maximum agreement subtree of two given evolutionary trees [9].

Let T_1 and T_2 be two evolutionary trees with n leaves each. Let d be the maximum degree of these trees. Previously, Kubicka, Kubicki and McMorris [18] gave an algorithm that can compute the number of leaves in a maximum agreement subtree of T_1 and T_2 in $O(n^{(\frac{1}{2}+\epsilon)\log n})$ time for $d = 2$. Steel and Warnow [20] gave the first polynomial-time algorithm. Their algorithm runs in $O(\min\{d!n^2, d^{2.5}n^2 \log n\})$ time if d is bounded by a constant and in $O(n^{4.5} \log n)$ time for general trees. Farach and Thorup [7] later reduced the time complexity of this algorithm to $O(n^2)$ for general trees. More recently, they gave an algorithm

* Supported in part by NSF Grant CCR-9101385.

[8] that runs in $O(n^{1.5}\log n)$ time for general trees. If d is bounded by a constant, this algorithm runs in $O(nc^{\sqrt{\log n}} + n\sqrt{d}\log n)$ time for some constant $c > 1$.

This paper presents an algorithm for computing a maximum agreement subtree in $O(n\log^2 n)$ time for d bounded by a constant. Since there is a lower bound of $\Omega(n\log n)$, our algorithm is within a $\log n$ factor of optimal. For general d, this algorithm runs in $O(nd^2\log d\log^2 n)$ time or alternatively in $O(nd\sqrt{d}\log^3 n)$ time. This algorithm employs new tree contraction techniques [17, 19]. With tree contraction, we can immediately obtain an $O(n\log^5 n)$-time algorithm for d bounded by a constant. Reducing the time bound to $O(n\log^2 n)$ requires additional techniques. We develop new results that are useful for bounding the time complexity of tree contraction algorithms. As in [7, 8, 20], we also explore the dynamic programming structure of the problem. We obtain some highly regular structural properties and combine these properties with the tree contraction techniques to reduce the time bound by a factor of $\log^2 n$. To remove the last $\log n$ factor, we incorporate some techniques that can compute maxima of multiple sets of sequences at multiple points, where the input sequences are in a compressed format.

We present tree contraction techniques in §2 and outline our algorithms in §3. The maximum agreement subtree problem is solved in §4 and §5 with a discussion of condensed sequence techniques in §4. Section §6 concludes this paper with an open problem.

2 New tree contraction techniques

Throughout this paper, all trees are rooted ones, and every nonempty tree path is a vertex-simple one from a vertex to a descendant. For a tree T and a vertex u, let T^u denote the subtree of T formed by u and all its descendants in T.

A key idea of our dynamic programming approach is to partition T_1 and T_2 into well-structured tree paths. We recursively solve our problem for T_1^x and T_2^y for all heads x and y of the tree paths in the partitions of T_1 and T_2, respectively. The partitioning is based on new tree contraction techniques developed in this section.

A tree is *homeomorphic* if every internal vertex of that tree has at least two children. Note that the size of a homeomorphic tree is less than twice its number of leaves. Let S be a tree that may or may not be homeomorphic. A *chain of* S is a tree path in S such that every vertex of the given path has at most one child in S. A *tube of* S is a maximal chain of S. A *root path of a tree* is a tree path whose head is the root of that tree; similarly, a *leaf path* is one ending at a leaf. A *leaf tube of* S is a tube that is also a leaf path. Let $\mathcal{L}(S)$ denote the set of leaf tubes in S. Let $\mathcal{R}(S) = S - \mathcal{L}(S)$, i.e., the subtree of S obtained by deleting from S all its leaf tubes. The operation $\mathcal{R}$ is called the *rake operation.*

Our dynamic programming approach iteratively rakes T_1 and T_2 until they become empty. The tubes obtained in the process form the desired partitions of T_1 and T_2. Our rake-based algorithms focus on certain sets of tubes described here. A *tube system* of a tree T is a set of nonempty tree paths $P_1, \cdots, P_m$ in

T such that (1) the paths P_i contain no leaves of T and (2) $T^{h_1}, \cdots, T^{h_m}$ are pairwise disjoint, where h_i is the head of P_i. Condition (1) is required here because our rake-based algorithms process leaves and non-leaf vertices differently. Condition (2) holds if and only if for all i and j, h_i is not an ancestor or descendant of h_j. We can iteratively rake T to obtain tube systems. The set of tubes obtained by the first rake, i.e., $\mathcal{L}(T)$, is not a tube system of T because $\mathcal{L}(T)$ simply consists of the leaves of T and thus violates Condition (1). Every further rake produces a tube system of T until T is raked to emtpy. Our rake-based algorithms only use these systems although there may be others.

We next develop a theorem to bound the time complexities of rake-based algorithms in this paper. For a tree path P in a tree T,

- $K(P,T)$ denotes the set of children of P's vertices in T, excluding P's vertices;
- $t(P)$ denotes the number of vertices in P;
- $b(P,T)$ denotes the number of leaves in T^h where h is the head of P.

(The symbol K stands for the word kids, t for top, and b for bottom.)

Given T, we recursively define a mapping Φ_T from the subtrees S of T to reals. If S is an empty tree, then $\Phi_T(S) = 0$. Otherwise,

$$\Phi_T(S) = \Phi_T(\mathcal{R}(S)) + \sum_{P \in \mathcal{L}(S)} b(P,T) \cdot \log(1 + t(P)).$$

(*Note.* All logarithmic functions log in this paper are in base 2.)

Theorem 1. *For all positive integers n and all n-leaf homeomorphic trees T, $\Phi_T(T) \leq n(1 + \log n)$.*

Proof. $\Phi_T(T)$ is maximized when T is a binary tree formed by attaching n leaves to a path of $n-1$ vertices. The proof is by induction. A naive proof would only show $\Phi_T(T) = O(n \log^2 n)$.

3 Comparing evolutionary trees

Formally, an *evolutionary tree* is a homeomorphic tree whose leaves are labeled by distinct labels. The *label set of an evolutionary tree* is the set of all the leaf labels of that tree.

The *homeomorphic version T' of a tree T* is the homeomorphic tree constructed from T as follows. Let $W = \{w \mid w \text{ is a leaf of } T \text{ or is the lowest common ancestor of two leaves}\}$. T' is the tree over W that preserves the ancestor-descendant relationship of T. Let T_1 and T_2 be two evolutionary trees with label sets L_1 and L_2, respectively.

- For a subset L_1' of L_1, $T_1||L_1'$ denotes the homeomorphic version of the tree constructed by deleting from T_1 all the leaves with labels outside L_1'.
- Let $T_1||T_2 = T_1||(L_1 \cap L_2)$.

- For a tree path P of T_1, $P||T_2$ denotes the tree path in $T_1||T_2$ formed by the vertices of P that remain in $T_1||T_2$.
- For a set $\mathcal{P}$ of tree paths $P_1, \cdots, P_m$ of T_1, $\mathcal{P}||T_2$ denotes the set of all $P_i||T_2$.

Formally, if L' is a maximum cardinality subset of $L_1 \cap L_2$ such that there exists a label-preserving tree isomorphism between $T_1||L'$ and $T_2||L'$, then $T_1||L'$ and $T_2||L'$ are called *maximum agreement subtrees of T_1 and T_2*.

- $\text{RR}(T_1, T_2)$ denotes the number of leaves in a maximum agreement subtree of T_1 and T_2.
- $\text{RA}(T_1, T_2)$ is the mapping from each vertex $v \in T_2||T_1$ to $\text{RR}(T_1, (T_2||T_1)^v)$, i.e., $\text{RA}(T_1, T_2)(v) = \text{RR}(T_1, (T_2||T_1)^v)$.

For a tree path Q of T_2, if Q is nonempty, let $H(Q, T_2)$ be the set of all vertices in Q and those in $K(Q, T_2)$. If Q is empty, let $H(Q, T_2)$ consist of the root of T_2, and thus, if both T_2 and Q are empty, $H(Q, T_2) = \emptyset$.

- For a set $\mathcal{Q}$ of tree paths $Q_1, \cdots, Q_m$ of T_2, let $\text{RP}(T_1, T_2, \mathcal{Q})$ be the mapping from $v \in \cup_{i=1}^m H(Q_i||T_1, T_2||T_1)$ to $\text{RR}(T_1, (T_2||T_1)^v)$, i.e., $\text{RP}(T_1, T_2, \mathcal{Q})(v) = \text{RR}(T_1, (T_2||T_1)^v)$. For simplicity, when $\mathcal{Q}$ consists of only one path Q, let $\text{RP}(T_1, T_2, Q)$ denote $\text{RP}(T_1, T_2, \mathcal{Q})$.

(The notations RR, RA and RP abbreviate the phrases root to root, root to all and root to path. We use RR to replace the notation MAST of previous work [7, 8, 20] for the sake of notational uniformity.)

Lemma 2. *Let T_1, T_2, T_3 be evolutionary trees.*

- $(T_1||T_2)||T_3 = T_1||(T_2||T_3)$.
- *If T_3 is a subtree of T_1, then $T_3||T_1 = T_1||T_3 = T_3$.*
- $\text{RR}(T_1, T_2) = \text{RR}(T_1||T_2, T_2) = \text{RR}(T_1, T_2||T_1) = \text{RR}(T_1||T_2, T_2||T_1)$.

Fact 1 ([7]) *Given an n-leaf evolutionary tree T and k disjoint sets $L_1, \cdots, L_k$ of leaf labels of T, the subtrees $T||L_1, \cdots, T||L_k$ can be computed in $O(n)$ time.*

Given T_1 and T_2, our main goal is to evaluate $\text{RR}(T_1, T_2)$ efficiently. Note that $\text{RR}(T_1, T_2) = \text{RR}(T_1||T_2, T_2||T_1)$ and that $T_1||T_2$ and $T_2||T_1$ can be computed in linear time. Thus, the remaining discussion assumes that T_1 and T_2 have the same label set. To evaluate $\text{RR}(T_1, T_2)$, we actually compute $\text{RA}(T_2, T_1)$ and divide the discussion among the five problems defined below. Each problem is named as a p-q case, where p and q are the numbers of tree paths in T_1 and T_2 contained in the input.

Problem 3. (one-one case)

Input:
1. T_1 and T_2;
2. root paths P of T_1 and Q of T_2 with no leaves from their respective trees;
3. $\text{RP}(T_1^u, T_2, Q)$ for all $u \in K(P, T_1)$;

4. $\text{RP}(T_2^v, T_1, P)$ for all $v \in K(Q, T_2)$.

Output: $\text{RP}(T_1, T_2, Q)$ and $\text{RP}(T_2, T_1, P)$.

The next problem generalizes Problem 3.

Problem 4. (many-one case)

Input:

1. T_1 and T_2;
2. a tube system $\mathcal{P} = \{P_1, \cdots, P_m\}$ of T_1 and a root path Q of T_2 with no leaf from T_2;
3. $\text{RP}(T_1^u, T_2, Q)$ for all P_i and $u \in K(P_i, T_1)$;
4. $\text{RP}(T_2^v, T_1, \mathcal{P})$ for all $v \in K(Q, T_2)$.

Output:

1. $\text{RP}(T_1^{h_i}, T_2, Q)$ for the head h_i of each P_i;
2. $\text{RP}(T_2, T_1, \mathcal{P})$.

Problem 5. (zero-one case)

Input:

1. T_1 and T_2;
2. a root path Q of T_2 with no leaf from T_2;
3. $\text{RA}(T_2^v, T_1)$ for all $v \in K(Q, T_2)$.

Output: $\text{RA}(T_2, T_1)$.

The next problem generalizes Problem 5.

Problem 6. (zero-many case)

Input:

1. T_1 and T_2;
2. a tube system $\mathcal{Q} = \{Q_1, \cdots, Q_m\}$ of T_2;
3. $\text{RA}(T_2^v, T_1)$ for all Q_i and $v \in K(Q_i, T_2)$.

Output: $\text{RA}(T_2^{h_i}, T_1)$ for the head h_i of each Q_i.

Our main goal is to evaluate $\text{RR}(T_1, T_2)$. It suffices to solve the next problem.

Problem 7. (zero-zero case)

Input: T_1 and T_2.

Output: $\text{RA}(T_2, T_1)$.

Our algorithms for these problems are called *One-One, Many-One, Zero-One, Zero-Many* and *Zero-Zero*, respectively. Each algorithm except One-One uses the preceding one in this list as a subroutine. These reductions are based on the rake operation defined in §2. We give One-One in §4 and the other four in §5.1-5.4.

These five algorithms assume that the input trees T_1 and T_2 have n leaves each and d is the maximum degree. We use integer sort and radix sort extensively to help achieve the desired time complexity. (For brevity, from here onwards, radix sort refers to both integer and radix sorts.) For this reason, we make the following *integer indexing assumptions*:

- An integer array of size $O(n)$ is allocated to each algorithm.
- The vertices of T_1 and T_2 are indexed by integers from $[1, O(n)]$.
- The leaf labels are indexed by integers from $[1, O(n)]$.

We call Zero-Zero only once to compare two given trees. Consequently, we may reasonably assume that the tree vertices are indexed with integers from $[1, O(n)]$. When we call Zero-Zero, we simply allocate an array of size $O(n)$. As for indexing the leaf labels, this paper considers only evolutionary trees whose leaf labels are drawn from a total order. Before we call Zero-Zero, we can sort the leaf labels and index them with integers from $[1, O(n)]$. This preprocessing takes $O(n \log n)$ time, which is well within our desired time complexity for Zero-Zero.

The other four algorithms are called more than once, and their integer indexing assumptions are maintained in slightly different situations from that for Zero-Zero. When an algorithm issues subroutine calls, it is responsible for maintaining the indexing assumptions for the callees. In certain cases, the caller uses radix sort to reindex the labels and the vertices of each callee's input trees. The caller also partitions its array into segments and allocates to each callee a segment in proportion to that callee's input size. The new indices and the array segments for subroutine calls can be computed in obvious manners within the desired time complexity of each caller. For brevity of presentation, such preprocessing steps are omitted in the descriptions of the five algorithms.

Some inputs to the algorithms are mappings. We represent a mapping f by the set of all pairs $(x, f(x))$. With this representation, the total size of the input mappings in an algorithm is $O(n)$. Since the input mappings have integer values at most n, this representation and the integer indexing assumptions together enable us to evaluate the input mappings at many points in a batch by means of radix sort. Other mappings that are produced within the algorithms are similarly evaluated. When these algorithms are detailed, it becomes evident that such evaluations can computed in straightforward manners in time linear in n and the number of points evaluated. The descriptions of these algorithms assume that the values of mappings are accessed by radix sort.

4 The one-one case

Our algorithm for Problem 3 makes extensive use of bisection-based dynamic programming and implicit computation in compressed formats. This problem generalizes the longest common subsequence problem [11], which has efficient dynamic programming solutions. A direct dynamic programming approach to our problem would recursively solve the problem with T_1^x and T_2^y in place of T_1 and T_2 for all vertices $x \in P$ and $y \in Q$. This approach may require solving $\Omega(n^2)$ subproblems. To improve the time complexity, observe that the number of leaves in a maximum agreement subtree of T_1^x and T_2^y can range only from 0 to n. Moreover, this number never increases when x moves from the root of T_1 along P to P's endpoint, and y remains fixed, or vice versa. Compared to the length of P, $\text{RR}(T_1^x, T_2^y)$ often assumes relatively few different values. Thus, to compute this number along P, it is useful to compute the locations at P where the number decreases. We can find those locations with a bisection scheme and use them to implicitly solve the $O(n^2)$ subproblems in certain compressed formats.

The following discussion focuses on the basic technique used in such implicit computation. The details of our dynamic programming technique are similar to those in [8, 20]. They are are omitted in this extended abstract and can be found in the full version of this paper [16].

For integers k_1 and k_2 with $k_1 \leq k_2$, let $[k_1, k_2] = \{k_1, \cdots, k_2\}$, i.e., the integer interval between k_1 and k_2. The *length* of an integer interval is the number of its integers. The *upper* and *lower halves* of an even length $[k_1, k_2]$ are $[k_1, \frac{k_1+k_2-1}{2}]$ and $[\frac{k_1+k_2+1}{2}, k_2]$, respectively. The *regular* integer intervals are defined recursively. For all integers $\alpha \geq 0$, $[1, 2^\alpha]$ is regular. The upper and lower halves of an even length regular interval are also regular. For example, $[1, 8]$ is regular. Its regular subintervals are $[1, 4]$, $[5, 8]$, $[1, 2]$, $[3, 4]$, $[5, 6]$, $[7, 8]$, and the singletons $[1, 1], [2, 2], \ldots, [8, 8]$.

A *normal sequence* is a nonincreasing sequence $\{f(j)\}_{j=1}^{l}$ of nonnegative numbers. A normal sequence is *nontrivial* if it has at least one nonzero term. For example, $5, 4, 4, 0$ is a nontrivial normal sequence, whereas $0, 0, 0$ is a trivial one.

Let $f_1, \cdots, f_k$ be k normal sequences of length l. An *interval query* for $f_1, \cdots, f_k$ is a pair $([k_1, k_2], j)$ where $[k_1, k_2] \subseteq [1, k]$ and $j \in [1, l]$. If $k_1 = k_2$, $([k_1, k_2], j)$ is also called a *point query*. The *value* of a query $([k_1, k_2], j)$ is $\max_{k_1 \leq i \leq k_2} f_i(j)$. A query $([k_1, k_2], j)$ is *regular* if $[k_1, k_2]$ is a regular integer interval. For example, let $f_1 = 5, 4, 4, 3, 2$; $f_2 = 8, 7, 4, 2, 0$; $f_3 = 9, 9, 5, 0, 0$. Then, f_1, f_2 and f_3 are normal sequences of length 5. Here, $k = 3$ and $l = 5$. Thus, $([1, 3], 2)$ is an interval query; its value is $\max\{f_1(2), f_2(2), f_3(2)\} = 9$. The pair $([1, 1], 3)$ is a point query; its value is $f_1(3) = 4$. The pair $([1, 2], 2)$ is a regular query; its values is $\max\{f_1(2), f_2(2)\} = 7$.

The *joint* of $f_1, \cdots, f_k$ is the normal sequence $\hat{f}$ also of length l where $\hat{f}(j) = \max\{f_1(j), \cdots, f_k(j)\}$. Continuing the above example, the joint of f_1, f_2, f_3 is $\hat{f} = 9, 9, 5, 3, 2$.

The *minimal condensed form* of a normal sequence $\{f(j)\}_{j=1}^{l}$ is the set of all pairs $(j, f(j))$ where $f(j) \neq 0$ and j is the largest index of any $f(j')$ with $f(j') = f(j)$. A *condensed form* is a set of pairs $(j, f(j))$ that includes the minimal condensed form. The *size* of a condensed form is the number of pairs in it. The *total size* of a collection of condensed forms is the sum of the sizes of those forms. Continuing the above example, the minimal condensed form of f_3 is $\{(2, 9), (3, 5)\}$; its size is 2. The set $\{(1, 9), (2, 9), (3, 5), (5, 0)\}$ is a condensed form of f_3; its size is 4. The total size of these two forms is 6.

Lemma 8. *Let $F_1, \cdots, F_k$ be sets of nontrivial normal sequences of length l. Let $\hat{f}_i$ be the joint of the sequences in F_i. Given a condensed form of each sequence in each F_i, we can compute the minimal condensed forms of all $\hat{f}_i$ in $O(l + s)$ time where s is the total size of the input forms.*

Lemma 9. *Let $f_1, \cdots, f_k$ be nontrivial normal sequences of length l. Given a condensed form of each f_i with a total size of s, we can evaluate m point queries* *in $O(m+l+s)$ time, and evaluate m_1 regular queries and m_2 irregular queries in a total of $O(m_1+(m_2+l+s)\log(k+1))$ time.*

Theorem 10. *One-One solves Problem 3 in $O(nd^2 \log d + n \log(p+1)\log(q+1))$ time or alternatively in $O(nd\sqrt{d}\log n + n\log(p+1)\log(q+1))$ time.*

Proof. By means of bisection-based dynamic programming and condensed forms.

5 The rake-based reductions

This section solves Problems 4 through 7.

5.1 The many-one case

The following algorithm is for Problem 4 and uses One-One as a subroutine. Note that Problem 4 is merely a multi-path version of Problem 3.

Algorithm Many-One;
begin

1. For all P_i, compute $T_{1,i} = T_1^{h_i}$, $T_{2,i} = T_2 || T_{1,i}$, and $Q_i = Q || T_{1,i}$;
2. For all empty Q_i, compute part of the output as follows:
   (a) Compute the root $\hat{v}$ of $T_{2,i}$ and $v \in K(Q, T_2)$ such that $\hat{v} \in T_2^v$;
   (b) $\text{RP}(T_1^{h_i}, T_2, Q)(\hat{v}) \leftarrow \text{RP}(T_2^v, T_1, \mathcal{P})(h_i)$; (*Note.* $H(Q_i, T_{2,i}) = \{\hat{v}\}$. This is part of the output.)
   (c) For all $x \in H(P_i, T_1)$, $\text{RP}(T_2, T_1, \mathcal{P})(x) \leftarrow \text{RP}(T_2^v, T_1, \mathcal{P})(x)$; (*Note.* This is part of the output.)
3. For all nonempty Q_i, compute the remaining output as follows: (*Note.* The many-one case is reduced to the one-one case with input $T_{1,i}$, $T_{2,i}$, P_i and Q_i.)
   (a) For all $u \in K(P_i, T_{1,i})$, $\text{RP}(T_{1,i}^u, T_{2,i}, Q_i) \leftarrow \text{RP}(T_1^u, T_2, Q)$;
   (b) For all $\hat{v} \in K(Q_i, T_{2,i})$, compute $\text{RP}(T_{2,i}^{\hat{v}}, T_{1,i}, P_i)$ as follows:
      i. Compute the vertex $v \in K(Q, T_2)$ such that $\hat{v} \in T_2^v$;
      ii. $\text{RP}(T_{2,i}^{\hat{v}}, T_{1,i}, P_i)(x) \leftarrow \text{RP}(T_2^v, T_1, \mathcal{P})(x)$ for all $x \in H(P_i, T_{1,i})$;
   (c) Compute $\text{RP}(T_{1,i}, T_{2,i}, Q_i)$ and $\text{RP}(T_{2,i}, T_{1,i}, P_i)$ by applying One-One to $T_{1,i}, T_{2,i}$, P_i, Q_i and the mappings computed at Steps 3a and 3b;
   (d) $\text{RP}(T_1^{h_i}, T_2, Q) \leftarrow \text{RP}(T_{1,i}, T_{2,i}, Q_i)$; (*Note.* This is part of the output.)
   (e) For all $x \in H(P_i, T_{1,i})$, $\text{RP}(T_2, T_1, \mathcal{P})(x) \leftarrow \text{RP}(T_{2,i}, T_{1,i}, P_i)(x)$; (*Note.* This is part of the output.)

end.

Theorem 11. *Many-One solves Problem 4 in $O(nd^2 \log d + \log(1+t(Q)) \cdot \sum_{i=1}^m b(P_i, T_1)\log(1+t(P_i)))$ time or alternatively in $O(nd\sqrt{d}\log n + \log(1+t(Q)) \cdot \sum_{i=1}^m b(P_i, T_1)\log(1+t(P_i)))$ time.*

Proof. Since T_1 and T_2 have the same label set, all $T_{2,i}$ are nonempty. To compute the output RP, there are two cases depending on whether Q_i is empty or nonempty. These cases are computed by Steps 2 and 3. The correctness of Many-One is then determined by that of Steps 2b, 2c, 3a, 3b, 3(b)ii, 3d and 3e. These steps can be verified using Lemma 2. As for the time complexity, these steps take $O(n)$ time using radix sort to evaluate RP. Step 1 uses Fact 1 and takes $O(n)$ time. Steps 2a and 3(b)i take $O(n)$ time using tree traversal and radix sort. As discussed in §3, Step 3c preprocesses the input of its One-One calls to maintain their integer indexing assumptions. We reindex the labels and vertices of $T_{1,i}$ and $T_{2,i}$ and pass the new indices to the calls. We also partition Many-One's $O(n)$-size array to allocate a segment of size $|T_{1,i}|$ to the call with input $T_{1,i}$. Since the total input size of the calls is $O(n)$, this preprocessing takes $O(n)$ time in an obvious manner. After this preprocessing, the running time of Step 3c dominates that of Many-One. The stated time bounds follow from Theorem 10 and the fact that Q_i is not longer than Q and the degrees of $T_{2,i}$ are at most d.

5.2 The zero-one case

The following algorithm is for Problem 5. It uses Many-One as a subroutine to recursively compare T_2 with the subtrees of T_1 rooted at the heads of the tubes obtained by iteratively raking T_1. The tubes obtained by the first rake are compared with T_2 first, and the tube obtained by the last rake is compared last.
Algorithm Zero-One;
begin

1. $S \leftarrow T_1$;
2. $LF \leftarrow \mathcal{L}(S)$; (*Note.* LF consists of the leaves of T_1.)
3. For all $x \in LF$, $\text{RA}(T_2, T_1)(x) \leftarrow 1$; (*Note.* This is part of the output.)
4. For all $u \in LF$, $\text{RP}(T_1^u, T_2, Q)(y) \leftarrow 1$, where y is the unique vertex of $T_2||T_1^u$; (*Note.* This is the base case of rake-based recursion.)
5. $S \leftarrow S - \mathcal{L}(S)$;
6. **while** S is not empty **do** the following steps:
   (a) Compute $\mathcal{L}(S) = \{P_1, \cdots, P_m\}$;
   (b) Gather the mappings $\text{RP}(T_1^u, T_2, Q)$ for all P_i and $u \in K(P_i, T_1)$; (*Note.* These mappings are either initialized at Step 4 or computed at previous iterations of Step 6d.)
   (c) $\text{RP}(T_2^v, T_1, \mathcal{L}(S))(x) \leftarrow \text{RA}(T_2^v, T_1)(x)$ for all $v \in K(Q, T_2)$ and $x \in \cup_{i=1}^m H(P_i, T_1)$;
   (d) Compute $\text{RP}(T_1^{h_i}, T_2, Q)$ for the head h_i of each P_i and $\text{RP}(T_2, T_1, \mathcal{L}(S))$ by applying Many-One to T_1, T_2, $\mathcal{L}(S)$, Q and the mappings obtained at Steps 6b and 6c; (*Note.* This is the recursion step of rake-based recursion.)
   (e) For all $x \in \cup_{i=1}^m K(P_i, T_1)$, $\text{RA}(T_2, T_1)(x) \leftarrow \text{RP}(T_2, T_1, \mathcal{L}(S))(x)$; (*Note.* This is part of the output.)
   (f) $S \leftarrow S - \mathcal{L}(S)$;

end.

Theorem 12. *Zero-One solves Problem 5 in* $O(nd^2 \log d \log n + n \log n \log(1 + t(Q)))$ *time or alternatively in* $O(nd\sqrt{d} \log^2 n + n \log n \log(1 + t(Q)))$ *time.*

Proof. The $\mathcal{L}(S)$ at Step 6a is a tube system. The heads of the tubes in $\mathcal{L}(S)$ become children of the tubes in future $\mathcal{L}(S)$. The vertices $u \in K(P_i, T_1)$ at Step 6b are either leaves of T_1 or heads of the tubes in previous $\mathcal{L}(S)$. These properties ensure the correctness of the rake-based recursion. The remaining correctness proof uses Lemma 2 to verify the correctness of Steps 3, 4, 6c and 6e. Steps 1-5, 6a, 6b and 6f are straightforward and take $O(n)$ time. Step 6c and 6e take $O(n)$ time using radix sort to access RP and RA. At Step 6d, to maintain the integer indexing assumptions for the call to Many-One, we simply pass to Many-One the indices of T_1 and T_2 and the whole array of Zero-One. Step 6d has the same time complexity as Zero-One. The desired time bounds follow from Theorems 1 and Theorem 11.

5.3 The zero-many case

Problem 6 is merely a multi-path version of Problem 5. Algorithm Zero-Many is for Problem 6 and uses Zero-One as a subroutine. It is similar to Many-One, and its details are omitted here.

Theorem 13. *Zero-Many solves Problem 6 in* $O(nd^2 \log d \log n + \log n \cdot \sum_{i=1}^{m} b(Q_i, T_2) \log(1 + t(Q_i)))$ *time or alternatively in* $O(nd\sqrt{d} \log^2 n + \log n \cdot \sum_{i=1}^{m} b(Q_i, T_2) \log(1 + t(Q_i)))$ *time.*

Proof. The proof is similar to that of Theorem 11. The time bounds follow from Theorem 12.

5.4 The zero-zero case

Algorithm Zero-Zero is for Problem 7. It is similar to Zero-One, and its details are omitted here. It uses Zero-Many as a subroutine to recursively compare T_1 with the subtrees of T_2 rooted at the heads of the tubes obtained by iteratively raking T_2. The tubes obtained by the first rake are compared with T_1 first, and the tube obtained by the last rake is compared last.

Theorem 14. *Zero-Zero solves Problem 7 in* $O(nd^2 \log d \log^2 n)$ *time or alternatively in* $O(nd\sqrt{d} \log^3 n)$ *time.*

Proof. The proof is similar to that of Theorem 12 and follows from Theorems 1 and 13.

6 Discussions

We answer the main problem of this paper with the following theorem and conclude with an open problem.

Theorem 15. *Let T_1 and T_2 be two evolutionary trees with n leaves each. Let d be their maximum degree. Given T_1 and T_2, a maximum agreement subtree of T_1 and T_2 can be computed in $O(nd^2 \log d \log^2 n)$ time or alternatively in $O(nd\sqrt{d} \log^3 n)$ time. Thus, if d is bounded by a constant, a maximum agreement subtree can be computed in $O(n \log^2 n)$ time.*

Proof. By Theorem 14, the algorithms in §5–4 compute $\mathrm{RR}(T_1, T_2)$ within the desired time bounds. With straightforward modifications, these algorithms can compute a maximum agreement subtree within the same time bounds.

The next lemma establishes a reduction from the longest common subsequence problem to that of computing a maximum agreement subtree.

Lemma 16. *Let $M_1 = x_1, \ldots, x_n$ and $M_2 = y_1, \ldots, y_n$ be two sequences. Assume that the symbols x_i are all distinct and so are the symbols y_j. Then, the problem of computing a longest common subsequence of M_1 and M_2 can be reduced in linear time to that of computing a maximum agreement subtree of two binary evolutionary trees.*

We can use Lemma 16 to derive lower complexity bounds for computing a maximum agreement subtree from known bounds for the longest common subsequence problem in various models of computation [1, 2, 11, 12, 21]. This paper assumes a comparison model where two labels x and y can be compared to determine whether x is smaller than y or $x = y$ or x is greater than y. Since the longest common subsequence problem in Lemma 16 requires $\Omega(n \log n)$ time in this model, the same bound holds for the problem of computing a maximum agreement subtree of two evolutionary trees where d is bounded by a constant. It would be significant to close the gap between this lower bound and the upper bound of $O(n \log^2 n)$ stated in Theorem 15. Recently, Farach, Przytycka and Thorup [6] independently developed an algorithm that runs in $O(n\sqrt{d} \log^3 n)$ time. Cole and Hariharan [4] gave an $O(n \log n)$-time algorithm for binary trees. It may be possible to close the gap by incorporating ideas used in those two results and this paper.

References

1. A. V. Aho, D. S. Hirschberg, and J. D. Ullman. Bounds on the complexity of the longest common subsequence problem. *Journal of the ACM*, 23(1):1–12, January 1976.
2. A. Apostolico and C. Guerra. The longest common subsequence problem revisited. *Algorithmica*, 2:315–336, 1987.

3. H. L. Bodlaender, M. R. Fellows, and T. J. Warnow. Two strikes against perfect phylogeny. In *Lecture Notes in Computer Science 623: Proceedings of the 19th International Colloquium on Automata, Languages, and Programming*, pages 273–283. Springer-Verlag, New York, NY, 1992.
4. R. Cole and R. Hariharan. An $O(n \log n)$ algorithm for the maximum agreement subtree problem for binary trees. In *Proceedings of the 7th Annual ACM-SIAM Symposium on Discrete Algorithms*, pages 323–332, 1996.
5. M. Farach, S. Kannan, and T. Warnow. A robust model for finding optimal evolutionary trees. *Algorithmica*, 13(1/2):155–179, 1995.
6. M. Farach, T. M. Przytycka, and M. Thorup. Computing the agreement of trees with bounded degrees. In *Lecture Notes in Computer Science 979: Proceedings of the Third Annual European Symposium on Algorithms*, pages 381–393, 1995.
7. M. Farach and M. Thorup. Fast comparison of evolutionary trees (extended abstract). In *Proceedings of the 5th Annual ACM-SIAM Symposium on Discrete Algorithms*, pages 481–488, 1994.
8. M. Farach and M. Thorup. Optimal evolutionary tree comparison by sparse dynamic programming (extended abstract). In *Proceedings of the 35th Annual IEEE Symposium on Foundations of Computer Science*, pages 770–779, 1994.
9. C. R. Finden and A. D. Gordon. Obtaining common pruned trees. *Journal of Classification*, 2:255–276, 1985.
10. D. Gusfield. Efficient algorithms for inferring evolutionary trees. *Networks*, 21:19–28, 1991.
11. D. S. Hirschberg. Algorithms for the longest common subsequence problem. *Journal of the ACM*, 24(4):664–675, 1977.
12. J. W. Hunt and T. G. Szymanski. A fast algorithm for computing longest common subsequences. *Communications of the ACM*, 20:350–353, 1977.
13. T. Jiang, E. L. Lawler, and L. Wang. Aligning sequences via an evolutionary tree: complexity and approximation. In *Proceedings of the 26th Annual ACM Symposium on Theory of Computing*, pages 760–769, 1994.
14. S. Kannan, E. Lawler, and T. Warnow. Determining the evolutionary tree. In *Proceedings of the 1st Annual ACM-SIAM Symposium on Discrete Algorithms*, pages 475–484, 1990. To appear in Journal of Algorithms.
15. S. K. Kannan and T. J. Warnow. Inferring evolutionary history from DNA sequences. *SIAM Journal on Computing*, 23(4):713–737, August 1994.
16. M. Y. Kao. Tree contractions and evolutionary trees. Submitted to SIAM Journal on Computing, 1996.
17. S. R. Kosaraju and A. L. Delcher. Optimal parallel evaluation of tree-structured computations by raking. In *Lecture Notes in Computer Science 319: Proceedings of the 3rd Aegean Workshop on Computing*, pages 101–110, 1988.
18. E. Kubicka, G. Kubicki, and F.R. McMorris. An algorithm to find agreement subtrees. *Journal of Classification*, 12(1):91–99, 1995.
19. G. L. Miller and J. H. Reif. Parallel tree contraction, part 1: Fundamentals. In *Advances in Computing Research: Randomness and Computation*, volume 5, pages 47–72. JAI Press, Greenwich, CT, 1989.
20. M. Steel and T. Warnow. Kaikoura tree theorems: Computing the maximum agreement subtree. *Information Processing Letters*, 48:77–82, 1993.
21. C. K. Wong and A. K. Chandra. Bounds for the string editing problem. *Journal of the ACM*, 23(1):13–16, January 1976.

Author Index

Lecture Notes in Computer Science

For information about Vols. 1–1122

please contact your bookseller or Springer-Verlag

Vol. 1123: L. Bougé, P. Fraigniaud, A. Mignotte, Y. Robert (Eds.), Euro-Par'96. Parallel Processing. Proceedings, 1996, Vol. I. XXXIII, 842 pages. 1996.

Vol. 1124: L. Bougé, P. Fraigniaud, A. Mignotte, Y. Robert (Eds.), Euro-Par'96. Parallel Processing. Proceedings, 1996, Vol. II. XXXIII, 926 pages. 1996.

Vol. 1125: J. von Wright, J. Grundy, J. Harrison (Eds.), Theorem Proving in Higher Order Logics. Proceedings, 1996. VIII, 447 pages. 1996.

Vol. 1126: J.J. Alferes, L. Moniz Pereira, E. Orlowska (Eds.), Logics in Artificial Intelligence. Proceedings, 1996. IX, 417 pages. 1996. (Subseries LNAI).

Vol. 1127: L. Böszörményi (Ed.), Parallel Computation. Proceedings, 1996. XI, 235 pages. 1996.

Vol. 1128: J. Calmet, C. Limongelli (Eds.), Design and Implementation of Symbolic Computation Systems. Proceedings, 1996. IX, 356 pages. 1996.

Vol. 1129: J. Launchbury, E. Meijer, T. Sheard (Eds.), Advanced Functional Programming. Proceedings, 1996. VII, 238 pages. 1996.

Vol. 1130: M. Haveraaen, O. Owe, O.-J. Dahl (Eds.), Recent Trends in Data Type Specification. Proceedings, 1995. VIII, 551 pages. 1996.

Vol. 1131: K.H. Höhne, R. Kikinis (Eds.), Visualization in Biomedical Computing. Proceedings, 1996. XII, 610 pages. 1996.

Vol. 1132: G.-R. Perrin, A. Darte (Eds.), The Data Parallel Programming Model. XV, 284 pages. 1996.

Vol. 1133: J.-Y. Chouinard, P. Fortier, T.A. Gulliver (Eds.), Information Theory and Applications II. Proceedings, 1995. XII, 309 pages. 1996.

Vol. 1134: R. Wagner, H. Thoma (Eds.), Database and Expert Systems Applications. Proceedings, 1996. XV, 921 pages. 1996.

Vol. 1135: B. Jonsson, J. Parrow (Eds.), Formal Techniques in Real-Time and Fault-Tolerant Systems. Proceedings, 1996. X, 479 pages. 1996.

Vol. 1136: J. Diaz, M. Serna (Eds.), Algorithms – ESA '96. Proceedings, 1996. XII, 566 pages. 1996.

Vol. 1137: G. Görz, S. Hölldobler (Eds.), KI-96: Advances in Artificial Intelligence. Proceedings, 1996. XI, 387 pages. 1996. (Subseries LNAI).

Vol. 1138: J. Calmet, J.A. Campbell, J. Pfalzgraf (Eds.), Artificial Intelligence and Symbolic Mathematical Computation. Proceedings, 1996. VIII, 381 pages. 1996.

Vol. 1139: M. Hanus, M. Rogriguez-Artalejo (Eds.), Algebraic and Logic Programming. Proceedings, 1996. VIII, 345 pages. 1996.

Vol. 1140: H. Kuchen, S. Doaitse Swierstra (Eds.), Programming Languages: Implementations, Logics, and Programs. Proceedings, 1996. XI, 479 pages. 1996.

Vol. 1141: H.-M. Voigt, W. Ebeling, I. Rechenberg, H.-P. Schwefel (Eds.), Parallel Problem Solving from Nature – PPSN IV. Proceedings, 1996. XVII, 1.050 pages. 1996.

Vol. 1142: R.W. Hartenstein, M. Glesner (Eds.), Field-Programmable Logic. Proceedings, 1996. X, 432 pages. 1996.

Vol. 1143: T.C. Fogarty (Ed.), Evolutionary Computing. Proceedings, 1996. VIII, 305 pages. 1996.

Vol. 1144: J. Ponce, A. Zisserman, M. Hebert (Eds.), Object Representation in Computer Vision. Proceedings, 1996. VIII, 403 pages. 1996.

Vol. 1145: R. Cousot, D.A. Schmidt (Eds.), Static Analysis. Proceedings, 1996. IX, 389 pages. 1996.

Vol. 1146: E. Bertino, H. Kurth, G. Martella, E. Montolivo (Eds.), Computer Security – ESORICS 96. Proceedings, 1996. X, 365 pages. 1996.

Vol. 1147: L. Miclet, C. de la Higuera (Eds.), Grammatical Inference: Learning Syntax from Sentences. Proceedings, 1996. VIII, 327 pages. 1996. (Subseries LNAI).

Vol. 1148: M.C. Lin, D. Manocha (Eds.), Applied Computational Geometry. Proceedings, 1996. VIII, 223 pages. 1996.

Vol. 1149: C. Montangero (Ed.), Software Process Technology. Proceedings, 1996. IX, 291 pages. 1996.

Vol. 1150: A. Hlawiczka, J.G. Silva, L. Simoncini (Eds.), Dependable Computing – EDCC-2. Proceedings, 1996. XVI, 440 pages. 1996.

Vol. 1151: Ö. Babaoğlu, K. Marzullo (Eds.), Distributed Algorithms. Proceedings, 1996. VIII, 381 pages. 1996.

Vol. 1152: T. Furuhashi, Y. Uchikawa (Eds.), Fuzzy Logic, Neural Networks, and Evolutionary Computation. Proceedings, 1995. VIII, 243 pages. 1996. (Subseries LNAI).

Vol. 1153: E. Burke, P. Ross (Eds.), Practice and Theory of Automated Timetabling. Proceedings, 1995. XIII, 381 pages. 1996.

Vol. 1154: D. Pedreschi, C. Zaniolo (Eds.), Logic in Databases. Proceedings, 1996. X, 497 pages. 1996.

Vol. 1155: J. Roberts, U. Mocci, J. Virtamo (Eds.), Broadbank Network Teletraffic. XXII, 584 pages. 1996.

Vol. 1156: A. Bode, J. Dongarra, T. Ludwig, V. Sunderam (Eds.), Parallel Virtual Machine – EuroPVM '96. Proceedings, 1996. XIV, 362 pages. 1996.

Vol. 1157: B. Thalheim (Ed.), Conceptual Modeling – ER '96. Proceedings, 1996. XII, 489 pages. 1996.

Vol. 1158: S. Berardi, M. Coppo (Eds.), Types for Proofs and Programs. Proceedings, 1995. X, 296 pages. 1996.

Vol. 1159: D.L. Borges, C.A.A. Kaestner (Eds.), Advances in Artificial Intelligence. Proceedings, 1996. XI, 243 pages. (Subseries LNAI).

Vol. 1160: S. Arikawa, A.K. Sharma (Eds.), Algorithmic Learning Theory. Proceedings, 1996. XVII, 337 pages. 1996. (Subseries LNAI).

Vol. 1161: O. Spaniol, C. Linnhoff-Popien, B. Meyer (Eds.), Trends in Distributed Systems. Proceedings, 1996. VIII, 289 pages. 1996.

Vol. 1162: D.G. Feitelson, L. Rudolph (Eds.), Job Scheduling Strategies for Parallel Processing. Proceedings, 1996. VIII, 291 pages. 1996.

Vol. 1163: K. Kim, T. Matsumoto (Eds.), Advances in Cryptology – ASIACRYPT '96. Proceedings, 1996. XII, 395 pages. 1996.

Vol. 1164: K. Berquist, A. Berquist (Eds.), Managing Information Highways. XIV, 417 pages. 1996.

Vol. 1165: J.-R. Abrial, E. Börger, H. Langmaack (Eds.), Formal Methods for Industrial Applications. VIII, 511 pages. 1996.

Vol. 1166: M. Srivas, A. Camilleri (Eds.), Formal Methods in Computer-Aided Design. Proceedings, 1996. IX, 470 pages. 1996.

Vol. 1167: I. Sommerville (Ed.), Software Configuration Management. VII, 291 pages. 1996.

Vol. 1168: I. Smith, B. Faltings (Eds.), Advances in Case-Based Reasoning. Proceedings, 1996. IX, 531 pages. 1996. (Subseries LNAI).

Vol. 1169: M. Broy, S. Merz, K. Spies (Eds.), Formal Systems Specification. XXIII, 541 pages. 1996.

Vol. 1170: M. Nagl (Ed.), Building Tightly Integrated Software Development Environments: The IPSEN Approach. IX, 709 pages. 1996.

Vol. 1171: A. Franz, Automatic Ambiguity Resolution in Natural Language Processing. XIX, 155 pages. 1996. (Subseries LNAI).

Vol. 1172: J. Pieprzyk, J. Seberry (Eds.), Information Security and Privacy. Proceedings, 1996. IX, 333 pages. 1996.

Vol. 1173: W. Rucklidge, Efficient Visual Recognition Using the Hausdorff Distance. XIII, 178 pages. 1996.

Vol. 1174: R. Anderson (Ed.), Information Hiding. Proceedings, 1996. VIII, 351 pages. 1996.

Vol. 1175: K.G. Jeffery, J. Král, M. Bartošek (Eds.), SOFSEM'96: Theory and Practice of Informatics. Proceedings, 1996. XII, 491 pages. 1996.

Vol. 1176: S. Miguet, A. Montanvert, S. Ubéda (Eds.), Discrete Geometry for Computer Imagery. Proceedings, 1996. XI, 349 pages. 1996.

Vol. 1177: J.P. Müller, The Design of Intelligent Agents. XV, 227 pages. 1996. (Subseries LNAI).

Vol. 1178: T. Asano, Y. Igarashi, H. Nagamochi, S. Miyano, S. Suri (Eds.), Algorithms and Computation. Proceedings, 1996. X, 448 pages. 1996.

Vol. 1179: J. Jaffar, R.H.C. Yap (Eds.), Concurrency and Parallelism, Programming, Networking, and Security. Proceedings, 1996. XIII, 394 pages. 1996.

Vol. 1180: V. Chandru, V. Vinay (Eds.), Foundations of Software Technology and Theoretical Computer Science. Proceedings, 1996. XI, 387 pages. 1996.

Vol. 1181: D. Bjørner, M. Broy, I.V. Pottosin (Eds.), Perspectives of System Informatics. Proceedings, 1996. XVII, 447 pages. 1996.

Vol. 1182: W. Hasan, Optimization of SQL Queries for Parallel Machines. XVIII, 133 pages. 1996.

Vol. 1183: A. Wierse, G.G. Grinstein, U. Lang (Eds.), Database Issues for Data Visualization. Proceedings, 1995. XIV, 219 pages. 1996.

Vol. 1184: J. Waśniewski, J. Dongarra, K. Madsen, D. Olesen (Eds.), Applied Parallel Computing. Proceedings, 1996. XIII, 722 pages. 1996.

Vol. 1185: G. Ventre, J. Domingo-Pascual, A. Danthine (Eds.), Multimedia Telecommunications and Applications. Proceedings, 1996. XII, 267 pages. 1996.

Vol. 1186: F. Afrati, P. Kolaitis (Eds.), Database Theory - ICDT'97. Proceedings, 1997. XIII, 477 pages. 1997.

Vol. 1187: K. Schlechta, Nonmonotonic Logics. IX, 243 pages. 1997. (Subseries LNAI).

Vol. 1188: T. Martin, A.L. Ralescu (Eds.), Fuzzy Logic in Artificial Intelligence. Proceedings, 1995. VIII, 272 pages. 1997. (Subseries LNAI).

Vol. 1189: M. Lomas (Ed.), Security Protocols. Proceedings, 1996. VIII, 203 pages. 1997.

Vol. 1190: S. North (Ed.), Graph Drawing. Proceedings, 1996. XI, 409 pages. 1997.

Vol. 1191: V. Gaede, A. Brodsky, O. Günther, D. Srivastava, V. Vianu, M. Wallace (Eds.), Constraint Databases and Applications. Proceedings, 1996. X, 345 pages. 1996.

Vol. 1192: M. Dam (Ed.), Analysis and Verification of Multiple-Agent Languages. Proceedings, 1996. VIII, 435 pages. 1997.

Vol. 1193: J.P. Müller, M.J. Wooldridge, N.R. Jennings (Eds.), Intelligent Agents III. XV, 401 pages. 1997. (Subseries LNAI).

Vol. 1196: L. Vulkov, J. Waśniewski, P. Yalamov (Eds.), Numerical Analysis and Its Applications. Proceedings, 1996. XIII, 608 pages. 1997.

Vol. 1197: F. d'Amore, P.G. Franciosa, A. Marchetti-Spaccamela (Eds.), Graph-Theoretic Concepts in Computer Science. Proceedings, 1996. XI, 410 pages. 1997.

Vol. 1198: H.S. Nwana, N. Azarmi (Eds.), Software Agents and Soft Computing: Towards Enhancing Machine Intelligence. XIV, 298 pages. 1997. (Subseries LNAI).

Vol. 1199: D.K. Panda, C.B. Stunkel (Eds.), Communication and Architectural Support for Network-Based Parallel Computing. Proceedings, 1997. X, 269 pages. 1997.

Vol. 1200: R. Reischuk, M. Morvan (Eds.), STACS 97. Proceedings, 1997. XIII, 614 pages. 1997.

Vol. 1201: O. Maler (Ed.), Hybrid and Real-Time Systems. Proceedings, 1997. IX, 417 pages. 1997.

Vol. 1202: P. Kandzia, M. Klusch (Eds.), Cooperative Information Agents. Proceedings, 1997. IX, 287 pages. 1997. (Subseries LNAI).

Vol. 1203: G. Bongiovanni, D.P. Bovet, G. Di Battista (Eds.), Algorithms and Complexity. Proceedings, 1997. VIII, 311 pages. 1997.

Vol. 1204: H. Mössenböck (Ed.), Modular Programming Languages. Proceedings, 1997. X, 379 pages. 1997.